Global Transmission Expansion

Recipes For Success

Global Transmission Expansion

Recipes For Success

Fiona Woolf

PennWell®

PennWell Corporation
1421 S. Sheridan Road
Tulsa, OK 74112 USA

1.800.752.9764
+1.918.831.9421
sales@pennwell.com
www.pennwell-store.com
www.pennwell.com

Cover Design by Clark Bell
Book Design by Wes Rowell

Library of Congress Cataloging-in-Publication Data
Woolf, Fiona, 1948-
Global transmission expansion : recipes for success / by Fiona Woolf.
p. cm.
ISBN 0-87814-862-0
1. Electric utilities--Prices. 2. Electric utilities--Costs. 3.
Electric power transmission. I. Title.
HD9685.A2 W66 2003
333,793'2--dc21

2002152359

Printed in the United States of America

1 2 3 4 5 07 06 05 04 03

For Nicholas, who gave me unending encouragement and support.

CONTENTS

FIGURES

ACRONYMS

AC	alternating current
ACCC	Australian Commerce and Competition Commission, a division of which is the regulator of the Australian National Electricity Market
ADR	alternative dispute resolution
APT	arbitrage pricing theory
APV	adjusted present value
ARR	auction revenue receipts
ATC	available transfer capability
BAPE	Bureau d'audiences publiques sur l'environnement
BATNA	best alternative to negotiated agreement
BSC	Balancing and Settlement Code
CA ISO	California ISO
CAPM	capital asset pricing model
CBT	cross border trade
CCGT	combined cycle gas turbine
CEER	Council of European Energy Regulators
CEGB	Central Electricity Generating Board
CfD	contract for differences
CPI	consumer price index
CRR	congestion revenue right
CSC	Cross Sound Cable
DC	direct current
DCF	discounted cash flow
DEA	data envelopment analysis
DG	distributed generation
DORV	depreciated optimized replacement value
DSM	demand side management
DSO	distribution system operator
DTLR	Department of Transport, Local Government and the Regions (U.K.)
EBIT	earnings before interest and tax

EC	European Commission
EdF	Electricité de France
EIR	environmental impact report
EMF	electromagnetic field
ENEL	the Italian electricity utility
ENRE	Ente National Regulador de la Electricidad (electricity regulator in Argentina)
EPSC	Electric Planning and Siting Commission
ETSO	European Transmission System Operator's Association
EU	European Union
FACTS	flexible AC transmission system
FDC	fully distributed cost
FERC	Federal Energy Regulatory Commission (U.S.)
FGR	flowgate right (a type of transmission right)
FTR	financial transmission right (includes fixed transmission right, transmission congestion contract, and point-to-point transmission right)
GWh	gigawatt hour
HVDC	high voltage direct current
ICRP	investment cost-related pricing
IIAS	integrated impact assessment statement
IIP	Information Incentives Project
IMA	independent market administrator or operator
IME	internal market in electricity
IMO	independent market operator
IPP	independent power project
ISO	independent system operator
ITC	independent transmission company
ITP	independent transmission provider
JAP	joint advisory panel
KPI	key performance indicators
kV	kilovolt
kW	kiloWatt

kWh	kiloWatt-hour
LMP	location-based or locational marginal pricing
LRMC	long-run marginal costs
MAAC	Mid-Atlantic Area Council
MCOS	marginal cost of service
MEA	modern equivalent asset
MEPA	Massachusetts Environmental Protection Act
MEU	municipal electricity utility
MPSC	Maryland Public Service Commission
MVA	megavolt-amperes
MVAR	megavolt-amperes-reactive or megavars
MVOS	marginal value of service
MW	megawatt
MWh	megawatt-hour
NECA	the Australian National Electricity Code Administrator
NEM	National Electricity Market (Australia)
NEMMCO	National Electricity Market Management Company Limited (Australia)
NEP	New England Power Company
NERA	National Economic Research Associates
NERC	North American Electricity Reliability Council
NETA	new electricity trading arrangements (introduced into England and Wales in 2001)
NGC	National Grid Company plc, the system operator and transmission owner of the transmission system of England and Wales
NIE	Northern Ireland Electricity plc
NIMBY	"not in my back yard"
NPV	net present value
NSP	network service provider
OASIS	open access same-time information system
OATT	open access transmission tariff
OEB	Ontario Energy Board
OEP	Office of Energy Projects

OFGEM	Gas and Electricity Markets Authority, (British electricity regulator)
OFWAT	Office of Water Regulation, (water regulator in England and Wales)
OPF	Office of the Provincial Facilitator
P_0	the initial or reset price on which a price cap operates
PBR	performance-based or incentive regulation
PG&E	Pacific Gas & Electric, (utility that owns much of the transmission system in northern California)
PJM	PJM Interconnection LLC, (Pennsylvania New Jersey Maryland power pool, the RTO of that power market)
PTDF	power transfer distribution factor
QC	Queen's Counsel
PUC	In the U.S., a public utility or public service commission (state regulator), also referred to as PSC
QNI	Queensland-New South Wales interconnector
REx	revealed expectation
RFP	request for proposal (in a competitive solicitation process)
ROA	return on assets
ROE	return on equity
ROI	return on investment
ROR	rate-of-return regulation
RPI	retail price index or index of consumer prices. It may also refer to any price index (e.g., CPI)
RTG	regional transmission group
RTO	regional transmission organization (may be an ISO or a transco)
SDG&E	San Diego Gas & Electric
SMES	superconducting magnetic energy storage

SNI	South Australia-New South Wales interconnector
SRMC	short-run marginal cost
SRO	self-regulating organization
TA	transmission (system) administrator
TCC	transmission congestion contract (see FTR)
TFO	transmission facility owner
TFP	total factor productivity
TLR	transmission loading relief
TNB	Tenega Nasional Berhad (Malaysian electricity utility)
TSO	transmission system operator
TSR	transmission system right
TSS	Transmission Services Scheme, i.e., the regulatory, uplift management incentive scheme that applied to NGC in its role as system operator that has since been replaced by the Balancing Services Activity Revenue Restriction
TX	transmission
UMIS	uplift management incentive scheme
UMIST	University of Manchester Institute of Science and Technology
VAR	kilovar (measure of reactive power)
WACC	weighted average cost of capital

ACKNOWLEDGEMENTS

Many people have helped to create these recipes. It is a case where too many cooks cannot spoil the broth. Above all, I should thank Professor William Hogan for making my senior fellowship at Harvard a reality. This book came about because of my ability to take sabbatical leave from my firm and to enjoy the richness and abundance of the many excellent resources that the Center for Business and Government at the John F. Kennedy School of Government made so freely available to me. I would like to thank Ira Jackson and his team, Dow Davies, Elizabeth Bulette, and all the other fellows who encouraged me, particularly Karen Yeung, Dan Bavly, Paul Hodges, Rina Spence, and Maria Christina Scharf. In Boston, David and Cathy Maister were a tower of strength.

Professor Hogan's contribution to this book, the figures, and to the body of knowledge and research upon which it is based has been enormous and I cannot thank him enough. Ashley Brown, Constance Burns, Steve Anderson, Stephen Bird, Ann Stewart, and Jo-Ann Mahoney provided every possible form of support, much friendship, and good humor in so many different ways. The assistance of Professors José Gómez-Ibáñez, Roger Porter, Jerold Kayden, and Cary Coglianese was freely given and gratefully received.

I am truly fortunate in knowing so many people whom I could turn to for information, thoughts, the ability to talk things through, comments on texts, as well as support and sympathy. I received excellent explanations from many eminent economists including Paul Joskow, Sally Hunt, Richard Tabors, Robert Wilson, David Newbery, Seabron Adamson, Sam Lovick, Thomas Olivier Léautier, Larry Ruff, Graham Shuttleworth, Jon Stern, Ralph Turvey, Oliver Letwin, Peter Bird, Richard O'Neill, Robin Cohen, Jack Neushloss, Alex Henney, Shmuel Oren, David Clement, Steve Taub, Eric Woychik, David Currie, Roy Shankar, and Michael Pollitt.

Daniel Kirschen, Goran Strbac, Phil Gall, Reinier Lock, Pam Kozlowski, David Maister, Nicholas Woolf, Alan Kidd, Denise Church, David Sheppard, Tim Russell, Peter Bird, Reinier Lock, and Eric Woychik gave me very valuable input and comments.

The idea for this book arose in Mexico City and I must thank José Luiz Aburto, Ruben Flores, Raul Nocedal, Marco Antonio González, Derek Woodhouse, and Dionisio Perez-Jácome for helping me to do some early thinking on transmission expansion.

On land use issues, my thanks go to Malcolm Grant for setting me off in the right direction. Lawrence Susskind gave me an enormous amount of enthusiasm for and information on consensus building, and his assistant Sossi Aroyan was most helpful. In New Zealand, Judge David Sheppard provided time and much kindness, and I must also thank Kenneth Palmer, Tom Fookes, Mike Taggart, Denise Church, John Turner, Gerard van Bohemen, and Bernard Hill for their time, thoughts, and ideas. In the U.K., I was greatly helped by my partners Tony Kitson and Chris Williams as well as Michael Fitzgerald Q.C. I would also like to thank Ashley Brown and Deidre Matthews for their special help. Alan Barringer gave me details of the saga of the Brighton-High Ridge line.

Kari Nyman, Mohinder Gulati, Eduardo Zolezzi, Clemencia Torres, Michael Klein, Bernard Tenenbaum, Phil Gray, and Jean-Pierre Charpentier of the World Bank were most helpful. Under the auspices of the World Bank, I had useful discussions with Donal O'Leary and Rolf Musielak of Siemens (who kindly provided me with insights into Siemens technological advances).

The people who have kept my feet on the ground (from the point of view of the operational, economic, and financial realities of implementation) include Jim Barker, Bill Dunn, Mark Rossi, Ian Grubb, Alan Kidd, David Jefferies, Richard Morse, David Owens, Kevin Kelly, Harvey Salgo, Ross Baldick, as well as Dave Nevius and Virginia Sulzberger at NERC.

The transmission owners and system operators of the world made a huge contribution. Many people in the National Grid Company plc and National Grid USA freely gave me time and excellent input in many ways, including Fiona Smith, Tim Tutton, Chris Spencer, Ian Davis, Tony Hill, Nick Winser, Brian Henderson, Masheed Rosenqvist, Jeff Scott, Charles Davies, Mike Metcalfe, Colin Ray, and Peter Bingham.

In Norway, I was greatly helped by Odd Hoelsaeter, Astrid Skojnborg-Brunt, Lars Kristian Vormendal, and Jovard Svoen.

In New Zealand, the assistance given by Bob Thomson, Allan Carvell, Howard Cattermole, Bruce Girdwood, Chandra Kumble, Peter Robertson, and Len Gould of Transpower was invaluable.

I am extremely grateful to Phil Gall of Transgrid in New South Wales for his unending and painstaking assistance and for enabling me to talk to many of his colleagues, including David Croft, Don Hutt, Geoff Holden, Norman Zip, Christopher FitzGerald, Ross Ranger, Colin Parker, Ashok Manglick, Garrie Chubb, and Chris Baker. At NEMMCO, I met with Charlie Macaulay and Tim George. In California, I was assisted by Stephen Metague, Terry Winter, and Elena Schmidt.

I received considerable input from Brian Lunn and Leslie Bryans in Northern Ireland and from Nora Coary, Dermot Byrne, Peter O'Shea, and Adele Sléator in the Republic of Ireland.

From Latin America, I was assisted by Uriel O'Farrell, Alberto Devoto, Sebastian Bernstein, and Vivianne Blanlot.

The TransÉnergie Group generously provided me with an array of valuable input on a wide number of issues in a variety of jurisdictions, for which I am extremely grateful and would like to thank Robert Benoit, Ray Coxe, José Rotger, and Juan Carlos Araneda especially, but also Francis Dupuis, Guillermo Espinosa Ihnen, and Stephane Pepin.

My thanks also go to Roslina Zainal, Zarina Bt. Nadzimuddin, and their colleagues at Tenaga Nasional Berhad in Malaysia and to Dolly Mogkatle and her team at ESKOM in South Africa.

PJM was very supportive of my efforts and gave me a wide variety of inputs and materials. I would like to thank Phil Harris, Bill Whitehead Rick Drom, Ken Laughlin, Joseph Bowring, Steve Herling, and Gerry Kook, in particular, for their many thoughts and assistance. Roberto Paliza kindly helped me with the Midwest ISO proposals.

I am grateful to TransAlta for the thoughts of Barry Chuddy, Cam Barton, Dennis Frehlich, and Scott Thon.

Dr. Karel Kodeda gave me invaluable legal and practical information with respect to the Czech Republic. Christopher Jones and

Michael Albers of the European Commission helped me to focus on expansion of European interconnectors and transmission systems.

Paolo Ricci, Sebastien Bumbolo, and Francesco Giorgianni of Terna and ENEL gave me a variety of thoughts. The same is true of Andrew Claxton, who helped me with flowgates and other insights from a trading perspective.

Other people who deserve an acknowledgement are Jeremiah Lambert (who helped me to find a publisher) and Michael Burt (who gave me very helpful guidance on computer software).

I must thank the partners and staff of CMS Cameron McKenna for making this book possible. Robert Derry-Evans pushed me into applying for the fellowship at Harvard in the first place. My own team provided both input and tolerance of my absence. In particular, I would like to thank Mark Moseley, Robert Lane, Mark Bartholomew, Amanda Seaton, Andrew Ivison, Robert Phillips, David Marks, Greg Williams, Joel Zipp, Amy Koch, Peter Budd, Rob Power, and Munir Hassan. The support that I had from Ada McGrath, George Daves, Fae Longman, Louise Clifford, Alison Humphries, Vera Farrants, Jo Darley, and Kate Hodgson and her team was outstanding. I could not have managed without Harry Dupre, who cheerfully and expeditiously responded to dozens of requests for information and citations.

Most authors express special gratitude to their secretaries, and I have been truly fortunate to have had Vanessa Dennison as my secretary for so many years. She has done much more for me than word processing. She has organized me and my life, and remained a cheerful and enthusiastic supporter throughout the arduous book writing process.

CHAPTER 1
Satisfying a Hunger

Hunger for Infrastructure

Society is hungry for all kinds of network infrastructure. It needs roads, railways, fiber optics, electricity transmission systems, gas pipelines, and energy distribution systems.

One of the great challenges society faces is how it goes about making and implementing decisions that will provide the additional infrastructure capacity it needs when the infrastructure is almost certain to have some adverse impact on local communities and the environment. If the existing infrastructure is inadequate to do the job, society bears significant cost, impacting all sectors of the economy. There is media outcry. Governments and regulators are severely criticized.

We use our infrastructure inefficiently. We put up with the cost of congestion on roads because we do not know how to fix the problem. We have not really focused on the cost of congestion on other networks until now. Happily, we have developed some ideas as to how to reduce congestion on electricity transmission systems, and they work.

This book could have been written generically on the expansion of all networks, but it focuses on electricity transmission systems because they are the most complex to deal with and have, to date, received the least attention.

We will use the term "expansion" to mean any transmission system reinforcement, whether it expands capacity or not. Some reinforcements are necessary to protect the system from damage, some are necessary in order to ensure that continuity and standards of service to consumers are maintained. Some reinforcements are undertaken in order to be able to control power flows and to increase the flexibility of the system to avoid the need to build new lines. This book is about all of these reinforcements, whatever their reason.

It is also about how society goes about building the transmission infrastructure it needs, while recognizing the impact on local interests and the environment. It is designed to be relevant whether or not the industry has been or ever will be restructured.

Transmission is equally important in all jurisdictions around the world. This book will refer a good deal to the experience gained and lessons learned in deregulated industries, where the issues come more sharply into focus because of the spotlight that the functional separation (unbundling) of transmission and system operations from other industry functions places on transmission. The lessons will be relevant to those jurisdictions having no intention of introducing competitive markets, because the problems they face will be the same, whatever the structure of the industry and in whatever environment electricity is bought and sold. Moreover, the book assumes that in every country where electricity is bought and sold, there is a trading system that can be loosely described as a market, even though the electricity industry may essentially provide a public service and may not characterize itself as trading in and delivering electricity as a commodity.

Organization of this Book and the Ingredients of this Chapter

The topic of electricity transmission expansion is like a cake with many different ingredients and flavors that requires both raising agents and setting agents to cook successfully. The recipe metaphor provides a means of describing and organizing the many issues needing to be dealt with and the problems needing to be solved, so that recipes can be created for workable transmission expansion schemes.

It also implies that there *are* solutions and that, if the recipes are followed, the final product will be palatable.

This chapter describes from a global perspective:

- what transmission is
- why transmission expansion is important
- why it is a mess
- why it is complex

In other words, it explains why we need to create recipes to expand transmission systems, why there are many different ingredients, and why several different preparation techniques are needed. The book will also show that we need a variety of recipes for different types of expansion.

In designing the recipes, we must take into account the objectives of a well-designed scheme to encourage transmission investment. We need to:

- provide electricity consumers what they want—reasonable reliability at an affordable price
- create incentives for market participants to utilize the transmission system efficiently
- create incentives to maximize transmission asset availability, capability, and flexibility
- create incentives for efficient investment in transmission by incumbent transmission owners and third parties
- create incentives to develop and introduce new technologies[1]

These objectives look obvious and they *can* be achieved, despite the tensions and trade-offs between them. The recipes to achieve them involve understanding the basic project finance and economic principles of cooking transmission expansion projects (chapter 2), looking first in the store cupboard to see whether in fact we need to expand the system and then selecting the right utensils (industry participants) and the right oven to implement the project on a merchant or regulated basis (chapter

3). We shall then choose the key ingredients from market design and transmission pricing principles (chapter 4) and add a raising agent in the form of market forces (chapter 5). We need to decide on the right regulatory approach as a setting agent (chapter 6) and choose the right regulatory ingredients as incentives (chapter 7). We shall look for an efficient preparation technique in coordinating the planning of transmission expansion (chapter 8). After that, we shall look at the two aspects of cooking that make all the difference—timing and presentation. The problems of obtaining siting and planning permissions and rights of way (chapter 9) can take as much time, if not more, than the rest of the steps so far described. However, those problems can be mitigated or solved by the way in which a transmission expansion project is presented to the local communities and landowners affected by it (chapter 10).

Finally, we will look at the resulting recipes and try to picture the outcomes in the way that a good recipe book would show them (chapter 11), observing that much depends on the taste and the presentation.

What is Transmission?

The transmission grid

Transmission is usually defined in terms of a system rather than a single line or facility. A common definition is that *transmission* is an interconnected group of lines and associated equipment for the movement or transfer of electricity at high voltage between points of supply and points at which it is transformed for delivery to customers or is delivered to other electric systems.

Transmission lines are like the highways or motorways where vehicles move at high speeds. They connect generation facilities to points of electricity demand (including the points where electricity distributors or retailers take the power and pass it on over distribution lines to consumers). Transmission lines may interconnect other transmission systems. In many jurisdictions, transmission lines are configured as an integrated network or mesh system that is often called the "grid," to reflect the network properties of the transmission lines and facilities.

Like all producers, generators would prefer to deliver at the power station gate and consumers would prefer to buy at their own premises. They leave the transmission and distribution of electricity to the professionals who specialize in transporting it to the consumers and transforming it to the right voltage. It falls to the system operator and transmission owners to provide several other vital services that are needed because of the laws of physics, the properties of transmission systems, and the fact that electricity cannot be stored. Together, they provide the security of supply and reliability on which consumers depend. They also provide vital services that enable trading systems and markets in electricity to function smoothly. Indeed, transmission systems become the overhead (and underground) marketplaces for trading electricity. Despite their importance, they represent the smallest proportion of the value of the electricity industry's assets (*e.g.*, 13% compared with 59% for generation and 28% for distribution in the U.S.).[2]

Transmission vs. distribution

The distinction between the terms "transmission" and "distribution" is usually seen as one of voltage, but it is often blurred. There are different definitions and a wide variety of practices as to the voltage levels at which the distinction is made. The Federal Energy Regulatory Commission (FERC) makes the distinction between transmission and local distribution using the seven indicators of local distribution:

- Local distribution facilities are normally in close proximity to retail customers
- Local distribution facilities are primarily radial in character
- Power flows into local distribution systems; it rarely, if ever, flows out
- When power enters a local distribution system, it is not consigned or transported on to some other market

- Power entering a local distribution system is consumed in a comparatively restricted geographical area
- Meters are based at the transmission-local distribution interface to measure flows into the local distribution system
- Local distribution systems will be of reduced voltage[3]

However, the distinction between transmission and distribution is something of an art rather than a science. In many jurisdictions that have undertaken restructuring, it has been necessary to give explicit power to the regulator to determine whether certain assets should be regarded as transmission or distribution facilities in certain circumstances. For our purposes, the exact distinction between transmission and distribution, as a technical matter, is not particularly relevant.

Importance of Transmission

Transmission provides many benefits

Adequate transmission capacity is needed both for market participants and the consumers that many of them serve (whether or not the industry has been deregulated) for a wide range of reasons:

- It provides the security of supply and reliability[4] that are fundamental needs of modern society. If the transmission capacity is adequate, generation in other regions can supply more backup power in the event of emergencies, reducing local reserve margins needed to maintain reliability. This will enable new generation investment closer to the load centers to be deferred and reduces the total aggregate amount of new generation needed across the region
- Congestion on a transmission system causes costs that are ultimately borne by consumers, because cheaper generation that might be located behind a bottleneck or constraint on the system cannot be used to serve them

- Adequate transmission capacity minimizes other indirect costs of transmission such as transmission losses, "must run" generation needed to support the system, and certain ancillary services in order to maintain reliability, which are borne by the consumer
- Transmission can be a key ingredient in helping to reduce energy prices (in both the spot and bilateral contract markets or trading systems) by allowing more market participants access to buy and sell at cheaper prices in high-priced areas. It allows consumers to buy power from lower cost producers, reduces price volatility, increases supply choices, and improves wholesale competition[5]
- Interregional transmission lines and interconnectors help to enlarge the size of a trading system or market by combining smaller markets that may suffer from illiquidity and high prices. They aggregate regional markets by interconnecting systems, thereby increasing the numbers of buyers and sellers, improving their liquidity and competitiveness. Because of this, they help to reduce the risk of energy price volatility and the costs of hedging that risk
- In systems that rely on hydropower, transmission can provide insurance against dry years by allowing increased imports of electricity
- It helps to alleviate market power in generation by opening up access to local markets for competitors
- It gives freedom to buyers and sellers to trade dynamically without the fear of their transactions being physically curtailed because of lack of transmission capacity

System operations

The power grid is under the control of a *system operator* that dispatches power generated across the grid in order to ensure that demand and supply balance in real time. Since electricity cannot be stored, it is necessary to maintain system security and supply adequacy so that the transmission grid is still able to transfer power when sudden failures of generation or transmission facilities occur. The commercial and technical rules of any electricity trading system must be designed with sufficient reserve capacity to allow for uninterrupted service when emergencies or unanticipated failures occur.

Neither the system operator nor the transmission system actually directs electricity to flow from one particular point to another. The grid is like a pool of water—inputs are indistinguishable once they are made, and no offtake of electricity can be associated with any particular input. Electricity will flow in accordance with the laws of physics across all available parallel paths.

The system operator instructs all but the smallest generators when and how much to generate. It decides the switching of the circuits to configure the transmission system in the most appropriate manner and must manage congestion (bottlenecks) on the system. It will dispatch generators in an economic "merit order" so that the least cost or lowest priced generating units are run to meet demand before the more expensive or higher priced units.

Reliability

Reliability is defined by the North American Electricity Reliability Council (NERC) as "the degree to which the performance of the elements of [the electrical] system results in power being delivered to consumers within accepted standards and in the amount desired." This definition marries the concept of adequacy with that of security. *Adequacy* is defined as "the ability of the system to supply the aggregate electric power and energy requirements of the consumers at all times." NERC defines *security* as "the ability of the system to withstand sudden disturbances."

The reason reliability is important in the planning and expansion of transmission systems is that the concept of reliability is at the heart of maintaining security of supply and continuity of service to consumers. Reliability means that transmission systems must be planned to survive certain critical system disturbances or contingencies by operating reliably within specified thermal, voltage, and stability limits.[6] For our purposes, we shall use the terms system security and reliability interchangeably. They imply that the system will remain intact even after outages or other equipment failures occur. This does not mean that planning and expanding the transmission system is carried out without regard to cost. It does not mean that a transmission system with adequate capacity to meet all demands that are placed upon it is the only ingredient of security of supply and continuity of service to consumers.

How congestion arises

Congestion on a transmission system is no different than congestion on a road network. The delay that is caused by bottlenecks creates cost to the drivers or owners of the vehicles and creates inefficiency and frustration for society. Congestion on a transmission system is created by the thermal or voltage limits of the transmission facilities that cannot be overloaded without damaging the equipment. The system operator must respect these limits. This will mean that some generating units will not be able to run because the power they would inject into the system would violate these limits. They are described as "constrained off." The system operator must, therefore, dispatch additional, more expensive generation to replace the generation that is constrained off the system as a result of these limitations, in order to maintain reliability and to ensure that supply and demand balance, moment by moment.

Transmission congestion, even on a single line, can increase energy prices dramatically. For example, a weak link in the U.S. between West Virginia and the Carolinas increased wholesale prices as much as 88% at times. The transmission link that transports electricity from upstate New York to New York City created congestion costs of $724.7 million in the summer of 2000, partly as a result of a nuclear power plant being withdrawn from service on the New York City side of the constraint. The congestion costs dropped to $64.6 million in the summer of 2001.[7]

Congestion costs take the industry by surprise

In the world of the vertically integrated utility, congestion costs are usually internalized and not separately identified. Many utilities will argue that there are no congestion costs, because the transmission lines are rarely operated at their thermal ratings. When the industry is vertically integrated, these costs are internalized. Generation and transmission costs are traded off and the focus is on the overall cost of delivered power.

However, as Figure 1-1 shows, congestion costs tend to appear and, indeed, are capable of increasing quite dramatically and unpredictably, particularly after restructuring, when a new wholesale market begins operations.[8] The fact that congestion costs can increase as a result of deregulation is perhaps not surprising, because market freedoms will result in changing patterns of power flows for which the transmission system was not designed. However, it is usually the scale of the congestion costs that takes market participants, system operators, transmission owners, and regulators by surprise. For example, in England and Wales, the previously hidden indirect costs of transmission rose after restructuring in 1990 to £571m in 1993. Figure 1-1 shows the level of congestion costs in the U.S. in the year 2000.

As Eric Hirst and Brendan Kirby point out:

> *Decisions on whether to build new transmission are complicated by uncertainties over the future costs of congestion. These uncertainties relate to load growth, the price responsiveness of load, fuel costs and, therefore, electricity prices, additions and retirements of generating capacity, and the location of those generators.*[9]

None of these factors has anything to do with the move to restructuring and deregulation of the electricity industry, but as we shall see, they can be exacerbated by it.

Exercise of market power

Transmission constraints essentially create multiple submarkets on the system, allowing some generators to exert local market power.

In larger markets, generators can set prices in the smaller submarkets for all sellers of energy. The cost of local market power is difficult to quantify because the excess profits or rents that generators with market power are able to realize are hidden in their prices. Traditional measures of market concentration are not helpful in this context.[10]

There is even greater potential for local market power in the supply of ancillary services that the system operator needs to maintain reliability. Generators that are constrained on to the system (*i.e.*, required to run in place of cheaper generation that cannot run because of the constraint) are usually described as "must-run" generation. They must generate at certain critical times in order for the system operator to be able to maintain reliability. They represent a clear example of local market power, simply because the system operator cannot manage without them at these times. These must-run generators usually provide reactive power to maintain voltage and reliability, but they can also generate real power to maintain the transmission capacity on some interfaces. They are local monopolies at the points on the system

	New England*	New York*	PJM*	California*
January 2000	$7.23	$71.43	$21.32	$6.60
February 2000	$10.96	$50.00	$6.72	$3.29
March 2000	$22.05	$21.43	$1.79	$5.64
April 2000	$20.11	$10.71	$18.56	$6.30
May 2000	$20.36	$96.43	$13.26	$5.29
June 2000	$13.10	$317.86	$26.30	$18.42
July 2000	$7.87	$185.71	$9.10	$44.00
August 2000	$10.75	$214.29	$3.70	$73.00
September 2000	$9.14	$71.43	$9.70	$24.00
October 2000	$8.62	$32.14	$3.70	$28.70
November 2000	$4.52	$39.29	$(6.80)	$72.40
December 2000	$10.85	$53.57	not reported	$103.40
Total	$145.56	$1,164.72	$107.35	$391.04

*$ million
Source: ISO-New England

Fig. 1-1 Congestion costs
(Source: ISO-New England)

where they are needed. They have to be remunerated by contractual rather than market mechanisms, because they are monopolies. Cheaper resources in one region or submarket cannot be used to meet demand or to maintain reliability in another region or submarket if transmission constraints are present.

Constraints may lead to inefficient generation investments

Constraints may cause inefficient generation investments. A generator could take into account transmission constraints in deciding whether to acquire generating capacity on the "right" side of the constraint. Indeed, an investor might be willing to pay a premium for constrained-on generation or deliberately to site power plants in constrained regions. It could even decide to retire some of its existing generating units in order to increase congestion on the network. The "counterflows" from those units might have helped to reduce congestion in the past by providing power to serve customers locally.

It is unrealistic to expect that local market power could be simply limited by market rules or reliability "must-run" contracts that give the system operator the right to call the unit to produce at a specific price, if that unit would not otherwise be economic. The generators will always be able to game the rules and contracts. They can exploit inconsistencies and loopholes in market rules and adopt strategic behavior (*e.g.*, by withholding available generating capacity from the market) in a manner that is anti-competitive and has the effect of driving up prices.[11] Expansion of the transmission system is the best way to deal with local market power.

Thomas-Olivier Léautier[12] points out that there is extensive literature that has examined the interaction between the availability of transmission capacity and market power in generation. Back in 1983, long before deregulation of the electricity industry had started in earnest, Paul Joskow and Richard Schmalensee warned against neglecting the transmission network, and showed numerically that transmission expansion reduces generators' market power. More recently, models of imperfect quantity competition in generation have been proposed[13] and models of imperfect price competition have also been described.[14] The impact of transmission constraints on the bidding behavior of generators has been studied empirically.[15] All authors conclude that generators are able to exert local market power when the transmission system is congested, with the result that prices will increase. Moreover, it has been demonstrated that imperfect competition in generation results in a transfer of profits or rents from owners of transmission assets to generators.[16]

Benefits of regional trading

Many jurisdictions have taken active steps to achieve more economic benefits for society by requiring or encouraging greater regional or national trading. They have strived to create large regional markets to overcome balkanization and weak links between systems, and to maximize liquidity (in terms of increased the numbers of buyers and sellers) and economies of scale. The main examples are Nordpool (Norway, Sweden, Finland, and parts of Denmark), the European Union, and Australia. However, the lack of interconnector capacity between the existing transmission systems has held back the development of fully competitive regional markets and inhibited the capture of economies of scale.

This is an issue that has been exercising regulators and governments in many parts of the world. More capacity to interconnect different transmission systems would greatly assist the development of the proposed "single market" in electricity in the European Union.[17] In Scandinavia, greater interconnection capacity would facilitate trading between and improve reliability in the Nordpool countries. Australia is very conscious of the need for interconnection capacity to facilitate the working of the National Electricity Market (NEM). Two new interconnectors have been built, and several more are being developed. In the U.S., FERC has taken a strong stand in favor of regional markets and would like to see the industry restructured into large regional markets (in addition to that already established in Texas).[18]

The benefits of larger regional markets that combine smaller less competitive markets are simple and compelling, but it takes interconnector and transmission capacity to realize the benefits. If a good recipe can be found and the preparation and implementation carefully handled, the results will be beneficial to the consumer. Workable markets and trading systems (whether competitive or not) can be created with significant numbers of buyers and sellers that improve liquidity. This will give industry participants more flexibility, increase efficiency, and put a downward pressure on prices for the benefit of consumers. The larger regional markets and trading systems will also enable the weaker parts of existing transmission systems that may already be constrained to be managed more effectively, so that power may flow to consumers across other routes.

Why Transmission Expansion Is a Mess

Transmission has been the poor orphan of the electricity industry. Regulators, government decision makers, utility management, and market designers have focused their primary efforts on generation, distribution, energy trading systems, and markets rather than transmission expansion. The reason for this may be that, typically, transmission accounts for only 4-11% of the total cost of electricity to the consumer.[19] Now, however, the issue has become urgent, not least because of the increased awareness of the costs of congestion and the important contribution that adequate transmission capacity makes. The unbundling of transmission and system operations on restructuring and deregulation has revealed this.

Decline in transmission investments

There has been little increase in transmission capacity in many countries over the last decade.[20] Indeed, the U.S. transmission capacity has not kept up with load growth and generating capacity additions. The transmission owners in Argentina and Norway have described their systems as being under stress. The siting of new generation in the north of England has necessitated the construction of a line using a new route corridor to bring power to the load centers in the south.

NERC has expressed concern that investment in transmission has fallen and is not meeting requirements:

> *The gap between the transmission expansion needs and proposed construction of transmission is widening. To support the reliability of the bulk power system, proper incentives must be developed to encourage transmission construction.*[21]

Figure 1-2 shows the existing capacity and planned investment in transmission at voltages of 230 kV and over.

Assessment period	Actual plant (pre-existing)	Planned additions	Change in planned additions from prior year	Total miles anticipated
1994-2003	150,953	10,400	-	161,353
1995-2004	150,826	8,851	(15%)	159,677
1996-2005	153,782	6,818	(23%)	160,600
1997-2006	150,510	5,834	(14%)	157,344
1998-2007	150,225	5,587	(4%)	155,812
1999-2008	155,691	6,155	10%	161,508
2000-2009	157,810	7,527	22%	165,682

230kV and over (circuit miles)

Fig. 1-2 New transmission investment (Source: NERC Reliability Assessments)

National Grid USA has told the Federal Energy Regulatory Commission (FERC): "There is more congestion on the grid and there are increasing violations of reliability rules, placing the transmission grid at prolonged risk of failure."[22]

Although the need for new transmission facilities is compelling in many jurisdictions, it is striking that so little expansion (apart from minor upgrades and the construction of facilities to connect or interconnect new generation and large consumers to the system) has been undertaken. There must be compelling reasons for investors to have been unenthusiastic about transmission expansion, despite the clear opportunities to invest.

Reasons for lack of enthusiasm to invest

There are many reasons that it has proved difficult to attract investment in transmission in a wide variety of different jurisdictions:

- There has been a perception of over-investment that has resulted in regulators discouraging further investment and reducing rates of return on transmission. Regulated depreciation rates over the full life of the assets (40 years) have also deterred investment
- There have been inadequate rewards, in comparison to the other allowed potential investments of transmission-owning utilities
- The structure of the industry, notably vertical integration and fragmentation of ownership of transmission assets among a number of owners of generation and distribution assets, has meant that less attention was paid to transmission by regulators, utility management, industry participants, and consumers. Its importance and value was little understood
- In the past, utilities focused only on meeting the needs of their own (native load) customers. They had no opportunity or little incentive to compete or to build plant in low-cost areas and transport it to high-cost regions. Any profits made by utilities that expanded their transmission capacity to accommodate the transactions of this nature would probably have been rebated to customers under rate-of-return regulation (representing a surplus over allowed costs). As a result, they had no incentive to expand their systems for this purpose
- There was a perception that a vertically integrated utility would lose market share if it built transmission allowing competitors access to its wholesale and retail customers

- The roles, responsibilities, risks, and costs associated with transmission system ownership and control may have been inappropriately assigned, resulting in responsibility or liability without the corresponding ability or incentives to invest. Conversely, the ability to invest or expand may have been allocated to an entity that has no responsibility or liability for direct or indirect transmission costs
- There has been a lack of attention paid to the level of congestion costs, and anticipated, future congestion costs, with the result that there was no perceived need to invest to reduce these costs
- Those transmission owners that were able to profit from congestion (because they were permitted to keep congestion rents) had an incentive to maintain, or even increase congestion, and, therefore, did not invest to reduce it
- There were uncertainties about the best method to deal with loop flows or parallel transfers of energy across the lines of more than one transmission owner which have deterred investment. Power flows scheduled by one transmission owner with ample transmission capacity may overload the lines of another transmission owner. The first transmission owner may be unwilling to expand transmission capacity because it needs no extra transmission capacity itself, and the second utility may be unwilling to invest in its own system because it will collect no revenues from the power flows scheduled by others
- Where the industry has been restructured, the resulting transmission business appeared to be one of passive ownership and asset management. Because the control of the assets would pass to the system operator, the business no longer represented an attractive investment
- Few regulatory regimes have managed to achieve the necessary balance between the interests of consumers and investors. They have failed to create financial incentives to maintain and invest in transmission to meet consumer and market participant needs. They have focused on incentives for transmission owners to achieve cost savings. On balance, all governments

and regulators tend toward short-term cost minimization in order to protect consumers as to prices, rather than taking a long-term view as to investment and quality of service. This has deterred potential investors

- Regulatory risk is an increasing concern to investors, lenders, and capital markets everywhere. Transmission has always been closely regulated because of its character as a natural monopoly. The risk takes the form of periodic reviews of the regulated asset base, prices, changing rules, expensive and lengthy proceedings and rate cases, and regulatory micromanagement. The concerns are increasing even in those countries having a long history of utility regulation
- It has become much more difficult and costly to obtain approvals for the siting of new transmission lines, planning permissions, and environmental consents, and the process is protracted. It has taken 10 years in England and 7 years in Norway to obtain the necessary consents and planning permissions. A similar process for the relatively short Brighton-High Ridge line outside Washington D.C. started in 1976 and was completed in 1992
- There are increasing difficulties, costs, and delays in obtaining rights-of-way, easements, and interests in land
- Objectors to transmission projects are better organized than they used to be. They include a broader range of interests, including environmentalists and organizations concerned with quality of life issues, who may not be concerned with the effect on property values
- Decision making with respect to permits, consents, and the expropriation of rights of way can be fragmented. This is frequently the case, where there are several state and federal agencies that must review and approve projects, making the whole process quite complex, lengthy, costly, and uncertain
- Government and regulatory policy conflicts (particularly conflicts between energy and environmental policies) create uncertainty and delay

- A contributing factor is parochialism and the desire for self-sufficiency that destroy incentives to integrate into larger regional trading systems. This can result in permits or consents not being granted or onerous conditions being imposed. It is common in cases where the local permitting or siting authority does not perceive a local need for an infrastructure project that may benefit a neighboring area or the region as a whole
- The necessary conditions described in chapter 5 that would encourage market participants to undertake transmission expansion on a purely merchant basis have not been put in place. As a result, a transmission expansion market is only beginning to emerge in Australia and North America. However, a market-based transmission expansion regime was quite successfully put in place in Argentina in 1994. Regimes that allow for both regulated and merchant transmission exist in Peru and Colombia

We now know that transmission is important and that transmission expansion is a mess. Why is it so complicated to fix it?

Why Transmission Expansion is Complicated

Most transmission systems are networks

Transmission systems are networks and, for the most part, power cannot be directed to flow on a particular path. This means that building a transmission line can actually increase network capacity by more than the capacity of the new line. It can also decrease capacity on some other parts of the system. Calculating the net amount of the capacity increase can be complex, because the pattern of congestion changes across various interfaces. In a world where transmission rights are granted in return for investment in new capacity, this issue becomes extremely important.

Transmission pricing is a key ingredient

Pricing is an important economic tool to ensure that resources are used efficiently. If transmission pricing is well designed, it will send signals to market participants and consumers to locate their plant and to use the existing transmission facilities efficiently so that the need to construct new facilities is minimized. The prices will also send signals when and where to invest.

A key ingredient of the market design relates to how costs are allocated and how energy and transmission services are priced. Industry participants will take a keen interest in anything, however small, that increases their risk or uncertainty or reduces profits. A market participant will object to free riders on facilities it has paid for—similarly it will be unwilling to pay additional charges in order to meet a transmission owner's revenue requirements that are inflated due to other users terminating their commitment to take service and by-passing the system. Transmission owners will also look for solutions to loop flows or parallel power transfers from other transmission systems (because power flows over parallel paths) that cause them economic harm or may limit the transfer capabilities of their systems.

Indirect costs of transmission

There are other, more indirect costs involved in using the transmission system, such as the costs of transmission system congestion, line losses, and ancillary services used by the system operator to maintain system security and reliability. Market participants will want these costs to be as small as possible and will be very concerned if the market rules result in their paying more than what they regard to be a fair share. It follows, therefore, that one of the primary reasons that transmission expansion is complicated is because of its interrelationship with trading arrangements, the market design, and, indeed, the structure of the industry. The interactions relate not simply to how the investment costs are recovered through transmission prices, but how maintenance and switching are carried out, how the effects on congestion and losses will be treated, and the entire transmission and operational planning process. The complicating factors include whether there is:

- an independent system operator or transmission company that is both a system operator and a transmission owner
- the allocation of functions and responsibilities between the system operator, transmission owners, and market participants creating incentives to expand the system
- a right of first refusal or obligation on the part of incumbent transmission owners to build new capacity additions
- a centrally-administered or market-based arrangement for managing congestion
- a separate scheme for providing, trading, or paying for ancillary services and losses that is market-based, centrally-administered, or some combination of the two
- a right to acquire rights-of-way or easements compulsorily that is available only to incumbent utilities

There is a further complication that generation and loads (demand) will operate primarily in competitive markets after deregulation, while transmission remains a regulated monopoly service. When a generator is selling its services in both a competitive energy market and a regulated ancillary services market, for example, it is impossible to determine unambiguously the appropriate allocation of costs between those markets and to establish appropriate incentives for efficient investments (or product substitution). The uncertainty may have resulted in under-investment or cross-subsidization.

Even without deregulation, expanding transmission was never easy

Transmission expansion has always been complicated. Investment in transmission necessarily involves high development costs and long lead times (permits, environmental consents, and rights-of-way have to be obtained). The visually intrusive nature of steel lattice transmission lines always provokes strong local opposition to their siting. The long asset lives (as long as 40 years), low regulated rates-of-return, and straight-line depreciation over the life of the assets make investment in transmission unappealing by comparison with other industries where the risks may be the same. The fact that regulators usually view transmission as a risk-free natural monopoly inviting a correspondingly low return (because it is allowed to recover all its costs from consumers and has no usage risk) has also deterred investors. There are large economies of scale, and plant and equipment that come in standard sizes (because they are expensive to manufacture to a unique specification). It is always difficult to anticipate the future needs of consumers, and complex advance planning for the construction of new facilities is necessary. It is not easy to ensure that transmission is planned and built as efficiently and economically as possible.

There is now a realization that transmission is not a risk-free, natural monopoly that should command a low return. It has always competed with generation and demand-side measures (where a consumer could agree to curtail its demand in order to relieve congestion). The spotlight is on the revenue and usage risk as users become free to build their own transmission facilities or purchase power from captive generation plant without using the transmission system. Moreover, innovation and the advent of new technologies involve taking risk.

Regulating transmission was never easy

There has always been a tension that regulators have found difficult to deal with between the need to encourage investment and the need to protect the consumer by ensuring that transmission service is provided efficiently, without overbuilding the system. Regulators have become increasingly critical of their own approach to the regulation of transmission. They have been criticized for encouraging over-investment in the past by guaranteeing a "fair" or "reasonable" rate of return on investment prudently incurred.[23] They have implicitly assumed over the last decade that the transmission systems were overbuilt as a result of this regulatory regime because there were no explicit effeciency incentives.

As a result, they have been under pressure to hold down the rates of return. However, a move has begun away from the traditional rate of return or cost of service regulation to what is described as performance-based or incentive regulation. Transmission owners will welcome this, for the most part. It will give a transmission owner the opportunity to take more risk, for which it would be able to earn an increased return. This can, however, create uncertainties about the rate of return that the regulator will allow with respect to a particular transmission expansion project. The transmission owner will not be sure whether an asset will remain in the regulated asset base for a period that is sufficiently long for the investor to realize an adequate return on the investment. This is a major issue in New Zealand and Australia.

Free riders

In all infrastructure development, there are some strong incentives for "free riding." An industry participant that would benefit from eliminating congestion on the transmission system might be better off waiting and hoping that someone else will expand the system first. It would save the cost of building the additional capacity. Transmission system investments may be too large for any individual investor to contemplate, and so consortia or coalitions of interested parties may be necessary. However, the free rider problem probably becomes worse when coalitions are needed, because the interested parties will hold off agreeing to participate in a coalition until they can see what deals other parties are able to make. As a result, everyone will delay reaching agreement as long as possible. In any event, coalitions are always difficult and time-consuming to put together.

Effects of deregulation

Experience in deregulated markets indicates that not only will the costs of congestion typically increase after restructuring, but also expansion planning becomes more difficult to carry out, coordinate, and implement. Competitive wholesale markets need a stronger and more flexible transmission system than before in order to handle the changing power flows that market forces will bring about. The transmission systems are placed under greater strain because they are used in ways that were not contemplated when they were designed. There may be pressure to relax planning and operating standards designed to maintain reliability. Transmission users may come and go at short notice in a competitive market. Transmission owners may fear that their investments may no longer produce revenues or that the regulator may force downward revaluation of their asset bases reflecting newly redundant facilities. This will adversely affect their balance sheets as well as their profit and loss accounts. Unbundled, stand-alone transmission companies incur increased risk from rapid technological innovation, competition, and regulatory changes as well as increased need for coordination.[24]

Achieving coordination after unbundling

Unbundling or separation of transmission and system operation functions of a vertically integrated electricity industry into separate divisions, business units, or corporations is undertaken in order to achieve:

- competition in some segments of the industry
- greater management focus on each of the separate functions
- increased efficiency
- cheaper prices
- better service to customers

However, all of these excellent benefits come at a price. There will always be a need for some degree of coordination. Indeed, it was because of the need for coordination that the functions were vertically integrated in the past.[25] The need to achieve coordination in the electricity industry is clear, given the lack of ability to store electricity and the importance of maintaining reliability and security of supply.

Encouraging coordination through regulation

The problem of achieving the necessary coordination is made more difficult by the need to regulate parts of the industry that retain the characteristics of a natural monopoly, such as the transmission system. This involves the creation of rules. Indeed, markets may, in fact, mean more rules.[26] They will certainly involve different rules. Therefore, it is necessary to put in place both market-based and regulatory strategies (including performance incentives) for encouraging coordination in vertically separated, regulated, private sector infrastructure. These rules will come in the form of both contracts and regulatory rules. Indeed, contracts and regulation can be substitutes for one another since they are all designed to protect consumers from opportunistic behavior by monopoly suppliers.

Vertical separation does not change the regulator's basic responsibility to protect the consumer, as to both prices and quality of service. However, the relationship between each of the newly separated firms and the consumer is often more complicated, because each provides an important, but different component providing the overall service the consumer needs and the regulator must protect. As a result, the regulator's traditional role may be more difficult, and new tasks and challenges will emerge. Moreover, contracts may be more difficult to negotiate if a third party, namely the regulator, is required to approve each one.

Another important issue raised by vertical separation is the extent to which the newly separated parts of the industry should be kept out of each others' business and, if not, the extent to which the regulator must prohibit or restrain certain behavior that would give one firm an unfair competitive advantage through vertical integration over another that is not vertically integrated. If one firm is engaged in two stages of production or service, but its competitors are engaged in only one, the firm engaged in both stages may use its monopoly at one stage to achieve an advantage over its competitors in the second stage. The new firms in the vertically separated industry may have been permitted to engage in more than one stage of production, transmission, distribution, or retail sales of electricity. This has been permitted on the basis of a regulatory prohibition on system operators participating in other specified industry-related activities, as well as governance arrangements to ensure independence and non-discriminatory access to the transmission system and the market.[27]

Increasingly, there is now the view that separation of transmission ownership from generation is necessary and not simply the separation of system operations from generation. Indeed, in Chile, steps were taken in 2000 to require Enersis, which had significant market power in generation, to divest its national high voltage transmission system that was acquired by TransÉnergie. Indeed, many countries that have deregulated their electricity industries have separated transmission ownership from all generation, including England and Wales, Australia, Argentina, Ontario, the Nordpool countries, and New Zealand. As we shall discuss in chapter 3, some of the jurisdictions such as Australia, Argentina, and Ontario have also separated system operations from transmission ownership.

Increased need for regulation

A further complication is that the regulator usually has closely regulated investments because service to consumers is often highly dependent upon the detail of investment decisions. This can involve what might be described as micro-regulation because the regulator may have looked at the design and pricing of services for particular consumers that are not able to protect themselves from the monopoly power of a transmission owner. Indeed, the regulator's skill in regulating individual services may be of critical importance since these services are all critical elements in providing what the consumer needs, namely, an integrated and efficient supply of electricity at the point of consumption at an affordable price.

Figure 1-3 shows the various participants and asset owners in a competitive wholesale market structure.

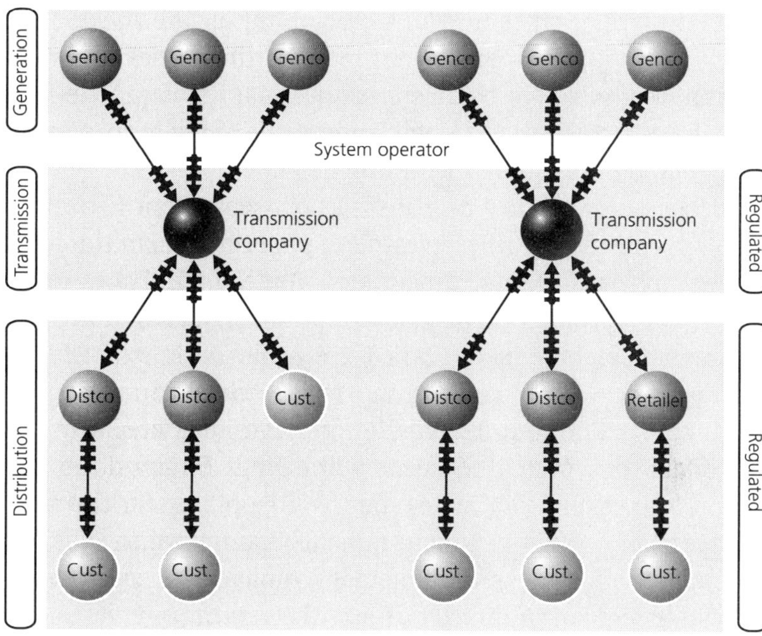

Fig. 1-3 Competitive wholesale electricity market structure

Most coordination of the electricity value chain is done through contracts or market rules enforced through contracts between the system operator, transmission owners, generators, and market participants. However, regulatory oversight of all these agreements is often carried out, not only to ensure that the contract and rules achieve the necessary degree of coordination, but also in order to assist the parties in actually achieving agreement. The regulator may well have to intervene in order to resolve disputes or to determine what these contracts should say.

There are advantages to contracts that are entered into without regulatory involvement. They can deal with detailed issues and situations in which the parties find themselves and also cover their technical needs. The contract parties may be able to think up innovative new solutions to problems they have not faced before when they were vertically integrated. They will have better industry knowledge than the regulator and will be able to identify critical or practical issues that the regulator will not. However, it takes time to negotiate these agreements, and it is costly. The negotiated approach also requires the parties to have sufficient bargaining power to be able to negotiate a fair deal. The regulator needs to be empowered and sufficiently well-informed to prevent one party from dominating the other at the negotiating table.

From the point of view of transmission expansion, unbundling of transmission means that the investment plans of transmission utilities affect industry participants, consumers, and the regulators in diverse ways. The appropriate level and nature of risks and rewards for a separated transmission company will have to be designed. The trading and market rules affecting planning will need to be redesigned, the terms for connection and use of the transmission system by users to accommodate their increased needs will have to be rewritten, and the regulation of transmission investment will be put under the spotlight. Every aspect of the way in which transmission is regulated as a whole will affect the willingness of investors to implement expansion projects, because regulation directly affects the return on investment.

Technical and market rules

Transmission systems were traditionally regarded as nothing more than an integral part of the delivery system between power producers and the consumers. Indeed, many transmission-owning utilities did not maintain separate accounts for their transmission businesses until restructuring forced them to unbundle the function. On deregulation, the transmission system becomes the marketplace to which market participants have open and non-discriminatory access. They are free to use it as they please within certain technical limitations. These limitations are normally to be found in the Grid Code (a set of technical rules relating to the planning of, connection to, and operation of the electric grid used in the UK and many other countries), technical rules, or operating protocols. They may relate to the design and operational capabilities of the system or to the software that administers the market or operates the system itself. The software implements the market design and is defined by it.

New concerns over transmission expansion

In the private sector, investment is driven by price signals and the profit motive. Electricity industry participants become increasingly concerned about how the regulated transmission business makes expansion decisions, if the transmission business incentives are not aligned with their own. The decisions of when and where to expand the system can impact the profitability of generators and loads in the energy market.

There is also a complex interaction between the perceived natural monopoly characteristics of the regulated transmission system and the freedom of industry participants to respond to market conditions and price signals by building transmission facilities themselves, should that prove a cheaper alternative to their paying for the direct and indirect costs of transmission. They may argue that it may be acceptable for a natural monopoly to exist, but there is no reason to add to it the monopoly right to build all the additional transmission facilities. This philosophy was adopted in Argentina, and many jurisdictions permit third parties to undertake what is now called market-based expansion or merchant transmission.

Regulated and market-based expansion

Regulating a monopoly is never easy. The regulator never has perfect information to do the job. After all, if we knew how to regulate monopolies, we would not be restructuring them. The introduction of market-based transmission expansion appears to be desirable. It should capture entrepreneurial innovation and financing and the other benefits of market forces. It should not need regulation; however, a market mechanism in transmission involves redesigning the way in which transmission is regulated as a function, not only to allow expansion to occur at market-based prices, but also to create a level playing field between incumbent or regulated transmission owners and merchant transmission investors. The regulatory regime needs to deal with the interfaces between these two approaches to transmission ownership and investment. Society would benefit if both market-based and regulated expansion were to flourish—these are the two ovens in which our recipes for transmission expansion must be successfully cooked.

Designing the Recipes for Expansion

We now know why transmission is important, why transmission expansion is a mess, and that it is complex. In order to design the recipes that we need to achieve efficient and palatable expansion, we must work through the complicating factors and choose the key ingredients before designing the preparation methods. Before doing that, we must look at the cooks and the kitchen equipment. First, we must analyze the motives of the investors and understand the economics of their investment decisions.

NOTES

1. There are new technologies that could make it easier to expand transmission systems. For a description of superconducting magnetic energy storage (SMES), flexible AC transmission system (FACTs) devices, high voltage DC (HVDC) systems, HVDC "light" transmission lines, and real-time ratings of transmission lines, see Eric Hirst and Brendan Kirby, "Key Transmission Planning Issues," *The Electricity Journal* (October 2001), pp. 66–67. The authors conclude that: "In spite of their wonderful attributes and recent declines in their costs, these technologies are generally too expensive to warrant widespread use. To date, they have been deployed in a few locations, primarily by utilities to improve the performance of their systems"

2. Thomas-Olivier Léautier, "Transmission Constraints and Imperfect Markets for Power," *Journal of Regulatory Economics*, 19:1 (2001), pp. 27-54

3. FERC, Promoting Wholesale Competition Through Open Access Non-Discriminatory Transmission Services by Public Utilities, Docket No. RM95-8-000, Order No. 888, FERC Statutes and Regulations P 31,036 (1996) at 31,771

4. Eric Hirst, "Transmission Crisis Looming," *Public Utilities Fortnightly* (September 15, 2000). The same level of reliability can be maintained with lower total investment

5. In some rare cases, transmission expansion can result in increased total payments by consumers that benefitted from prices that were low due to the local generators' export constraints.

6. NERC, "Transmission Expansion: Issues and Recommendations," Report to the NERC Planning Committee of the Transmission Adequacy Issues Task Force (approved September 28, 2001)

7. *Wall Street Journal*, Thursday, December 20, 2001 quoting FERC "Electric Transmission Constraint Study," EX02-7-000 (December 19, 2001)

8. When security-constrained dispatch is used to measure congestion, the congestion costs can be as high as 63% of the fixed cost of the network. Jacqueline Boucher, Benoit Ghilain, and Yves Smeers, "Security-Constrained Dispatch Gives Financially and Economically Significant Nodal Prices," *The Electricity Journal* (November 1998), p. 53

9. Eric Hirst and Brendan Kirby, "Key Transmission Planning Issues," *The Electricity Journal* (October 2001), p. 61

10. The Herfindahl-Hirshman Index is the main example of a measure of market power

11. For a description of local market power in California, see Thomas-Olivier Nasser "The Hidden Value of Transmission Assets," *The Electricity Journal* (June 1999), pp. 69-78

12. Thomas-Olivier Léautier, "Transmission Constraints and Imperfect Markets for Power," *Journal of Regulatory Economics*, 19:1 (2001), pp. 27-54

13. Shmuel S. Oren, "Economic Inefficiency of Passive Transmission Rights in Congested Electricity Systems with Competitive Generation," *The Energy Journal,* 18(1) (1997), pp. 63-84. Hung-po Chao and Stephen C. Peck, "An Institutional Design for an Electricity Contract Market with Central Dispatch," *The Energy Journal*, 18(1) (January 1997), pp. 85-110. Judith Cardell, Carrie Cullen Hitt, and William W. Hogan, "Market Power and Strategic Interaction in Electricity Networks," *Resource and Energy Economics* (October 1997), pp. 109-137. Severin Borenstein, James Bushnell, and Steven Stoft, "The Competitive Effects of Transmission Capacity in a Deregulated Electricity Industry," POWER Working Paper PWP-040R, University of California Energy Institute (1998) available at www.ucei.berkeley.edu/ucei

14. James Bushnell, "Transmission Rights and Market Power," *The Electricity Journal,* (October 1999), pp. 77-59. Paul L. Joskow and Jean Tirole "Transmission Rights and Market Power on Electric Power Networks," *RAND Journal of Economics,* vol. 31, no. 3 (Autumn 2000), pp. 450-487

15. Catherine Wolfram, "Strategic Bidding in a Multi-Unit Auction: An Empirical Analysis of Bids to Supply Electricity in England and Wales," *RAND Journal of Economics* 29 (Winter 1998), pp. 703-725. James Bushnell, "Transmission Rights and Market Power," *The Electricity Journal* (October 1999), pp. 77-85

16. James Bushnell, "Transmission Rights and Market Power," *The Electricity Journal* (October 1999), pp. 77-85. Paul L. Joskow and Jean Tirole "Transmission Rights and Market Power on Electric Power Networks," *RAND Journal of Economics,* vol. 31, no. 3 (Autumn 2000), pp. 450-487. They show that generators benefit from a reduction in transmission capacity. Catherine Wolfram, "Strategic Bidding in a Multi-Unit Auction: An Empirical Analysis of

Bids to Supply Electricity in England and Wales," *RAND Journal of Economics* 29 (Winter 1998), pp. 703-725. She finds, among other results, that in England and Wales, "constrained-on" units bid significantly higher markups than those not constrained on. James Bushnell and Frank Wolak in "Regulation and the Leverage of Local Market Power in the California Electricity Market," POWER Working Paper PWP-070, University of California Energy Institute (1999) find that in California, regulatory measures created to curb generators' local market power in fact led to higher energy prices during the summer of 1998. They conclude that competitive wholesale markets need stronger and more flexible transmission systems

17. Commission of the European Communities, Directive Concerning Common Rules for the Internal Market In Electricity, 96/92/EC, Official Journal L 027, 30/01/1997 p. 0020-0029

18. FERC Notice of Proposed Rulemaking–Continued Development of an Open and More Competitive Market in Bulk Power Sales of Electricity–Regional Transmission Organizations (RTOs), Docket No. RM99-2-000, 64 FR 31,390 (June 10, 1999), FERC Statutes and Regulation P 32,541 (1999)

19. There is evidence that in the U.S., historically, transmission has represented approximately 11% of the average cost or of power delivered to the consumer. See PA Consulting Group Report: "The Future of Electric Transmission in the United States: A Vision for Transmission as a Vibrant, Stand-Alone, For-Profit Business" (January 2001), pp. 3-4 and Appendix B

20. Eric Hirst, "Expanding U.S. Transmission Capacity," *EPRI Publication* (July 2001). In the U.S., transmission capacity has increased slowly from 1978 to 1998 at a lower rate than peak demand growth. To the extent that transmission is built to serve growing loads, peak demand is an appropriate normalizing factor for transmission capacity. The overall decline in normalized transmission capacity for the U.S. was 16% between 1989 and 1998. Utility projections of future transmission investment show further declines and expected transmission capacity additions are less than expected load growth

21. The North American Electric Reliability Council, "Reliability Assessment 2000: The Reliability of Bulk Electric Systems in North America," 33 (2000)

22. Response of National Grid USA to questions posed by the Commission, FERC Docket No. EL01–80–000, p.2. See also Prepared Testimony of David N. Cook,

General Counsel, North American Electric Reliability Council, before the United States Senate Committee on Energy and Natural Resources (July 25, 2001), pp. 4–7

23. For a discussion of the problem of rewards and incentives, see Alfred E. Kahn, *The Economics of Regulation* (Cambridge, Massachusetts: The MIT Press, 1970-71), vol. I, pp. 53-54

24. Michael Cragg, William Lehr, and Ron Rudkin, "Assessing the Cost of Capital for a Standalone Transmission Company," *The Electricity Journal* (January/February 2000), pp. 81-82. See also Shimon Awerbuch, Leonard S. Hyman, and Andrew Vesey, *Unlocking the Benefits of Restructuring: A Blueprint for Transmission* (Vienna, Virginia: Public Utilities Reports, Inc., November 1999), chapters 1, 2, and 9

25. It is possible to overdo the vertical separation and make it almost impossible to achieve the necessary coordination that results in services to society being provided efficiently or safely, *e.g.,* there is a widespread view that the railways in the UK have been "unbundled" too much, with the result that prices have increased and accidents have occurred. José A. Gómez–Ibáñez, "Regulating Coordination: The Promise and Problems of Vertically Unbundling Private Infrastructure," Discussion Paper, Taubman Center for State and Local Government, Harvard University, Cambridge, Massachusetts (December 1999)

26. Steven K. Vogel, *Freer Markets, More Rules: Regulatory Reform in Advanced Industrialized Countries* (Ithaca, New York: Cornell University Press, 1996)

27. James Barker, Bernard Tenenbaum, and Fiona Woolf, "The Governance of Power Pools and System Operators," *World Bank, Technical Paper No. 382* (1997))

CHAPTER 2
The Basic Principles

Organization of the Chapter

Most cookery writers devote a section of their cookbooks to the basic techniques or principles that underlie many of their recipes. This chapter is designed to explain the main principles on which the recipes and the discussion in the rest of this book are based. After looking at the elements of corporate and project finance that motivate investment and that should be taken into account by regulators in assessing the cost of capital of a regulated transmission owner, we shall discuss some key aspects of the economics of transmission systems. Finally, we will explain the relevant principles of the effect of regulation and market design on transmission systems. This chapter need not detain the advanced reader.

Principles of Corporate and Project Finance

All chefs like to know for whom they are cooking. The transmission expansion regime that is ultimately implemented must be attractive to the potential investors, their managements, and shareholders.

Corporations

It is an obvious point that the investors, whether they are entrepreneurs, market participants, or existing transmission owners, are likely to be organized as corporations. This means that they are legally distinct from their shareholders (whether public or private), and they have limited liability. Hence, the shareholders cannot be held responsible for the liabilities of the corporation. The shareholders are not usually the managers (although the managers may be minority shareholders). The main decision maker with respect to investment decisions within the management team of the corporation is likely to be the financial manager or chief financial officer (who may be an accountant), rather than the chief operating officer (who might be an engineer). He/she will focus on the fact that shareholders will want the management to increase the value of the corporation's stock and to pay dividends.

Net present value and the opportunity cost of capital

The role of the chief financial officer is to enable the management and board of the company to make the best possible decisions regarding both capital expenditure, on the one hand, and the raising of the finance (both equity and debt), on the other. Capital expenditure budgeting is a key activity. Several capital expenditure projects may compete for limited finance. For the present discussion, however, we assume that all investments can be equity-financed, so that decisions to spend money can be separated from decisions to raise money. Under this assumption, the chief financial officer would probably start by valuing alternative capital investment opportunities by a four-step procedure:

- The project's incremental after-tax cash flow would be forecast
- The project risk would be assessed
- The opportunity cost of capital would be estimated
- The net present value (NPV) of the project, using a discounted cash-flow formula would be calculated

The *net present value* is the present value plus any immediate cash flow. The *present value*, by contrast, is the forecasted cash flow discounted at the appropriate rate, usually the opportunity cost of capital.

The project with the highest NPV would be chosen. Calculating the present value simply involves discounting future cash flows at the appropriate rate, usually called the opportunity cost of capital, hurdle rate, or discount rate. The discount rate is determined by rates of return prevailing in capital markets. If the future cash flow is not subject to risk of fluctuation or volatility, then the discount rate could be the interest rate of debt. If the size of the future cash flow were uncertain, then the expected cash flow would be discounted at the expected rate of return offered by equivalent-risk securities.

Cash flows are discounted for the reason that a dollar today is worth more than a dollar tomorrow, and because a safe dollar is worth more than a risky one. Formulae for present values and NPVs are mathematical expressions of these ideas. The capital market is the market where safe and risky future cash flows are traded. By calculating the present value of an asset or investment, the chief financial officer is, in effect, estimating how much investors will pay for the asset or be willing to invest, given alternative investments available to them in the capital markets.

It is argued that the concept of NPV allows efficient separation of ownership and management of the corporation.[1] If management invests only in assets with positive NPVs, it will be serving the interests of each one of the shareholders regardless of his/her needs and preferences. Investors will focus on the certainty of the non-diversifiable future cash flows and the level of risk associated with them in making their investment decisions. As a technical matter, uncertainty can be separated into diversifiable risk (uncorrelated with the market) and non-diversifiable (correlated) risk. The former risk has a price of zero. The latter risk is what leads to higher discount rates.

Application of the discounted cash flow model to rates of return

When regulators allow utilities to earn a "fair rate of return," they tend to base the rate on that which is offered by securities having the same risk as the utility's common stock or ordinary shares. Small variations in estimates of the utility's cost of equity capital can have a substantial effect on the prices charged to consumers and on the utility's profits. As a result, both utilities and regulators devote considerable resources to estimating the cost of equity capital. This is important, because the value of a stock is equal to the stream of cash payments discounted at the rate of return investors expect to receive on comparable securities. Common stocks or ordinary shares do not have a fixed maturity; their cash payments consist, in concept, of an infinite stream of dividends. Therefore, the present value is derived from the stream of the anticipated dividends applying the appropriate discount rate, but taking into account the fact that investors do not purchase common stocks or ordinary shares only to receive dividends. Rather, they invest for both dividends and capital gains.

Risk, return, and the opportunity cost of capital

Returns to investors vary according to the risks they bear. The riskiest securities are common stocks or ordinary shares. The stock market can provide a benchmark for the opportunity cost of capital. If a project of average risk is being evaluated, the discount rate will be the expected return on the average common stock or ordinary shares. Historical evidence suggests that it is about 8-9% above the risk-free rate of interest, but many financial managers and economists opt for a lower figure.[2]

In investment terms, risk means that the future returns are uncertain. The relevant measure of risk is that of a portfolio of investments, because most investors do not put all their eggs into one basket. The standard deviation of the market portfolio generally represented by

Standard and Poor's Composite Index is around 20% per annum. A security's contribution to the risk of a well-diversified portfolio depends on how the security is liable to be affected by a general market decline. This sensitivity to market movements is known as "beta." Beta measures the amount that investors expect the stock price to change for each additional 1% change in market prices. The average beta of all stocks is 1.0. A stock with a beta greater than 1.0 is unusually sensitive to market movements. A stock with a beta below 1.0 is unusually insensitive to market movements. The correlation of uncertainty with diversifiable and non-diversifiable risk mentioned earlier also provides a good measure.

Capital asset pricing model

The marginal contribution of a stock to the overall risk of the market portfolio is measured by beta. This is the fundamental idea behind the capital asset pricing model (CAPM), which holds that each security's expected risk premium should increase in proportion to its beta.

The capital asset pricing theory is the best-known model of risk and return and is used by several regulators (including the UK regulator, OFGEM). It is said to be plausible, but far from perfect. Actual returns are related to beta over the long run, but the relationship is not as strong as the model predicts, and since the mid-1960s other factors seem to explain returns better. Stocks of small companies and stocks with high book values relative to market prices appear to have risks not captured by the model.

The CAPM has also been criticized for its strong simplifying assumptions. There are other theories such as the consumption capital asset pricing model and the arbitrage pricing theory. Each of these different models of risk and return has its advantages and disadvantages. However, all financial economists agree on two basic ideas:

- Investors require extra return for taking on risk
- They appear to be concerned predominantly with the risk they cannot eliminate by diversification into other business activities

Project risk

Each project should be evaluated at its own opportunity cost of capital. Obviously, the true cost of capital depends upon the use to which it is put. It is not appropriate to use the company cost of capital to discount the forecasted cash flows of new projects unless the project has the same risk as the company's existing business. It is the project risk that must be taken into account in determining the cost of capital for a particular investment (the covariance of the project return relative to the return on the existing portfolio of the investor). The company cost of capital may be appropriate for average risk projects and as a starting point for estimating discount rates for safer or riskier projects.

Cash flow forecasts will be adjusted to give due weight to bad outcomes as well as good outcomes. The chance of bad outcomes adding to the project risk is assessed. For regulated projects, the focus of attention will also be on the regulatory risk and the ability of the regulator to affect future cash flows from the project. This is discussed in more detail later and in chapter 6. For transmission projects that are not regulated, the focus will be on market and usage risk and the likelihood of market rule changes adversely affecting cash flows as well as market inefficiencies (*e.g.*, as a result of market power abuse).

Detailed risk assessment

Because of the critical nature of the cash flow forecasts in the NPV calculation and in the decision to make the investment, the financial manager or chief financial officer and his/her team will want to look closely at the details of the project and will not take it as a "black box."[3] This does not simply mean that they will carry out sensitivity analyses, break-even analyses, and Monte Carlo simulations to look at all the variables individually and in combination. They will look at all the details of the project and assess the many risks carefully, to ensure that they can be managed and mitigated.

They will look at insurance and risk management strategies such as hedging instruments, options, futures, and other derivatives. The way in which investors will look at the project will not be very different from that of their project lenders that may be financing the project on a limited recourse basis. Limited recourse means that the lenders will be looking primarily or wholly to the project revenues as a source of debt service and repayment of principal. Investors will be in it for the long haul.

However, investors will not simply assess the negatives of the project. They will try to ensure that the odds are stacked in their favor by enabling them to expand in areas in which the company has a comparative advantage, so they can generate additional profits. They will avoid expansion when competitive advantages are absent and additional profits are unlikely. They will not enter into a project without checking whether entry or expansion by competitors will drive future prices down.

Weighted average cost of capital

Modigliani and Miller showed that (assuming no taxes or financial market imperfections) the cost of capital does not depend on financing.[4] In other words, the weighted average of the expected returns to investors equals the opportunity cost of capital, regardless of the debt ratio. However, every dollar spent has to be raised somehow. The financing issues have to be taken into account if the investment is not being undertaken with internally generated funds or retained earnings.

The NPV can be calculated by discounting at the discount rate (typically adjusted downward to account for the value of interest tax shields) or at the opportunity cost of capital. The present value of the financing issues that have to be taken into account is added or subtracted. This is called adjusted present value or APV. The most commonly used adjusted discount rate is the after-tax weighted average cost of capital (WACC).

Weighted average cost of capital = $r_D(1-T_c \frac{D}{V} + r_E\frac{E}{V}$

where:

D	=	outstanding debt
E	=	number of outstanding shares x price per share
V	=	D + E
T_c	=	the marginal corporate tax rate
r_D	=	borrowing rate (as a percentage)
r_E	=	estimated return on share capital (as a percentage)

The formula really only works for projects that have the same risk profile as the existing business of the company, but the WACC can be used as a benchmark rate to be adjusted for differences in business risk or financing.[5]

Return on assets or return on investment

The management team will often measure performance of the company by the ratio of income to total assets. *Income* is usually defined as earnings before interest and taxes (EBIT). This ratio is also known as the return on assets (ROA) or return on investment (ROI).

$$\text{Return on assets} = \frac{(EBIT - tax)}{(Average\ total\ assets)}$$

where:

EBIT = Earnings before interest and taxes

The management team will want to compare the return that the company is earning with the opportunity cost of capital. The assets will be shown at net book value in the books of the company, reflecting original cost less any depreciation. The actual values of the assets may be less, but a low ROA does not necessarily imply that those assets could be better utilized elsewhere. Nor would a high ROA mean that it is possible to buy the same assets today and obtain a high return.

In a competitive industry, firms can, in the long run, expect to earn only their cost of capital. Therefore, a high ROA is sometimes regarded as an indication that the firm is charging excessive prices. When a regulator tries to determine whether the utility is charging a fair price, it will compare the ROA and its cost of capital.

Another measure that is commonly used with respect to the performance of a corporation is its return on equity (ROE). The calculation divides the earnings available for ordinary shareholders or common stockholders by average equity.

Raising debt for transmission projects

There is relatively little experience of lending for investment in new transmission assets (particularly on a project finance basis), but the issues that are specific to transmission are distinctive. The risks affecting the raising of finance are important, because, ultimately, they will affect the cost of the debt and equity for the expansion and, therefore, its economic feasibility and cost to the consumer. Some investors may be able to use internally-generated funds or retained earnings. They will essentially be engaging their balance sheets and raising finance as if transmission expansion were part of their core businesses.

However, in some situations, investors may prefer not to engage their balance sheets and to finance the expansion on a limited or non-recourse basis. This means that the banks would rely on the cash flows generated by the project for the payment of interest and the repayment of the loans. In this case, the "bankability" of the project (*i.e.*, whether the banks are prepared to lend on the basis of the project revenues and risks) will dictate whether or not it proceeds. Risk assessment then becomes a critical part of the project preparation or "cooking" process. Project finance tends to cost more than on-balance sheet financing, which will be an issue for the regulator to take into account if the expansion is being carried out on a regulated rather than a market basis. The reason for this is that the lenders are being asked to take the project risk, with recourse only to the project revenues as a source of repayment of principal and payment of interest. Typically, they will have very limited or no recourse to the developers or investors for funding. The project company will be thinly capitalized with no balance sheet strength and no other revenue-earning ability to finance unforeseen costs or delays. This increases risk for the lenders, for which they will command a premium.

Managing and mitigating the risks

In theory, lenders should prefer a project that does not involve a significant element of usage and price risk because of the need to create maximum certainty as to the revenue-earning capability and cash flows of the project in order to service and ultimately repay the debt on a timely basis. If the project company is not able to enter into long-term agreements to secure the revenues, the lenders will have to assess future market prices, which will be difficult. They may prefer a project that is remunerated on a regulated basis. However, they will be extremely concerned about regulatory risk (which we will discuss more fully in chapter 6), notwithstanding that there should be less price uncertainty and usage risk.

Regulatory risk is the expression describing the uncertainties of the regulatory regime to which the project company and other transmission owners will be subject. These uncertainties come about because the regulators are empowered to exercise discretion and change the regulatory rules on little or no notice. Regulators set or approve tariffs, price caps, or revenue caps, and approve the contents of the regulated asset base, by reference to which the permitted revenues of investors in a regulated transmission project are calculated. Regulators have considerable powers to affect earnings and to increase costs because they also regulate the standard of service.

For example, if there is a danger that the value of the project assets will be written down in the future or taken out of the asset base by the regulator, the revenues upon which they rely could be adversely affected. Investors will also want to be assured that the rate of return will be reasonable throughout the life of the assets, that the review periods are sufficiently long, and that there are no unexpected changes once the investment is approved. Ideally, they would like to see the development of the transmission equivalent of an independent power project, where the revenues are secured by means of a very long-term power purchase agreement with no reopeners. The equivalent contract in the context of transmission would be one that granted or reserved long-term rights to the capacity of the project, securing sufficient revenues to cover debt service and repayment of principal.

Financing issues specific to transmission projects

The particular concerns of lenders in relation to transmission (as opposed to other forms of infrastructure) projects will focus on:

- the design of the system (mesh, radial) being built and its interconnections with existing systems and plant
- the level of redundancy in the system design to enable the project company to meet performance targets
- the design of performance targets and their enforcement through incentives and penalties
- the ability of the transmission owner to meet its performance targets utilizing the project as a whole, rather than having performance measured by reference to individual circuits or discrete parts of the project being built
- cross-border political and tax risk (central, local, stamp duty) in the case of interconnector projects between different jurisdictions
- market rules risk (particularly the risk of changes relating to transmission, congestion, and losses)
- usage risk and creditworthiness of the entities paying for physical or financial rights to the project over the life of the project
- nonpayment risk that may be affected by the way in which payment is collected (*i.e.,* whether it is through the independent system operator (ISO) under the market rules or directly under

transmission tariffs, use of system agreements, or connection agreements with transmission users)
- regulatory risk in all its forms, but with particular reference to the review periods, the ability to write down the value of assets in a regulatory asset base, depreciation rates, rates of return, cost of capital, ability to change rules with or without notice, and the degree of micromanagement
- the risk of the introduction of new technologies and whether the chosen technology is both proven and the best available
- force majeure risk and liability for circumstances outside the complete control of the investor as well as circumstances that are substantially influenced by the system operator or other market participants
- the risks and potential for delay associated with obtaining consents and permits (including environmental protection measures)
- the risk and potential for delay in obtaining rights-of-way and land

Of the several general principles that apply to the development of any major infrastructure project, these are two principles that are particularly true in relation to transmission projects:

- The risks are very low once the project has been built and is operational
- Delay is usually bad for investors, developers, and lenders

Economic Principles of Electricity Transmission Systems

Networks, controllable lines, and loop flows

Life would be simple if transmission systems consisted entirely of controllable, radial lines over which power flowed from generators to consumers along a direct path consisting of lines and valves. The power flows could, in effect, be labeled, directed, and tracked. The valves could be closed if a consumer did not pay. However, transmission systems, for the most part, are mesh or reticulated networks, and power flows along the path of least impedance. It cannot be directed to a particular consumer. This creates interactions called network externalities. They can be very significant. The power will divide and flow across every parallel path between a generator to consumer. These flows are called "parallel" or "loop" flows. They occur because every time a new parallel route is opened up between two points, the electrical impedance between those two points is reduced. Power will flow across each possible path in inverse proportion to its impedance (although phase shifting allows some directional control by altering impedances). Even where there is a line that is controllable that is part of a larger system, the impact on the controllable line cannot be fully separated from the pattern of use of the rest of the system. Loop flows are calculated as the difference between scheduled (intended) and actual power flows between two points on the system. For this reason, loop flows are sometimes referred to as "inadvertent" transfers of power.

When transmission systems are interconnected with one another, the situation becomes even more complicated. Open access to transmission systems and the introduction of electricity markets that expand to cover wider areas can add to the complexities. There are technologies that allow the system operator to control power flows. They are still relatively expensive and are used sparingly at critical points on the system. However, most electricity transmission systems are virtually impossible to operate independently of interconnected networks that belong to different transmission owners. This has a direct impact on transmission expansion.

For example, if Transmission Owner A (TO-A) agrees to transmit electricity from a particular generator to a particular customer located on its system, some of that electricity will flow over its lines. As Figure 2-1 shows, if its system is interconnected with the system of Transmission Owner B (TO-B), some of the electricity may also flow over the lines of TO-B, because it will flow over all available parallel paths seeking routes where the impedance is lower. The electricity flowing over the lines of TO-B will increase its costs and its transmission losses (*i.e.*, energy lost in the process of transmission).

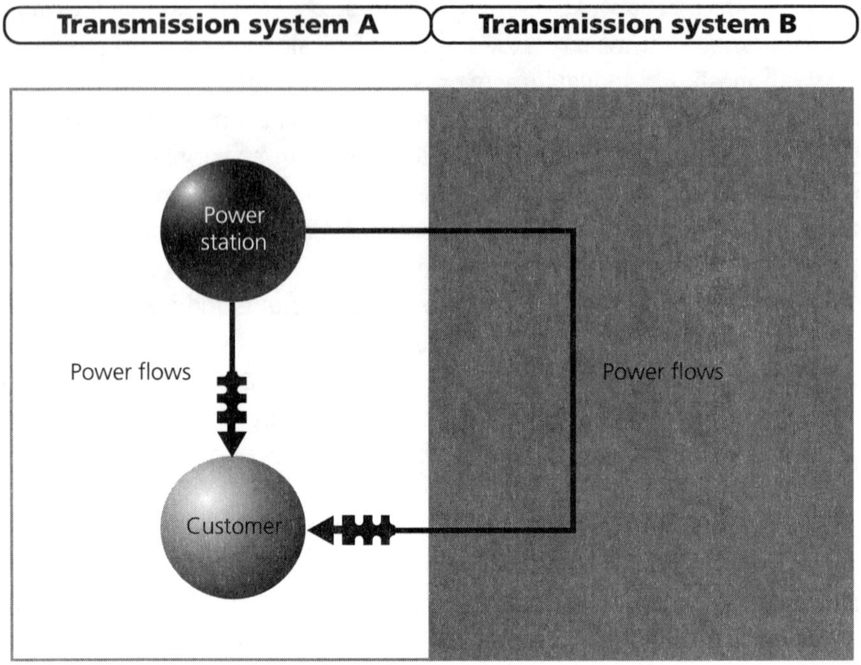

Fig. 2-1 Loop flow

Because of loop flow, a transmission owner does not capture fully the benefits of expanding its own capacity, nor does it bear the full costs of increasing its contractual obligations to transmit power. For example, if TO-B were to invest to increase its capacity, the amount that it would cost TO-A to deliver electricity between the original generator and customer would decrease. The amount of electricity that TO-A agreed to transmit would go over TO-B's expanded line, reducing the losses and the other costs incurred by TO-A. In short, TO-A benefits from TO-B's expansion without having to pay for it. Therefore, it is easy to understand why loop flow effects make congestion and pricing issues as well as transmission expansion more complicated.[6]

Different pricing schemes can create perverse incentives to build uneconomic transmission that reduces total welfare. For example, the building of a weak parallel line between a low cost and a high cost region would split the power flows between the two parallel lines. Flows on the original line would have to be reduced to keep the new line energized at its minimum level. A high cost generator in the high cost region at the end of the original line might be prepared to make the investment in order to lower the power transfers and raise prices in the high cost region. Even though the total welfare and economic efficiency would be reduced, the profit would justify the investment.

Marginal costing

Marginal cost is the cost of producing one more unit (or the cost that would be saved by producing one less unit). The concept of marginal cost pricing with respect to large infrastructure projects comes from a proposition that it would be both excessive and inefficient to charge the entirety of the investment costs to users in proportion to consumption.[7] Indeed, it would deter usage. It would be more appropriate to charge only the marginal cost for actual use and to recover investment costs in some other way. In the context of a transmission system, most of the costs represent fixed investments, and if the transmission system were priced by taking all the investment costs and dividing them by a number representing total usage, the resulting charge per unit of use would be much higher than the marginal costs of usage. This concept had a significant impact in many sectors and was commonly reflected in a two-part tariff, taking the form of a fixed fee with lower usage payments, which enhanced efficiency and increased sales.

Because the investment costs are by far the largest part of the costs of providing transmission service, any approach to pricing based upon marginal costing principles is unlikely to yield sufficient revenues to remunerate the transmission owner for its investment costs both in relation to new assets and sunk investment. This poses a difficult question—which users should pay the charges necessary to yield the revenues required to cover the investment, and more importantly, how those charges should be allocated across users of the system. The answer is not easy to construct and is always contentious, because all users will want to pay as little as possible, as we shall discuss in chapter 4.

Marginal costing methodology is, however, commonly applied in the electricity industry to derive the cost of transmission service. The concept of marginal cost embraces all costs imposed on the system by an increment of use by one user, assessed on the basis of the demand placed on the system by all other users.

There are two types of marginal costs that are commonly referred to:

- short-run marginal costs (SRMC)—the costs of increasing (or decreasing) output to meet an increment (or decrement) of demand without building new capacity. This can also be the price necessary to reduce demand so that it may be met by existing capacity. The SRMCs of a transmission system are taken to be the energy costs of constraints and transmission losses
- long-run marginal costs (LRMC)—the costs of accommodating an increment of output by increasing the physical transmission capacity that would include the cost of building new facilities as well as of any remaining losses

In general, it is assumed that in an optimally-dispatched system, the SRMC of transmission between two points is equal to the difference between the marginal costs of generation at each of those points. The system will be optimally dispatched if generation can be dispatched by the system operator in economic order, *i.e.*, a lower cost or priced generating unit is dispatched up to the full capacity offered at that cost or price before a more expensive or higher-priced unit (assuming no constraints). Transmission costs are then calculated from the estimate derived from a power flow model of the marginal cost of generation at all points on the network.[8]

Economies of scale, "lumpiness," and free riders

Economies of scale. Transmission investments may have large economies of scale.[9] Once the land and rights-of-way have been acquired, it becomes economic to build a high capacity line rather than a low capacity line. There may not be much difference in the cost of lines of different capacities. For example, the cost per MW-mile of a 500 kilovolt (kV) transmission line could be about half that of a 230 kV line. High-voltage lines also require less land per MW mile than lower voltage lines.[10] It cost Northern Ireland Electricity only an additional 20% to achieve a 100% increase in capacity in the new Moyle Interconnector between Scotland and Northern Ireland.

It obviously makes no sense to build lines to match short-term needs. Building the line with a higher capacity than is needed will typically reduce costs in the long term by avoiding the need to build several smaller lines that would be much more expensive. This is true particularly if a second line would involve acquiring more land and rights-of-way that would inevitably attract opposition associated with obtaining the necessary permits and consents. However, building for the long term involves uncertainty, and a series of smaller projects could be optimal. This is because there is some value in delaying a major investment until the future needs become clearer and, in the meantime, building smaller system enhancements to meet certain needs.

"Lumpiness." Transmission investments tend to be "lumpy" in that capacity is constructed in discrete blocks or "lumps" rather than in continually increasing amounts. For this reason, as well as because of economies of scale, these investments may be made in advance of need, but in anticipation of increased demand. As we have discussed, unless it is optimal to build a series of smaller projects, the investment could involve a major addition to the asset base of the utility within a relatively short period of time on which it will expect to earn a reasonable rate of return. This could involve a significant increase in prices that users must pay. In other words, the scale of any individual expansion project could cause associated lumpiness in the investor's revenue requirement that might disturb the smoothness or predictability of the progression of tariff increases over time that consumers look for.

Transmission facilities tend to come in standard sizes, *e.g.*, it is much easier to build a 345 kV line or a 500 kV line rather than a 389 kV line. A large transmission line may also affect system operations on a small system differently than several smaller lines. For example, if the large line becomes the largest single contingency on a transmission system, the reserve requirements may have to be increased to cover the contingency of the loss or outage of the line.

Free riders. If, as result of the investment, congestion is eliminated, cheap but previously isolated generators will be able to sell their power into the market at the same prices as everyone else. This means that there may be some strong incentives for "free riding." The cheap generator, benefiting from the elimination of congestion, may be better off waiting and hoping that someone else will come along and expand the transmission system first. That would save it the cost of building facilities to obtain access to high priced markets. However, if everyone thinks in this way, nothing will ever be built.

Transmission investments are likely to be bigger in scale than any individual market participant is prepared to undertake. Coalitions, consortia, or joint ventures of interested parties may be necessary. The complications of putting coalitions together probably exacerbate the free-riding problem. This is because no potential beneficiary will want to commit to bear part of the cost until all potential consortium members have agreed on their shares. They will all have an incentive to delay reaching agreement for as long as possible in the hope of minimizing the costs they bear.

Economies of scale, lumpiness, and the need for advance construction in order to meet anticipated future demand, in effect, create common costs. In the past, these costs were recovered from all transmission users in proportion to the MW of transmission service they took, largely because no individual user or group of users would want to bear a disproportionate part of these common costs. It is necessary to find a group of customers or a mechanism to ensure that the common costs can be recovered. The key ingredient is the power to persuade or compel those who benefit to pay. Without this, utilities may be unwilling to expand their systems or investors to invest. This thinking has resulted in the belief (discussed in chapter 5) that projects involving significant economies of scale could only be undertaken by regulated utilities able to recover the common costs from all consumers, because the regulator has decided that they are for the public good.

If the total common cost is spread over all consumers, it may cause some marginal users to bypass the system, in which case the cost will be recovered from the remaining users. This will increase their total share of the cost and could create even more bypass. As a result, it is desirable to develop a mechanism to ensure that the common costs are allocated to consumers in a way that does not distort investment decisions. In other words, they should be charged to beneficiaries, if they can be identified. The methodology developed in Argentina described in chapter 6 attempts to identify the beneficiaries of the expansion that should be willing to bear a share of the costs equating to the benefits they receive. However, it is not necessarily the case that they will be the users most willing to bear the common cost. The so-called Ramsey pricing principles would indicate that these costs should be borne by those users and consumers representing the most inelastic part of the demand.[11]

Direct costs of transmission

The direct costs of transmission are essentially the capital, operating, and maintenance costs of the transmission business. The operating costs, however, will need to be reallocated as the industry is unbundled and operating functions are differently assigned, particularly if system control or system operations are separated from transmission asset ownership and management. System operations also are highly interrelated with the indirect costs of transmission.

Indirect costs of transmission

The traditional approach of regulators to transmission focused on the direct capital, operating, and maintenance costs and paid little attention to the indirect costs of transmission, notably congestion, transmission losses, certain ancillary services, and the so-called "must run" generation costs. These indirect costs are associated with system operations. They are costs the system operator necessarily incurs (and which the consumer will ultimately bear) in order to maintain security of supply (because demand and supply must balance in real time) and system reliability.

In simple terms, the cost of transmission losses is the cost of generating the energy that is lost in the process of transmission. The cost of congestion is caused by limits on the transfer capability of particular transmission lines and facilities (which operate like bottlenecks on a road network) that result in the system operator having to dispatch generation to serve load that may not be as economic (cheap) as generation located more remotely that could have run to meet the demand if the constraint or bottleneck had not occurred. Certain ancillary services and "must run" generation output are required in order to enable the system operator to keep the supply and demand in balance and to maintain system reliability. The costs of congestion and losses are the costs upon which the principles of marginal costing discussed earlier are based. They represent the short run marginal costs of transmission.

It is clear that the direct and indirect costs of transmission are closely interlinked and that there are trade-offs between them. For example, if transmission line maintenance is carried out using live line working that would increase the cost of the maintenance by comparison with conventional techniques, the line could remain in service, thereby reducing or avoiding the costs of congestion that would otherwise be created if the line were taken out of service for maintenance using conventional techniques. Another example relates to transmission losses and their relationship to capital investment. Certain types of technologies or methods of creating additional transmission capacity may be more expensive in terms of their capital cost, but may reduce losses.

Nature of congestion costs

Where there is a constraint that inhibits the transfer of power from a specified export point to a specified import point, more expensive generation may have to be run on the import side of the constraint if there is a shortage of low-cost generation able to serve load on the import side or a shortage of generation that is usable by the system operator to maintain reliability. Figure 2-2 gives an example of how costs can rise in the event of a constraint that results in the running of more expensive generation to meet load.

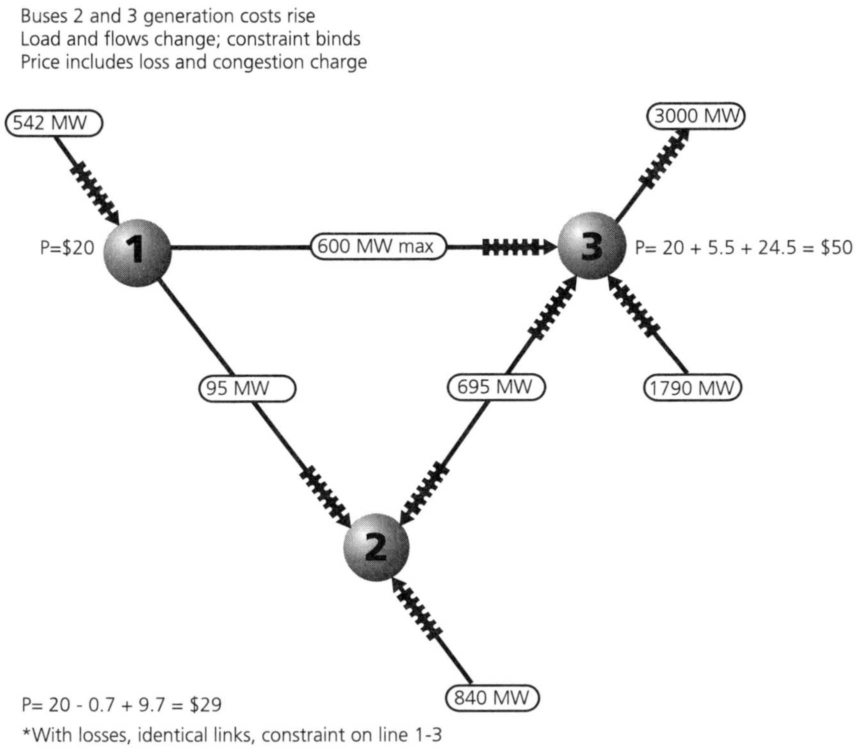

Fig. 2-2 Constraints with out-of-merit costs

The congestion will result in a difference in generation prices between the two locations and may impose a higher marginal cost in certain locations. Power will flow over the transmission line from a low-cost to a high-cost location. If lines are congested in periods of high demand, not all the power that could be generated in a low-cost region will be utilized, and some of the cheaper generation will be "constrained off." The higher-cost generation used to meet demand would be "constrained on", because, in the absence of the constraint, it would not run. The marginal cost in the two locations differs because of transmission congestion. The difference between these two costs—net of marginal transmission losses (see later discussion)—is the congestion rent.

In the case of a meshed network where the interactions between one part of the system and another are numerous, with loop flows and multiple contingencies affecting the thermal limits or voltage limits of the transmission facilities, the calculation of the difference in the marginal cost between two locations will be more complicated. A contingency is a foreseeable event such as the possible loss of a single infeed that will affect the ability of the system operator to maintain reliability. The system is constrained by the need to operate in a manner that will preserve reliable operation if the event occurs. If congestion is present on the network, the marginal costs are very likely to differ. It is easy to construct examples where congestion leads to marginal costs that differ by more than 100% across different locations.

If there were no constraint or congestion, any lower-cost power generated on one side of the transmission path or interface could be sold at the higher price on the other side of the path. The difference between these electricity prices is essentially the cost of transportation between the locations on either side of the potentially-constrained transmission path and reflects the cost of congestion and transmission losses. The "economic rent" (profit in excess of marginal cost) that this difference reflects would represent a windfall or bonus to generators able to utilize the path. This thinking has led to the design of transmission pricing called "locational" or "location-based" marginal pricing (LMP) that will recover these economic rents from those generators or wholesale sellers that are able to benefit from trade across a constrained interface.

Losses

Transmission losses are essentially losses of energy resulting from resistance in the wires that is encountered in the course of conducting electricity. The resistance generates heat in the wires that causes energy to be lost. Losses are proportional to the square of the power delivered to the load on a direct current (DC) line and approximately so in the case of an alternating current (AC) line. The greater the distance the electricity is transmitted and the nearer the line is loaded to its maximum capacity, the higher the losses. Therefore, losses are influenced by the location of generation and loads and the transactions they enter into. In the U.S., transmission and distribution losses can amount to 7-12% of energy produced.[12]

Losses are the equivalent of fuel in other transportation systems. They are part of the marginal cost of transporting power in the same way that gasoline is part of the marginal cost of road transportation. The system operator has to make up the energy lost in this way by dispatching additional generation to compensate for the losses. The marginal cost of delivering power to different locations differs at least by the marginal effect of losses on the system.

Losses can be accommodated in the market design by applying the principles of economic dispatch; the market equilibrium price would change accordingly. This approach would yield different marginal costs and different prices depending on the location, but the basic market design would be preserved. The replacement of losses is regarded as a cost of providing transmission.[13]

There are various loss management and allocation schemes including:

- loss factors (generator meter multipliers) applied to generators at their locations on the system that allocate the cost of the losses to them in the settlement or billing process (with the possibility of self-provision of energy to cover losses)
- nodal losses calculated every hour (as in Argentina), calculated on a year-ahead basis as in Sweden, or calculated six times a year as in Norway
- counter-trading (as in Finland) by the system operator, to purchase energy to cover the losses with the costs recovered from

all market participants in the form of an uplift or additional charge (rolled into the transmission charge) made to all wholesale buyers (at a single price per MWh)

- scaled marginal losses (scaled back so that the marginal losses equal the total average losses) recovered through an uplift or the transmission charges

The schemes should be designed on the principle that those who cause the losses because of their locations on the system should pay for them. This will send locational signals to generators and loads that will encourage efficient use of the transmission system. The technology utilized to expand the system can reduce losses (*e.g.*, by increasing conductivity).

Locational marginal pricing schemes can cover losses as well as congestion by including marginal losses in nodal prices. However, the average losses are approximately half of the marginal losses that are calculated under nodal prices. If marginal costs are increasing, then marginal and average costs cannot be the same. This means that an economic rent or profit in excess of marginal cost is unavoidably created by accounting for losses in nodal prices in the same way as congestion rent.[14] The rent can be used to defray fixed transmission costs and reduce access charges, in the same way as congestion rents.

The Effect of Regulation and Market Design on Transmission Systems

The effect of regulation

In theory, where an industry functions in accordance with perfect competition, there is little need for regulation. The perfect paradigm would consist of a large number of sellers and buyers. Buyers would be fully informed of the products, the prices, and their implications. Unfortunately, the theoretical, perfect world of competition is seldom present in the real world. Many industries are dominated by a small number of large firms, and in the electricity industry there will be natural monopolies such as the transmission and distribution systems, where it is economically efficient to allow only one provider to serve consumers. Consumers may not fully understand the consequences of their actions. Governments respond to these market failures by controlling behavior through price incentives (taxes) and regulation.

Regulation is designed to protect consumers served by monopolists with respect to prices and quality of service. Ideally, it should create most of the benefits for the consumer that would have been captured by market forces. Regulation attempts to induce behavior in monopolists that is similar to the behavior that would be induced if there were a market. Where a monopolist provides essential services that facilitate the working of a market (such as a transmission owner with respect to the energy market), the regulatory regime will need to reflect the industry structure as well as the design and state of development of the market.

Regulation has a significant impact on investment, and investors take the risks associated with regulation closely into account. Because of the natural monopoly characteristics of most transmission systems, regulation is seen as the necessary constraint within which transmission investment must operate. It is not something that can be ignored or resisted, but must be well understood and managed by investors. Most regulated transmission companies focus effort in maintaining good relations with their regulators, stay in strong regulatory compliance, and have staff skilled in regulatory affairs. The reason for this is that the regulatory regime will not simply touch on every factor that is taken into account in making the investment, but will probably control many of them. For example, the regulatory rules will affect the level of profits and the rate of return the investors can earn, the asset base and how it is valued, the allowed revenues and permitted prices, the investment, operations and maintenance costs that can be passed through to consumers, the cost of capital, and the gearing (debt: equity ratio) of the utility.

In the case of natural monopolies, the regulator may well appear to be defining or managing many aspects of the utility's business. The level of detail varies according to whether behavior is controlled through incentives or rules. In some cases, the regulation is so intrusive that it is described as micromanagement. It gives rise to concerns over what is called regulatory risk, namely, the risk that the regulator may exercise its powers and discretions or that the regulatory rules and incentives may be changed unpredictably, in a manner that affects investments adversely. The subject of regulation is discussed in more detail in chapters 6 and 7.

Regulatory incentives

The behavior of utilities to achieve specific outcomes may be controlled through enforceable rules or by means of economic incentives. Incentives are attractive, because they appear to overcome the inherent difficulties of creating both the rules and their means of enforcement. It is difficult to frame rules to meet all eventualities over time and to create both the commitment and flexibility that the parties need. If the rules are poorly drafted or designed, they can create opportunities for behavior by one party that unfairly disadvantages another. The greater the complexity of the rules, the greater the number and scope of these opportunities. Monopolists and market participants with market power will tend to have more ability to "game" the rules than consumers and small market participants. In other words, a utility may live by the letter rather than the spirit of the rule, but if there is a loophole or a defect in the way that the rules are framed, it will rapidly find and exploit all possibilities that are created.

Enforcement of regulatory rules

Finding an effective method of enforcement of regulatory rules that acts as a deterrent to the parties to break or ignore the rules is also difficult. Recourse to the courts or to arbitration can be slow and expensive. Fines and penalties can be difficult to set—at one extreme they can be so low that it is cheaper for the utility to pay them than to comply with the rule. At the other end of the spectrum, penalties may punish the parties unfairly and act as a deterrent to investment.

Difficult to establish incentives

Incentives essentially leave it to the utility to respond commercially to the incentive to behave in a manner that achieves the objective of the incentive. However, incentives are not without their problems. It is not easy to establish an incentive that provides an adequate reward to encourage investment at the same time as providing an incentive for operational efficiency and innovation, particularly with a view to minimizing the need for investment. It is not clear what standards of performance are available that will separate results attributable to good or bad management from those attributable to other factors. It may be attractive to punish firms that perform poorly with a reduced rate of return but punishing them in this way and lowering the rate of return, may affect the ability of firms to attract the funding for investment in the capital markets at favorable rates that they need to perform better.

Returns must be earned

Regulators that have allowed earnings well in excess of the cost of capital of the utility may in some cases have succeeded in bringing about the necessary expansion of capacity, but it is not necessarily the case that utilities have been offered the optimum amount of incentive for undertaking risky investments. Any restriction on aggregate earnings may have some immeasurable effect of discouraging efficient investments. If the incentive results in profits that, with the benefit of hindsight, are regarded as excessive by the regulator, they may be clawed back when the incentive is reviewed.

This regulatory behavior may discourage further investment. Alfred Khan explains it succinctly:

> *Merely permitting all regulated companies, as a matter of course, to earn rates in excess of the cost of capital does not supply the answer; there has to be some means of seeing to it that those supranormal returns are* earned, *some means, for example, of identifying the companies that have been unusually enterprising or efficient and offering higher profits to them while denying them to others.*[15]

The Averch–Johnson effect

Incentives to encourage investment can give rise to distortions that produce inefficient results. Harvey Averch, Leland Johnson, and Stanislaw Wellisz, among others, have pointed out that a utility may be induced by an incentive to make investments, the social benefits of which fall short of their social costs. This is because the investments would expand the asset base on which the utility is entitled to a rate of return in excess of its cost of capital. If the net revenues that are generated by incremental investments fall short of yielding the allowed rate of return, the utility can recover any shortfall by raising its rates or prices. As a result, the utility may adopt an excessively capital-intensive technology and take on additional business, if necessary, at unremunerative rates.[16]

Regulatory incentives have been successful

By appealing to the profit motive to achieve the desired outcomes, regulators have been able to devise explicit incentives for dynamic and efficient performance in an attempt to induce managements to invest and to overcome the inertia of monopoly and, indeed, of regulation itself. Incentives were already present in the form of "regulatory lag" (meaning the periods between intervention, approval, or review by the regulator) that enabled the regulated firm to keep any cost savings or productivity gains it was able to make during the periods between regulatory reviews or approvals. Incentives also took the form of the legally required "reasonableness" or "fairness" of the rate of return.

Economists had observed that regulatory lag tended to offer the same sort of automatic stimuli as were evident in the unregulated sectors of the economy where market forces were evident. The guaranteed reasonableness of the rate of return tended to stimulate investment. These observations led to proposals that these devices be institutionalized and explicitly adopted as regulatory policy to provide certainty to the utilities that they could expect to receive the rewards and suffer the penalties these devices provided.[17] Despite views to the contrary,[18] the institutionalized regulatory approach to review periods and allowing the utility to earn and, indeed, increase what was regarded as a reasonable rate of return led to the adoption of an approach to regulation that is now described as performance-based or incentive regulation. Chapters 7 and 8 discuss in more detail the optimal design of regulatory incentive schemes and performance standards.

Allocation of responsibilities and liabilities

In an unbundled electricity industry characterized by new players such as ISOs, transcos, transmission owners, transmission administrators, generators, and retailers, the allocation of functions, rights, obligations, and risks has not been considered from an investment perspective. It is important to ensure that signals and incentives that are intended to be created are not destroyed by an inappropriate allocation of cost, risk, and powers to invest. It would be unfortunate if those who are responsible for or bear the costs of transmission (notably the indirect costs) are not in a position to propose or invest in the needed expansion to reduce those costs and vice versa.

The responsibilities of the system operator to manage the indirect costs of transmission or to plan expansion may not be accompanied by the necessary expropriation or eminent domain powers to obtain the necessary rights-of-way or the financial ability to invest in order to expand the system. Conversely, the utilities, transmission owners, and market participants that are in a position to invest in transmission facilities may not be responsible for some or all of these indirect costs. As a result, they will face no signals to invest and will have little or no incentive to expand the system.

It is important in creating incentives for efficient expansion to ensure that the costs and benefits of the investment should both be borne and enjoyed by the same entity or class of industry participants and that the cost of denying access to the transmission system is borne by the same entity as the cost of providing access. Sally Hunt and Graham Shuttleworth point out that:

> *There is nothing wrong with not providing access: It may be cheaper to reimburse for lost profits than to put in a new line. ... it may be cheaper to pay for access denied than to pay overtime to get a transmission line back into service quickly.*[19]

The responsibilities and powers of any of the parties to take action should not be separated from the liability for the consequences of that action. In other words, liability for an act for omission should obviously fall on the party responsible for it. A party on whom liability falls will naturally be resistant to making further investment in an environment of an industry structure or regulatory or market rules that simply opens up additional exposure to liability for actions for which it has little or no responsibility or control. This is a particular problem in cases where there is a system of penalties or reductions in income for transmission outages or reductions in capacity.

Locational signals and incentives

As we have observed already, it is important to create incentives to ensure that the existing transmission system is used as efficiently as possible. Therefore, it is socially beneficial to develop a set of locational price signals and incentives that encourage new generation and loads to connect at points on the transmission system where there is relatively light utilization and spare capacity. This will reduce the need for new capacity for which the consumer will ultimately have to pay. In the context of a competitive electricity market, locational price differences define the opportunity cost of short-term transmission, which, as we saw earlier in this chapter, will influence investment decisions.

In the past, the costs of transmission may have been shared equally between all network users, regardless of their location on the system (perhaps as a matter of public policy). This is referred to as "postage stamp" pricing and it allows users to use the entire interconnected system at a single $/kWh price. Where the system belongs to multiple transmission owners, they will each levy their own charges on users of their assets. A user wishing to conclude a transaction across more than one system will have to pay more than one charge; hence, the adoption in the U.S. of the expression "pancaking," because it involves layering one charge on top of another. The uniform cost allocation inherent in postage stamp pricing is distinct from "license plate" access charges that require users to pay the charges of the local transmission owner that serves them, but then enables the users to use the assets of other transmission owners to gain access to the entire interconnected system.

Whether or not competition is introduced, it is important to make sure that prices are aligned with cost causation to the greatest possible extent, so that those who cause costs or utilize the system most bear their fair share of the costs. This will ensure that the market works efficiently and that the economic signals and incentives in the markets elicit the desired outcomes for society.[20]

If a new generator or load can be provided with incentives to locate at a point on the system that opposes the dominant power flow, congestion and the need for transmission expansion could actually be reduced, as could transmission losses. Not only could the need for new investment be reduced, but also the cost imposed on the system could be negative. Generators and loads could be paid to locate at points on the system that were beneficial to the system as a whole and, therefore, to all users.

Congestion pricing

Congestion pricing is used as a mechanism to allocate limited transfer capabilities over constrained interfaces to those market participants that value the ability to transfer power across the interface most highly. It represents a market-based approach using locational marginal pricing. It could be regarded as a scheme involving the payment by a market participant of a "bottleneck fee" for the use of a congested interface. The fee is usually referred to as a transmission usage charge and is equal to the energy price difference between the two ends of the transaction (nodes). This ensures that the system is used efficiently, because prices increase as the demand for electricity transmission increases (as in most markets). By contrast, the "postage stamp" and "license-plate" pricing that we have discussed do not take into account congestion or constraints on transmission systems.

Both postage stamp and license-plate pricing will result in allocation and use of transmission capacity that are likely to be inefficient. Available capacity will not necessarily be allocated to the user that places the highest value on the transfer capability of the constrained path or interface. The beauty of congestion pricing is that it properly integrates the energy market with the allocation and pricing of scarce transmission. These issues are discussed in more detail in chapter 4.

Fred Schweppe et al. created a mechanism to set an energy price for each location that also accounts for transmission congestion.[21] This mechanism was accompanied by a system of financial property rights that could, among other things, provide hedges against congestion costs.[22] The mechanism involves economic dispatch based on the least-cost equilibrium or market clearing prices subject to the usual system security limits or constraints. Generators would offer their minimum acceptable prices at their locations. Consumers could also bid, with dispatchable demand and price bids used to set the maximum prices that would be paid at the consumer's location. The security-constrained economic dispatch process would produce the corresponding prices at each location, incorporating the combined effect of generation, congestion, and losses. Market participants would see a single price, which would be the short-run marginal cost price of power at their own location. The difference in locational spot prices defines the opportunity cost of short-term transmission usage.

Where locational marginal pricing is in place, a power purchase agreement will not be sufficient to provide a complete long-term price hedge. Transmission congestion can produce significant movements in price that are different depending on location. If sellers and buyers are located far from each other, transmission congestion might confront the buyer with a high locational price and leave the seller with a low locational price. The seller needs something more to provide the natural back-to-back hedge on fluctuations of the short-run marginal price.

In the presence of congestion, revenues collected from buyers will substantially exceed the payments that need to be made to sellers. The difference is the congestion rent that arises because of constraints on the transmission system. If the system operator were allowed to keep the congestion revenue, it would create incentives for it to manipulate the dispatch and prevent transmission expansion in order to generate even greater congestion rents.

A system of long-run transmission congestion contracts operating in parallel with long-run generation contracts provides a solution to both these problems, namely, providing a price hedge against locational congestion differentials and removing the incentive for the system operator to distort both dispatch and expansion decisions.

Under a locational pricing system incorporating transmission congestion contracts, market participants will pay locational prices that include transmission opportunity costs. The transmission congestion contract holders receive the congestion payments, and the system operator keeps nothing. The transmission owner receives the regulated fixed charges for the existing system—the holder of the transmission congestion contract receives compensation for the additional cost that it will have incurred because of the congestion—the user pays the true opportunity cost.

Congestion management in the absence of congestion pricing

There are three main congestion management techniques used in the absence of congestion pricing:

- centrally-administered redispatch of more expensive generation to meet demand where the costs are charged at a single price per MWh to all users
- market splitting (where the constraint is used to split the market into separate markets on either side of the constraint)
- transmission loading relief (TLR) defined by the North American Electricity Reliability Council (NERC), which consists of a complex set of prescribed curtailment and other actions depending upon system conditions (broadly speaking, the more serious the system condition, the more likely and extensive the required curtailment of the intended transaction)

The main issue is the allocation of the associated costs and the resulting impacts on the efficiency of the market and the use of the transmission system. It is unlikely that any incentives to use the system efficiently will be created if the costs are socialized or rolled in (*i.e.*, a single price per kWh is charged to all users). If the costs are paid on a user-specific basis by reference to the use of the system or the value that the user places on a congested path, the signals and incentives for efficient use are bound to be stronger. Market splitting may increase volatility of energy prices, because the resultant markets will be smaller and less liquid. It may also create opportunities for the exercise of market power. The TLR procedures are widely regarded as artificial, complex, and unsatisfactory.[23]

Special considerations with respect to interconnector expansion between markets

A profit-motivated organization may detect an attractive opportunity to build an interconnector between different markets or systems in order to profit from the opportunity to arbitrage the price differences between different markets. It may invest for its own account or in joint venture with others, such as generators who see the opportunity to export low-cost power to a higher-price market or distributors wishing to hedge the volatility of their wholesale energy costs. The market conditions may be so compelling that, even in the short term, it is economic to invest in order to exploit the price differences or to sell physical or financial transmission rights to the interconnector to a variety of different market participants.

Different interests of industry participants

It is important, however, to analyze the different interest incentives that system operators, transmission owners, and market participants may face with respect to transmission or interconnection expansion between different markets or regions. For example, some market participants could be keen to see an increase in interconnector capacity between neighboring regions that would facilitate additional energy exports and imports. Generators in low cost regions and loads in high cost regions will have similar interests, and vice versa. The system operator might like to see the increase in interconnector capacity to assist it in maintaining system reliability. A transmission owner would see any interconnector that it might build as a profit opportunity. However, it is likely that the rights to the link would be more valuable to certain participants at one end of the interconnector at certain times. For competitive reasons, market participants and transmission owners at the other end of the interconnector or line might have concerns over the increased competition in their own market that the expansion might facilitate. These different interests have resulted in delays in developing interconnector projects and creating the necessary investor coalitions and consortia.

Regulated and market-based development

Interconnectors, like transmission lines, can be developed on a market-based or a regulated basis. In Australia, in principle, it is easier to develop an interconnector on a market basis, because it is not necessary to go through a formal regulatory test to justify the costs and benefits to the market and ultimately to those users and consumers who will be forced to pay for it if it is implemented on a regulated basis. The result has been parallel development of interconnectors on both bases, as we shall discuss in chapter 5.[24]

Minimizing barriers to interconnector development

It is important for the regional policy setters such as the European Commission (EC) or the Federal Energy Regulatory Commission (FERC), as well as local regulators, to consider carefully whether markets should be organized and transmission systems regulated to create incentives for or, at the very least, to minimize barriers to interconnector development. The lack of interconnector capacity may be a barrier to the type of wider regional trading that these regional or federal organizations hope to encourage.

Need for regulatory coordination

Regulators should not discourage market-based interconnection expansion, and indeed would do well to allow incumbent, regulated transmission owners to develop interconnectors on a merchant basis. In this way, they could achieve the needed development of interconnector capacity more quickly and efficiently. One of the reasons for this is that if the interconnector is built between two jurisdictions on a regulated basis with different regulators, regulatory regimes, and legal systems, it could take time for the regulators to develop a coordinated approach. This would be necessary to avoid regulating the same asset twice under different rules. Where the interconnector investor is allowed to charge market-based prices, the regulatory oversight and coordination issues become simpler. The regulators on either side of the border should have the same interests in focusing only on anti-competitive practices, non-discriminatory access, and abuse of market power (although regulators around the world have shown a remarkable diversity in anti-trust or competition law enforcement and protecting the national champions).

When Will Investment Occur?

In drawing together these basic principles, we need to answer the question "so, when will investment occur?" If there is one over-arching principle, it is that a decision to expand a transmission system will be taken when there is a clear signal that the marginal benefits (reduction in the costs of congestion and losses) are greater than the marginal cost of building new capacity.

However, there are a number of other conditions that may well apply:

- Investment will occur if the cost of capital for the investment is less than the return on the investment (over a reasonable payback period, less than the life of the asset)
- The value of additional power to be transported should be greater than the cost of the line
- Investment in transmission must be more cost-effective than transporting the fuel to a generating station located near a load center.

Investment will be less likely to occur if:

- there is a fear the investment will be stranded (*i.e.,* non-revenue earning) at a point in the future that is earlier than the point at which the return on the asset will exceed the cost of capital to build it
- there is uncertainty as to the market design and the changes to the market rules that may be made as the market settles down, particularly in the areas of congestion and loss management and transmission pricing
- the regulatory risk is perceived to be significant, either because of the need to redesign the regulatory regime or because there is a lack of credible commitment on the part of the regulator to stable regulatory rules and not to exercise its discretion to make changes unpredictably

- there is insufficient transparency and presence of adverse bias (*i.e.*, the lack of a level playing field)
- generally, there is insufficient certainty that the investment will be adequately rewarded and that the associated risks can effectively be managed

Finally, it is useful to remember that the goal of transmission expansion is not to eliminate congestion costs altogether, because it will be cheaper to tolerate a certain amount of congestion than to eliminate it completely. For the purposes of creating recipes for success in transmission expansion, our goal is to overcome the obstacles to transmission investment rather than to create an infinite system. However, congestion costs do provide signals as to the need for new transmission facilities, and, as we saw in chapter 1, congestion costs can explode quite suddenly, at the expense of electricity consumers.

NOTES

1. For a readable account of these general principles and their application, see Richard A. Brealey and Stewart C. Myers, *Principles of Corporate Finance* (International Edition: McGraw Hill, 2000), chapter 5. The authors regard the NPV methodology as superior to other analytical approaches to valuing investment opportunities, including the internal (or discounted cash-flow) rate of return. They do not think that there is a truly satisfactory way of defining the true rate of return of a long-lived asset, but that the best available concept is the internal rate of return (which is defined as the rate of discount that makes NPV zero). They think that it can also be a misleading measure unless it is calculated and used properly

2. Richard A. Brealey and Stewart C. Myers, *Principles of Corporate Finance* (International Edition: McGraw Hill, 2000), chapter 7

3. Richard A. Brealey and Stewart C. Myers, *Principles of Corporate Finance* (International Edition: McGraw Hill, 2000), chapter 10

4. F. Modigliani and M. H. Miller, "The Cost of Capital, Corporation Finance and the Theory of Investment," *American Economic Review* 48 (June 1958), pp. 261-297

5. Richard A. Brealey and Stewart C. Myers, *Principles of Corporate Finance* (International Edition: McGraw Hill, 2000), chapter 19. This contains a description of how to use the APV and WACC formulae and the issues that need to be taken into account in applying them to an NPV calculation in order to take into account what they describe as financing "side-effects." At the end of the summary on p. 572, they provide this explanation: "This chapter is almost 100 percent theory. The theory is difficult. If you think you understand all the formulas, assumptions, and relationships on the first reading, we suggest psychiatric assistance. We can, however, offer one simple, bullet-proof, easy-to-remember rule: Discount safe, nominal cash flow at the after-tax borrowing rate"

6. For a clear account of loop flow effects, see Timothy J. Brennan, Karen L. Palmer, Raymond J. Kopp, Alan J. Krupnick, Vito Stagliano, and Dallas Burtraw, *A Shock To The System: Restructuring America's Electricity Industry* (Washington, DC: Resources for the Future, 1996)

7. Harold Hotelling, "The General Welfare in Relation to Problems of Taxation and of Railway Utility Rates," *Econometrica* 6 (1938), pp. 242-269. See also Alfred E. Kahn, *The Economics of Regulation: Principles and Institutions* (Cambridge, Massachusetts: The MIT Press, 1988), vol. I, chapter 3

8. William W. Hogan, "Electricity Transmission and Emerging Competition," John F. Kennedy School of Government, Harvard University, prepared for the Public Utility Research Center Annual Conference (April 27, 1995). Explicit pricing of transmission can be ignored if the system operator can signal the economic value of electricity at every node on the network. Available on Professor Hogan's website, **www.whogan.com**

9. Ross Baldick and Edward Kahn, "Network Costs and the Regulation of Wholesale Competition in Electric Power," *Journal of Regulatory Economics* (1993), vol. 5, pp. 37-38

10. Eric Hirst and Brendan Kirby, "Key Transmission Planning Issues," *The Electricity Journal* (October 2001), p. 61

11. Frank Ramsey, "A Contribution to the Theory of Taxation," *Economic Journal* (March 1927). Marcel Boiteux, "On the Management of Public Monopolies Subject To Budgetary Constraints," *Journal of Economic Theory* 3 (1970), pp. 219-240 (first published in French in 1956). See also W. Vip Viscusi, John M. Vernon, and Joseph E. Harrington, Jr., *Economics of Regulation and Antitrust* (Cambridge, Massachusetts: The MIT Press, 1998), pp. 365-372

12. Source: U.S. Department of Energy, Energy Information Association, "Mitigating Greenhouse Gas Emissions: Voluntary Reporting" (October 1997)

13. See Steven Stoft, *Power System Economics: Designing Markets for Electricity* Piscataway, New Jersey 0 8855-1331: IEEE Press/John Wiley & Sons Inc. (2002)

14. For a more detailed explanation refer to the information booklet from Transpower New Zealand Limited, "Transmission Rentals" (December 1999). See also Scott M. Harvey and William W. Hogan, "Loss Hedging Financial Transmission Rights" (January 15 2002), available on Professor Hogan's website, **www.whogan.com**

15. Alfred E. Kahn, *The Economics of Regulation: Principles and Institutions* (Cambridge, Massachusetts: The MIT Press, 1988), vol. I, pp. 53-54

16. Harvey Averch and Leland L Johnson, "Behavior of the Firm under Regulatory Constraint," *American Economic Review* LII (December 1962), pp. 1,052-1,069. Stanislaw Welliz, "Regulation of Natural Gas Pipeline Companies: An Economic Analysis," *Journal of Political Economy* LXXI (February 1963), pp. 30-43

17. William J. Baumol, "Reasonable Rules for Rate Regulation: Plausible Policies for an Imperfect World," in Almarin Phillips and Oliver E. Williamson, *Prices: Issues in Theory, Practice and Public Policy* (Philadelphia: University of Pennsylvania Press, 1967), pp. 108-123. Stephen J. Littlechild, "The Regulation of British Telecommunications' Profitability," Report to the Department of Trade and Industry (1983)

18. Alfred E. Kahn, *The Economics of Regulation: Principles and Institutions* (Cambridge, Massachusetts: The MIT Press, 1988), vol. II, pp. 59-60

19. Sally Hunt and Graham Shuttleworth, *Competition and Choice in Electricity* (London: John Wiley & Sons, 1996), p. 230

20. Sally Hunt and Graham Shuttleworth, *Competition and Choice in Electricity* (London: John Wiley & Sons, 1996), pp. 228-229

21. Fred C. Schweppe, Michael C. Caramanis, Richard D. Tabors, and Roger E. Bohn, *Spot Pricing of Electricity* (Norwell, Massachusetts: Kluwer Academic Publisher, 1988)

22. William W. Hogan, "Contract Networks for Electric Power Transmission," *Journal of Regulatory Economics* 4 (1992), pp. 211-242 available on Professor Hogan's website, **www.whogan.com**

23. William W. Hogan, "Competitive Electricity Markets and Transmission Line Loading Relief," presentation to the Harvard Electricity Policy Group (September 10-11, 1998) available on Professor Hogan's website, **www.whogan.com**

24. The issues at the interface of regulated and market-based transmission project development are discussed in chapter 5

CHAPTER 3

The Store Cupboard, the Cooks, and the Oven

Ingredients of this Chapter

We must start by looking in the store cupboard and the fridge before we go shopping for new transmission facilities. We must also think about who the cooks will be and the oven we will use. In this chapter, we will be looking at the starting points of the transmission expansion process. Firstly, we must look at what potential there is to use the existing assets more effectively and flexibly to maximize throughput and increase the transfer capability of the key parts of the transmission system without significant investment. We need to think about the industry participants and the potential investors, the industry structure, how the responsibilities are allocated, and the rights and obligations of the players. Finally, we must consider the environment in which transmission expansion is to take place. We must consider whether the expansion will be market-based or whether it will be developed in the traditional regulated environment.

Getting More Out of the System to Minimize Investment

Spare capacity may be hidden in existing assets

At an early stage, we should look at the contents of the store cupboard before going out to buy new transmission lines. There may be spare capacity hidden in the existing assets that can be unlocked by a modest investment in new maintenance practices and technologies or the upgrading of existing facilities. It is clearly beneficial to society to implement schemes that result in more utilizable capacity being created on an existing network from which users would benefit, without incurring the costs and delays of new construction, particularly where the creation of a new route or corridor might be involved.

Traditional approaches have not worked

Traditional operational planning and outage coordination procedures may not have been geared, in the past, to maximizing transmission system availability, capability, or flexibility. Nor is it likely that the traditional regulatory regimes created any incentives to do so. In a deregulated environment there may be perverse incentives. A system operator may be reluctant to give releases and clearances for maintenance in an active market where the transmission capacity is barely adequate, because the need to minimize congestion and facilitate trading in the market will be regarded as crucial. As a result, the system is placed at greater risk of forced outages because of lack of maintenance or overloading. Argentina has been an example of this.

Maintenance standards create weak incentives

Some countries have put in place either broad-based or very detailed maintenance standards in an effort to maximize availability. The detailed approach runs the danger of inducing a minimum approach on the part of the transmission owner, which will focus only on achieving the specified standards, rather than an overall increase in availability of the key circuits and flexibility.[1] Moreover, this approach may result in a transfer of risk to the standard setter, if the standards prove to be inadequate for the purpose. In other words, if the transmission owner meets the standards, but the availability is still inadequate, it cannot be blamed.

Experience has shown that the standards may be easier to set where new technology is used and where the time required for maintenance can be predicted with more certainty. However, establishing the likely level of forced outages on an older system could be quite difficult, as could the time for replacement and repair because historic performance may be no guide where an asset could suddenly break down. As we will observe in chapter 8, broad economic standards reflecting cost targets work better than detailed technical standards.

Place the risk with the transmission owner

It is better to place the whole of the maintenance risk with the transmission owner, which is in the best position to manage it. However, the transmission owner may have no reason to make an effort to maximize availability, especially if it is not subject to specific performance incentives in this area and it is simply reimbursed its actual maintenance costs. In this case, it might well have the perverse incentive to save on staffing and repair costs and even bring circuits back from maintenance later than planned (as was the case in Alberta for a while), unless there are penalties imposed upon it in these circumstances. However, carrots work better than sticks. The transmission owner needs to be rewarded to make a special effort to manage the maintenance risk and invest in its system in order to maximize availability and flexibility. The cost of doing so will be very small by comparison to the benefits to consumers in terms of cost savings with respect to congestion and expansion.

Incentives also need to be put in place to ensure that, to the extent possible, outages are taken at times when the value of transmission transfer capability is at its lowest (*e.g.*, at the weekends) and to ensure that maintenance is focused on the key circuits that have the greatest value to market participants. However, this involves input from the system operator.

Split responsibilities

The roles, responsibilities, and risks with respect to maintenance and planning may have been split between an ISO, a transmission owner, and possibly also a transmission administrator (as in Alberta). To create the necessary incentives, it is important that the liability (if any) for lack of circuit availability, outages, and poor maintenance (which may take the form of compensation to those market participants that did not receive service) are properly aligned with the responsibilities for these activities and risks. Moreover, liability may fall on the transmission owner, notwithstanding that certain factors are beyond its control, perhaps because they are the responsibility of the ISO or transmission administrator. The transmission owners in Argentina have suffered in this way.

Benefits of incentive schemes

With the exception of the transmission services (system operator) incentive schemes (TSS) that have applied to National Grid Company plc (NGC) in England and Wales (described in more detail in the supplement to chapter 7), there have been very few incentive schemes that have been created, either in regulatory regimes or through market rules to maximize the availability, capability, and flexibility of the most used circuits and facilities of the transmission system. The result of an incentive scheme could be relatively cheap and quick enhancements to create needed transmission capacity that has the highest value to industry participants and consumers. Giving a transmission owner more risk and more reward in this way will derive a benefit to the consumer that far exceeds the cost.

NGC response to the TSS

The TSS (typically set for a two-year period) has created an incentive to reduce certain elements (notably congestion, balancing, and certain ancillary services) of a market price "uplift" that is charged to all wholesale buyers by setting a cost target. It is rewarded for beating the target and suffers a financial loss for failing to do so. NGC responded to the incentive scheme by adopting a whole new approach to transmission maintenance and expansion. The approach involved minor, but efficient upgrades of various transmission assets, particularly those that were key to the efficient working of the competitive wholesale market. From a maintenance perspective, there were better communications and control, better monitoring of the system, and live-line working. Live-line working is much used in England and Wales and even more so in Argentina, where NGC has a major transmission investment.

NGC claims that it has increased the capacity of its network by at least 22% since 1994 in this way and reduced system operations costs by improving business processes, spending on some operating costs to save other costs, and investing generally in minor capital expenditure schemes. For example, it:

- introduced multidisciplinary team working, based maintenance planning on asset condition rather than on fixed maintenance cycles, and changed the management culture to focus on reducing the costs to gain the benefits
- invested in accelerated maintenance including live-line working, double and triple shifts, working weekdays and weekends when economic
- developed software to analyze the drivers of the costs that it was incentivized to reduce
- installed intertripping schemes and on-line monitoring of transmission equipment to optimize plant ratings (*e.g.,* by monitoring actual cable temperatures)
- developed both generation and demand reduction contracts to mitigate bidding against constraints
- reviewed the risk assessment of the ratings of transmission equipment to increase transmission capacity in strategic areas of the system where possible
- sometimes advanced investments in the transmission system and undertook investments that were above the minimum required by the traditional planning standards, in order to reduce costs, and undertook investments such as reconductoring, reten-

sioning and regreasing lines, and installing static VAR compensators, capacitators, power flow stabilizers, quadrature boosters, gap conductors, and flexible alternating current (AC) transmission system (FACTS) devices to increase system capacity[2]
- used a modular construction technique that allowed voltage support equipment to be moved quickly and economically from one location to another to meet the changing needs of the system

Generally, the benefits arose from reductions in transmission constraints, but sometimes it proved possible to hold reserve more cheaply. There were, in effect, economies of scope between system operations and transmission asset management. As Nick Winser explained to FERC: "In short ...it's aggressive ...And it works."[3] Figure 3-1 needs no explanation.

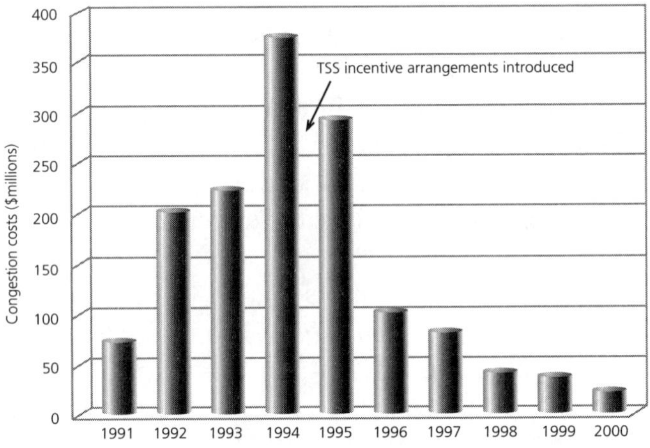

Fig. 3-1 Congestion costs in England and Wales

The TSS incentive arrangements essentially apply to a variety of aspects of system and market operations. They can be distinguished from the price cap (RPI–X) regulatory incentive scheme described in more detail in chapter 7 that applies to the direct costs of transmission. If the indirect costs (particularly congestion costs) are not subject to a mechanism designed to incentivize the transmission owner or system operator to minimize them, perverse incentives may be created that lead to an increase in indirect costs.

Incentives in market rules

The market design can include performance-based incentives to encourage efficient levels of maintenance and circuit availability by a transmission owner. In a balanced performance incentive, the transmission owner could be rewarded by a higher return (if it met superior maintenance and availability targets) and be penalized if it met lower targets.

In markets where financial transmission rights (FTR) are in place, the penalty would be set at a level sufficient to fund the shortfall in the revenues necessary to pay congestion rents to the holders of the outstanding FTRs, in the event that transmission outages were attributable to the transmission owner and resulted in the system operator collecting insufficient congestion revenues to fund the FTRs fully. This would give the transmission owner a market-driven incentive to maximize availability at times and in ways that would minimize the impact on congestion costs. We shall return to this in chapter 4.

The Industry Structure

Another starting point will be the chosen structure of the industry. There are many jurisdictions that still retain vertically integrated utilities that may or may not be state-owned. The diversity of industry structures that are in place around the world is enormous. It is hard to find two that are the same. The allocation of functions and responsibilities to similar business units or departments of a vertically integrated utility and to the various separated entities in a restructured industry is quite different.

System operators

Industries that remain vertically integrated often have system operations divisions or units that report to the chief operating officer, but are otherwise separate or distinct from the rest of the functions of the utility. When the industry is unbundled, the system operations staff and functions may be transferred to a separate division of the utility with independent management only, which is all that the European Union (EU) internal market in electricity (IME) directive (96/92/EC) originally required.[4] However, they may be transferred to an entirely separate company that may or may not own transmission assets. Indeed, as we shall observe in the supplement to chapter 5, the draft amendments to the IME directive would require full corporate separation of the transmission system operator (TSO) and the distribution system operator (DSO).[5] Many economists would go much further. As Sally Hurt put it, "Transmission must be separate from generation."[6] The corporate unbundling of an ISO from a utility that continues to own both generation and transmission will not deal with the potential exercise of market power adequately.

Unbundling usually signals the introduction of a competitive market when the independence of the system operator becomes a key ingredient. The industry structure must not allow any incentive to be created for the system operator to favor affiliated generation, distribution, and retail businesses. The system operator should not have any interest in the trading of energy in the market or the use of the transmission and distribution facilities. Access to these facilities must be made available on a completely nondiscriminatory basis. However, NGC (a transco, *i.e.*, combined system operator and transmission owner) has an interest in the trading in the England and Wales market in order to balance the system and to maintain reliability, but this is carefully regulated.

Independence of system operators

The necessary degree of independence of the system operator is usually achieved through a combination of a carefully constructed governance structure,[7] lack of electricity industry ownership interests on the part of the management and staff, and corporate separation to avoid its common ownership with other market participants. The question then arises as to who should own the system operator, because it is, essentially, there for a neutral purpose to serve the needs of market participants, consumers, and the public good. The most common solution, where the system operator is not to be combined with a transmission owner, is to avoid the question entirely and use a not-for-profit vehicle (usually a corporate entity) that has no shareholders. The character of each of these entities varies as it will depend upon local company and tax laws as to what vehicle is most suitable. The system operator usually recovers its costs without making a profit, under a budgeting exercise, translated into annual charges that are adjusted periodically to reflect operational outturn.

A blueprint for a system operator has been lacking

There is no universal blueprint for the role and responsibilities of the system operator.[8] Many of them are also market administrators and some are not. Most are transmission administrators that are responsible for long-term and operational planning up to and including the day-ahead stage. Some system operators simply coordinate long-term planning, leaving the detail to transmission owners or developers. Some are responsible for specifying the maintenance of the transmission system (not just giving clearance to take facilities out of service). In the Republic of Ireland, the ISO even has a "client engineer" role to supervise the implementation of transmission projects.

The acronym "ISO" means an independent system operator of the type we have discussed, and an "RTO" is a regional transmission organization (a term primarily used in the U.S.) that may be a system operator or a combined system operator and transmission company (transco). FERC also uses the expression "independent transmission provider" (ITP), which has another definition that is explained later (see Glossary), which could also be an ISO or a transco.

Transmission administrator

Alberta has had a transmission administrator that was separate from the system operator (the Pool). The dispatch and real-time functions were carried out by the Pool. The tariff and rule management function, transmission planning, procurement of ancillary services, and outage coordination for transmission maintenance were all handled by the transmission administrator, which acted as an interface with the transmission owners. The transmission administrator was hired under a management contract to carry out these functions.

Transmission owners

In some jurisdictions, such as England and Wales, Norway, Sweden, Finland, and New Zealand, there are transmission companies that are combined system operators and transmission owners. In the U.S. they are referred to as "transcos." In other countries, such as the U.S., Australia, and Argentina, there are transmission companies that have been unbundled from generation, distribution, and retail sales functions and that have been also been separated from the system operations function. These companies are referred to as independent transmission companies (ITC), indicating that they are not affiliated with other industry participants or network service providers (NSP). They exist alongside and under the system control of an ISO.

NGC and Stattnet were the earliest example of transcos. The question of separating the system operations functions from transmission ownership has been debated in England and Wales, and the regulator has decided that, for the time being, it is content for NGC to continue with the combined functions. However, the Scottish system of wholesale pricing is likely to change, and a market for the whole of Great Britain will be created. The regulator has proposed that a single Scottish system operator be formed, as a joint venture between Scottish Power and NGC. It has also indicated that there would be advantages to having a single system operator for the whole of Great Britain. This would be likely to entail the separation of system operations from transmission ownership for NGC.

ISO vs. Transco debate

There has been a heated debate as to the relative merits of ISOs and transcos. ISOs have been criticized, because they are usually nonprofit organizations requiring complex governance mechanisms, involving stakeholder committees, independent directors or governors, or both.[9] In some cases, there have been difficulties in negotiating, performing, and enforcing the contracts under which the ISO can operate transmission facilities owned and used by others. Another criticism of ISOs has been that they inhibit the provision of "joined up" transmission service.[10] It is provided partly by the ISO and partly by the transmission owner, and there may not be clarity as to the responsibility of either to expand the system or to operate efficiently (although this should not be the case if the market rules and agreements are well thought through).

ISO lacks balance sheet strength

An ISO is unlikely to have any funds available with which to meet its liabilities, without including a contingency or special fund in its budget that would have to be paid for by market participants. On the face of it, this might act as a disincentive to comply with the market and technical rules to the fullest possible extent. One of the major concerns has been over the lack of balance sheet strength of the ISO. An ISO would not easily be able to raise finance for transmission investment as easily or cheaply as a transco. A not-for-profit entity would not be suitable for this because investors would be looking to make a profit or return on their investment.

ISOs are easier to put in place

In practice, there is good experience of ISOs. They have many advantages in their own right. They can provide very professional, focused, and efficient service in terms of both system operations and market administration. Their establishment causes the least asset-ownership disturbance. This means that they gain acceptance more easily and are simpler to put in place than transcos, where complex tax and financing issues arise. Moreover, it is clear that they can add value, through their public interest focus and absence of conflict of interest, in many different ways. PJM Interconnection LLC is a leading example of an efficient and innovative ISO that adds value both to its members and to society. Its mission is as an information company, but it provides much more than an abundance of information on which a market will flourish. It acts as a consensus builder, an anticipator of market needs, and an innovator.

Potential efficiency losses

The issue of whether an ISO can be as efficient as a transco is now being put into perspective. As Paul Joskow explains:

> *There are good reasons to believe that there are potential efficiency losses (i.e. higher costs) associated with separating ownership of transmission assets from the responsibility of operating these assets. Indeed, this kind of separation, ownership and control is extremely unusual in any other sector of the economy. Moreover, the separation of ownership from operations makes it more difficult to develop and apply good incentive regulatory mechanisms to the transmission owner since it divides decisions that affect the direct and indirect costs of transmission between two organisations.*[11]

Trade-offs between the direct and indirect costs of transmission

To recap, the indirect costs of transmission include the costs of congestion management transmission losses, some of the costs of ancillary services, and "must-run" generation. Inefficient generation location decisions and excessive cost and delays in connecting new generators to the system also need to be taken into account. These are the costs of reinforcing the transmission system that would not otherwise have been incurred had the generator located at a point on the system where there was adequate transmission capacity and reinforcements were either not necessary or where fewer reinforcements were needed. These indirect transmission costs can be significant if there are no incentives on the part of transmission owners or system operators to mitigate them, as we have already seen.

There are important trade-offs between the indirect costs of transmission and improving the capacity and performance of the transmission system (the direct cost). Where there is a separate ISO, there is a danger that it will not have the incentive or the ability to manage these trade-offs by a combination of system operations, maintenance, asset management, and investment techniques. In effect, there are economies of scope between system operations and transmission asset ownership and management that can be captured for the benefit of the consumer.

Advantages of the transco model

The transco model is admired for a number of reasons:

- It has the necessary information with respect to maintenance and management of the existing assets, the financial muscle, and it can be incentivized to expand the transmission network[12]
- There are operational efficiencies to be achieved in terms of asset management and operations by using the same entity to control, operate, maintain, manage, and expand the transmission system
- There has been operational success in utilizing transcos to manage and control efficiently both the direct and indirect costs of transmission

The Italian government has reversed its decision to create an ISO by merging Terna, the transmission-owning subsidiary of the Italian electricity utility (ENEL), with Gestore Rete Transmissione Nazionale, the ISO (with the agreement of ENEL).[13]

Conflicts of interest on the part of transcos

The main concern with transcos is that there is an inherent bias that may conflict with the independence that a system operator needs to carry out its functions effectively and on a totally impartial basis. In other words, a for-profit transco will have an inherent tendency to favor transmission investment solutions in carrying out its functions as a system operator. It may overlook better or more cost-effective non-transmission solutions. This bias may be apparent in both planning and operational time scales. For example, a transco may overlook generation or demand-side solutions that are more cost-effective than a transmission investment. It may prefer to install reactive capacitance on its system rather than to purchase reactive power from a generator more cheaply. Other potential conflicts with the system operations functions include coordinating transmission maintenance outages, defining available transfer capability, and expansion planning, where a transco may be inclined to favor its own transmission business over other transmission owners.

In practice, this bias has not proved to be as problematic as originally feared. Because of the high degree of scrutiny of system and market operations and transmission expansion plans by the regulator and market participants, any potential bias can be identified and deterred. In operational time scales, any bias may well be picked up through an operational audit or through the same regulatory and market participant scrutiny. However, there is no doubt that the bias can exist. It may not result in deliberate actions, but may find expression in what the transco does, simply because transmission solutions involving its own assets are what it knows best. This issue is obviously more of a concern in jurisdictions where there are a number of potentially competing transmission owners or where new entrants must be attracted to expand the system. In Norway, there are a number of transmission owners, but they have leased their facilities to Statnett, which is a transco for the entire system.

Boiling the arguments down

The arguments in favor of the ISO seemed to boil down to the ease of its implementation, professionalism, impartiality, and focus on doing a good job purely in the public interest. It is easy to police. There has been satisfaction with the performance of ISOs (*e.g.*, PJM undertakes a survey each year of what it terms "customer satisfaction" and has scored very highly). There has been distaste for a "black box" monopoly transco that is potentially difficult to control and requires strong regulation to deal with any tendency to favor transmission expansion solutions and its own profit motivation. The arguments in favor of transcos are that it will have the balance sheet strength to invest in transmission, and, because of its profit-motivation, incentives can be created for it to perform more efficiently, to increase the availability and flexibility of the existing system and to invest. By contrast, an ISO is given control over transmission assets owned by others and may have neither the power nor the financial incentive to ensure that they are efficiently used and not congested.

The scope for creating incentives in "not-for-profit" ISOs

Although not-for-profit ISOs are subject to budgetary constraints and close monitoring by market participants through stakeholder committees and independent boards, there may be less scope for creating effective incentives through a system of performance-related staff remuneration than is the case with for-profit companies that are always fully responsible to their shareholders for the financial consequences of their actions. However, for-profit private transmission monopolies do require and respond to regulatory incentives to achieve efficiencies and to constrain monopoly behavior. It is more difficult to create similar incentives to apply effectively to a not-for-profit, monopoly ISO. It does have budgetary constraints, but, ultimately, if it exceeds its budget, it has the ability to pass on all of its costs to market participants. Although incentives could be placed in the compensation arrangements for ISO management and staff to replicate the characteristics of incentive regulation applied to a private company with equity at risk, there is a limit to how much financial risk the management and staff will be prepared to take. This risk could deter applications for employment by ISOs of high-calibre candidates who would be in a position to work in a less risky environment elsewhere. In any event, there is an argument that the incentives on the ISO (*e.g.,* to deliver the required regulatory or policy outcomes) should be different from those on the management and staff (*e.g.,* to manage the ISO efficiently to deliver the policy or regulatory outcomes).

FERC's position on ISOs and transcos

FERC's Order 2000 embodies the concept of regional transmission organizations (RTO), organizations that would plan and control regional transmission systems.[14] In an order providing guidance issued on November 7, 2001, FERC specified that RTOs are to be responsible for eight critical wholesale market functions:

- congestion management
- ancillary services
- administration of a balancing market
- open access same-time information system (OASIS) administration, including total transmission capacity and available transmission capacity calculations
- security coordination
- market monitoring
- regional transmission facility planning
- tariff administration and design[15]

RTOs must exhibit the characteristics of independence, scope and regional configurations, operational authority, and short-term reliability. Operational control of transmission facilities is to be ceded to the RTO by the member transmission owners. Otherwise, the nature of RTO is not restricted by Order 2000. Both transcos and ISOs, whether for-profit or not-for-profit, would meet its requirements. For the purposes of transmission expansion, Order 2000 assumes that the RTO is responsible for the transmission planning process.

In its proposed standard market design, FERC intends to require all public utilities to meet the definition of an ITP, turn their operations over to an RTO that meets the definition of an ITP, or contract with an entity that meets the definition of an ITP. FERC is quite prescriptive in its definition of an ITP. It is a public utility that owns, controls, or operates facilities used for the transmission of electric energy in interstate commerce, that administers the day-ahead and real-time and ancillary services markets in connection with its provision of transmission services pursuant to the standard market design tariff, and that is independent (*i.e.*, has no financial interest, either directly or through an affiliate, in any market participant in the region in which it provides transmission services or in neighboring regions).[16]

Although FERC has allowed common ownership of generation and transmission in deregulated markets where an ISO is created, the view is increasingly taken in the U.S. that complete separation of transmission ownership from generation is desirable. This would prevent generators from gaining inside knowledge of transmission constraints and outages that they could use to their advantage in the energy and transmission rights markets.

The transco vs. independent transmission company (ITC) debate

In countries such as the U.S. where there are multiple transmission owners, the debate has moved on. The possibility of competition between multiple transmission companies to capture the same economic value in reducing congestion is becoming a reality. The discussion has focused on the public interest duties of the system operator, leading to a proposal that structures other than a transco would provide a better answer. As John Chandley and William Hogan point out:

> *A critical issue is whether it is appropriate to allow an entity with legitimate but undeniably* private *interests in running a transmission business and pursuing grid investments as a for-profit enterprise to be responsible for performing the* public *interest responsibilities of the RTO in coordinating markets and related system operations.*[17]

What is now driving this debate are the objectives of:

- achieving competitive markets in energy
- transmission expansion
- the reduction of congestion costs

Hybrid proposals

In order to meet these concerns, there have been hybrid proposals to carve out several core functions of the transco, (while retaining some elements of system operations). The carve-out could allow the transco to take advantage of at least some of the trade-offs between transmission ownership and system operations mentioned earlier.[18] The examples, which are somewhat vague, because they depend on agreements to be reached between system operators and transmission owners, include:

- independent entities that share RTO functions on an agreed basis, namely an ISO and one or more ITCs that carry out transmission planning, but where the ISO approves the plans (the New England proposal)[19]
- an RTO with one or more ITCs that may perform some RTO functions related to transmission (the Midwest ISO proposal)[20]
- an independent market operator and a transco that does not carry out market administration functions[21]

In an effort to remove any potential conflict between the system operator's public interest functions and the private business interests of the transmission owner in the transco model, further proposals have been made that an ITC would be the best model, leaving the system operator as a full RTO, able to carry out all of its duties in an unbiased manner.

ITCs

The stand-alone, profit-motivated transmission business of an ITC would own and maintain transmission assets and operate them under the direction of an ISO or RTO. It could conduct both of the regulated and merchant transmission businesses (described later in this chapter), seeking out profitable investments in either environment. It would also be eligible for performance-based regulatory incentives (described in chapter 7).[22] The roles of the RTO and ITC in the wholesale market are explained in Figure 3-2.

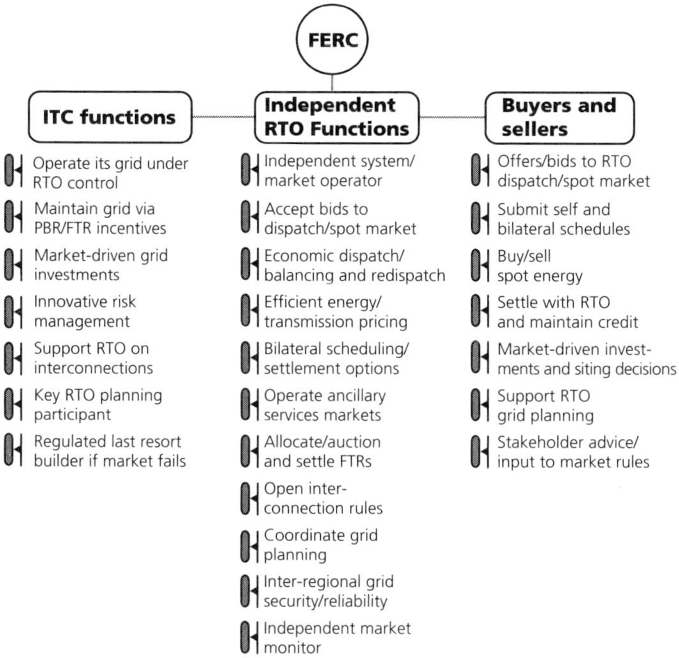

Fig. 3-2 Roles of RTO and ITC in the wholesale market (Source: John D. Chandley and William W. Hogan, "Independent Transmission Companies in a Regional Transmission Organization," December 2001)

Alex Henney has proposed the concept of a transmission services business that involves reducing the involvement of the ISO in grid access and enabling the transmission owners to provide a variety of different services in a contestable way that minimizes the need for regulation. The essential monopoly components in transmission would be regulated. The transmission services business is based on the idea that a market participant should be able to buy a variety of different services from a transmission owner, including some ancillary services, hedges, and different levels of service that command different levels of compensation if the service is not provided.[23]

Evaluation of the options

It is clear that ISOs, transcos, and ITCs can work well in the right circumstances because there has been good practical experience with all of these structural solutions around the world for a number of years. Figure 3-3 analyzes the advantages and disadvantages of the three options. Although an ISO could be a for-profit entity, the assumption, for the purposes of the following evaluation, is that all ISOs are not-for-profit entities because market participants and the regulator will be reluctant to allow a private company to make a profit out of what they regard as a public service. Of course, if some or all of the functions are contracted out to third parties for any reason, the fee could include a profit element. This might be justified on the grounds that it still represented the most cost-effective way to obtain the necessary services.

Advantages	**Transco**	**ITC**	**ISO**
In a good position to maximize economies of scope (*i.e.*, trade-offs between system operations and asset ownership to increase capacity)	✓	✗	✗
Can propose finance and construct efficient and innovative transmission investment	✓	✓	✗
Easy to regulate with a reasonable amount of information	✗	✓	✓
Performance-based (incentive) regulation easy to implement	✓	✓	✗
No bias in system operations (congestion management, ancillary services, etc.)	✗	✓	✓
No bias in planning coordination and selecting competing solutions to meet a transmission need	✗	✓	✓[1]
Willing to provide planning information	✓[2]	✗	✓
Good knowledge of performance of all existing assets to create efficient expansion proposals	✓	✗	✓
TOTAL SCORE (number of advantages)	5	5	5

1 Although the ISO will be independent and should have no bias, it will tend to favor solutions that improve its own ability to maintain reliability with minimum effort. We assume safeguards can be put in place in the market rules or regulatory regime to deal with this.

2 The assumption is that the transco will make information available, even if it would prefer to keep some of it confidential from other potential transmission developers, because, as system operator, it will see the need to do so.

Fig. 3-3 Evaluation of advantages of transcos, ITCs, and ISOs

No two industries are the same

The analysis in the table is obviously simplistic (and we could qualify some of the ticks and crosses, if not all of them), but it does support the view that the decision as to which is the best structure to adopt in all circumstances is finely balanced. However, no two electricity industries are the same. Therefore, we must look at the circumstances in which the electricity industry finds itself. We must also look at the transmission system itself in order to analyze which solution might be the most appropriate in view of the particular history, environment, needs, and objectives of the industry, the regulators, and consumers.

The evaluation of the three alternatives against possible circumstances, needs, and objectives is set out in Figure 3-4.

Circumstances, needs, and objectives	Transco	ITC	ISO
There are multiple potentially competing transmission owners	✗	✓	✗
Significant potential to maximize economies of and asset ownership to increase capacity	✓	✗	✗
There is a need to create competitive market in energy	✓[1]	✓	✓
There is a need to plan new transmission and evaluate competing options without bias	✗	✓	✓
There is a need to reduce congestion costs	✓	✓	✓
There is a need to attract merchant and non-incumbent transmission owner investment	✗	✓	✓
There is a need to encourage incumbent transmission owners to invest on a merchant or regulated basis (raising their own finance)	✓	✓	✓
There is a need for new lines using new route corridors (where the endorsement of an ISO that is not the developer will add credibility with environmental and permitting agencies)	✗	✓	✓
There is a need to minimize regulatory burden and overcome lack of information of the regulator	✗	✓	✓
There is a need to create incentives to improve operational efficiency	✓	✓	✗

1 This has been achieved using a transco in several countries, but it will involve tighter regulation and operational auditing than in the case of an ISO.

Fig. 3-4 Evaluation of circumstances, needs, and objectives

Conclusions as to transcos and ITCs

There are several conclusions that can be drawn from this analysis.

Transcos. A transco is an attractive model if there is a need for expansion and scope for significantly increasing the availability and flexibility of the existing system by creating incentives for more efficient maintenance and for modest investment to reconductor or upgrade it. It may be difficult to regulate a potentially large monopoly that will always have significantly better information from the regulator, but England and Wales have provided the model we looked at earlier in this chapter (described in more detail in chapter 7), and it works. Moreover, there is a theoretical economic analysis to support the regulatory approach.[24] NGC's performance has been extremely impressive for over a decade. It can fairly claim to have delivered high throughput, lowered congestion, increased flexibility, efficient connections of new generation, high availability, and reliability with an excellent safety record at low cost.[25] The performance of other transcos, such as Statnett and Transpower, has also been impressive.

The model works particularly well where there is only one transmission system and no pressing need to attract third party, non-incumbent investment in transmission. The transco model is also appropriate where the primary objective is to encourage the incumbent transmission owner to invest, where it is able to raise its own finance without government guarantees.

If there is to be no deregulation, the creation of a transco can add value. As Hogan has pointed out, to capture thc trade-offs between system operations and asset management, a large monopoly must be created that internalizes all the costs. However, the whole point of deregulation is to move away from regulated monopolies.[26]

ITCs. If there are multiple, potentially-competing transmission owners, a model involving independent transmission companies and an ISO will be the answer. A transco will lack credibility because of the potential bias of the system operator to favor its own transmission business, both in operational time scales and in the planning coordination process. The lack of bias in this model overcomes concerns that a transco might not manage real-time operations, congestion, imbalances, and the definition and administration of transmission rights on an entirely non-discriminatory basis. These elements are key ingredients of a workable, competitive market.

The ITC-ISO model would overcome fears that a process for evaluating competing solutions to a transmission need might not be fairly handled by a transco if it were allowed to submit its own proposal. There is a less obvious advantage to a transco where expansion involves building lines along new route corridors. These projects are likely to counter significant objections that will be voiced before environmental and permitting agencies. If an ISO has included a project in the outcome of a planning coordination process, it gives a degree of credibility and impartial endorsement of the project if the developer is independent from the system operator. The third-party developer should not find it difficult to reach agreement with a transco with respect to control of its facility by the system operator, because the transco will be subject to regulatory scrutiny. It will, however, be concerned about giving operational and planning information to a system operator it sees as a potential competitor, because the system operator also owns transmission.

Creating a national grid model

The success in regulating NGC and fears over market failure in transmission provoke the thought that it could be a better answer, in those countries where transmission ownership is fragmented, to create a much larger regional or national transmission company. The idea has been debated in both the U.S. and Australia.

The task of creating a national grid is not as daunting as it may seem. It need not involve forced divestiture of transmission assets. In Norway, the national grid was completed by Stattnet leasing the transmission assets that it did not own. In Australia, the concept of placing all the transmission assets in the National Electricity Market under the management and control of a single national grid company (pursuant to contracts) has been discussed.

The advantages are that weak links in and between the systems could be internalized, the so-called "seams" issues, or differences between systems could be minimized and economies of scale could more easily be captured. Although the resulting grid company would represent an even larger monopoly or "black box" for the regulator to deal with, the NGC regulatory experience indicates that it could be regulated to produce satisfactory results. Fragmentation of ownership is usually accompanied by fragmentation of regulation and governance, which is inefficient (particularly if different regulators have very different objectives and techniques). If market failure in transmission investment is predictable, for example, where the primary need is to improve the availability of existing capacity or to undertake major strengthening of the capacity between parts of highly reticulated AC systems, a national or regional grid company could provide a good answer.

The Allocation of Responsibilities

Impact of market design

We have looked at the industry structure as a starting point. It can obviously affect the responsibilities of the various industry participants. The impact of the design of the trading arrangements should not be underestimated. A competitive wholesale market may or may not have been introduced. Retail competition or consumer choice may or may not exist (in whole or in part). More importantly, market design affects the allocation of the indirect (short-run marginal) costs that each of the industry participants bears and, therefore, their incentives to invest in transmission to reduce those costs.

Diversity of responsibilities

There are many functions that can be allocated to system operators and transmission companies that relate to what happens between the power station fence and the consumer's meter.[27] Figure 3-5 shows how the main functional responsibilities have been allocated in some of the restructured industries. There is an ongoing debate as to the best allocation for the purposes of establishing an efficient market design, particularly what the core roles are of an ISO or RTO.[28] The multiplicity of industry structures, market designs, and also the diverse terminology are remarkable. Our recipes for the successful expansion of the transmission system must take account of the varying industry structures, the extent of deregulation (or otherwise), and the different roles and responsibilities of the players.

Responsibility	PJM	Argentina	Nordpool	Australia	England & Wales
System operations	RTO	ISO	Transcos	National ISO (NEMMCO),state ISOs and Transcos	Transco
Maintenance	TOs	TOs	Transcos	NSPs[1]	Transco
Planning	RTO (with TOs and stakeholders)	Market	Transcos	NSPs, ISOs and NEMMCO	Transco
Expansion	TOs and market	Market	Transcos (primarily)	NSPs and market	Transco
Congestion management	Market	Market	Transcos (zonal and postage stamp)	NEMMCO (zonal)	Transco (uplift)
Losses	RTO	Market	*	Market	Transco
Ancillary services	RTO	ISO	Transcos	NEMMCO, ISOs and Transcos	Transco
Spot market administration	RTO	ISO	Company owned by Transcos	NEMMCO	Transco
Technical and spot market rule administration and change	RTO	Government	Nordpool governance	NECA[2]	Governance structure
Billing and collection of transmission charges	RTO and TOs	ISO	Transcos	ISOs and NSPs	Transco
Oasis, ATC, etc.	RTO	ISO	Transcos	NEMMCO and NSPs	Transco
Market monitoring	RTO and regulator	Regulator/ Government	Regulators, competition authorities	NECA, Regulators	Regulator
Inter-regional coordination	RTO	ISO	Transcos	NEMMCO	Transco

1 NSP is a network service provider (transmission owner).
2 NECA is the National Electricity Code Administrator.
* Various arrangements are in place including nodal loss factors, socialized costs, and costs rolled into transmission prices.

Fig. 3-5 Who is responsible for what?

Following the general principle explained in chapter 2, a signal will be created that it is economic to build, once the future level of short-run costs incurred when using existing capacity exceeds the long-run costs of building additional capacity (taking into account the losses incurred when using it). If those who are in a position to expand the system do not bear any of these costs, they will receive no signals and have no incentive to invest. If those who bear the indirect costs of transmission are not able, individually or jointly, to invest in transmission, they will be unable to alleviate or hedge those costs. Figure 3-6 shows how these costs are initially born in the markets we have just looked at (before they are passed on).

	PJM	Argentina	Nordpool	Australia	England & Wales
Congestion	Market	Market	Transcos (zonal and postage stamp)	NEMMCO (zonal)	Transco (uplift)
Losses	RTO	Market	Transcos	Market	Transco

Fig. 3-6 Who bears the cost of congestion and losses?

The simple conclusion is that where transmission owners or market participants (users) bear these costs, we need to find recipes for transmission expansion encouraging them to invest. Where the costs need to be recovered by way of an uplift charged at a single price per MW to all users (or buyers), we need to create incentives to minimize the costs. In the alternative, the costs should be borne by market participants or transmission owners able to invest to reduce them.

Other Starting Points

Inherent powers, rights, and obligations

Another starting point that may be unlikely to change, because it is entrenched in the legal framework or in the regulatory regime, is the position of the incumbent transmission owner that may have either an obligation or the right to expand the transmission system.[29] For example, NGC has an obligation under its license[30] to connect any generator or load to its system that requests transmission service and must expand its system to accommodate these requests. The obligation may or may not be accompanied by a monopoly right to build. Many of the jurisdictions that have deregulated their electricity industries would, in theory, allow third parties to build transmission facilities. However, the design of the market, the transmission pricing, and the regulatory regime might not create incentives to third parties to do so. England and Wales would be an example of this.

Transmission owners may have the inherent right, as owners, to upgrade their own facilities, but they may not have a formal right or right of first refusal to do so. In Argentina, the incumbent transmission owners have limited rights to enhance their own systems, even to protect themselves from penalties for lack of transmission service. In the course of the market design process, the transmission owners may argue for a right of refusal, but this is not usually thought to be appropriate, because of concerns over monopoly power and the deterrent effect on beneficial market-based expansion by third parties. In the U.S., FERC turned down two proposals from GridSouth and ISO New England that would have given incumbent transmission owners a right of first refusal to build.[31]

Rights of expropriation or eminent domain

Another utensil and a key piece of equipment is the power to expropriate land and rights-of-way (eminent domain) through compulsory purchase. Usually these powers are derived from legislation and may be given to all public utilities of a specified class or character. They may be awarded on a project-specific basis by a regulator or an agency that approves the need for the project. An important issue that we shall look at in chapter 9 is whether these powers can be obtained by third parties, such as those market participants proposing an expansion project on a merchant basis. At the time of deregulation, many jurisdictions have rewritten the law in order to enable the appropriate regulator or agency to grant these powers to market participants or third parties on a project-specific basis or as inherent powers that enable the holder of the powers to fulfill its obligations as a transmission owner.

If these powers are not made available where they are needed, the developer of the transmission project will have no alternative but to attempt to acquire the land and rights-of-way that it needs by negotiation with landowners, which may prove to be a more time-consuming and expensive exercise than would be the case if it had the necessary powers. Ultimately, the costs may be borne by the consumers.

Tax-exempt bonds

Another starting point is the nature of the existing funding of incumbent transmission owners. Some utilities are able to finance investments with tax-exempt bonds or other instruments, the tax-exempt status of which is contingent upon the facilities that have been financed in this way being used to serve local customers or upon other conditions. An example of this is in the U.S. where San Diego Gas and Electric (SDG&E) has used tax-exempt local furnishing bonds to finance facilities to serve its native load customers. Instruments such as these that have been used to finance transmission facilities will have to be taken into account in the design of both the transmission expansion scheme and the market rules.

In the case of SDG&E, all of the costs associated with the loss of tax-exempt status, including the costs of defeasing, redeeming, and refinancing tax-exempt bonds, are considered costs of providing transmission service. The customer that takes service, knowing that the service will result in the loss of tax-exempt status, is responsible for the costs involved and the transmission owner can include in its tariff a provision permitting it to recover the costs. If the customer is not willing to pay the costs, SDG&E does not have to provide the requested service. The question will then arise as to whether the requested service would, in fact, have jeopardized the ability of the utility to serve its local customers or whether it would shift costs on to other customers, such as the costs of reinforcing the system following the connection or interconnection of new generation or other transmission facilities.

Government ownership

Several significant unbundled transmission owners in countries are still in government ownership, even where the industry has been de-regulated or partially de-regulated (*e.g.,* New Zealand, New South Wales, Republic of Ireland, Norway, Netherlands, India). A specific regulatory regime administered by the government or a regulatory agency for transmission has been defined (except in New Zealand). The difficulties for these government-owned transmission owners is that they may well wish to respond as commercial companies to incentives and signals to invest. They may be encouraged to make strategic investments by the government either as shareholder or as policy setter (such as investments in interconnectors between regions). Powergrid, in India, does not invest for profit or in response to its regulatory incentives. It is driven by a desire to provide adequate transmission service to the whole of India and by engineering excellence. Typically, all state-owned transmission companies will try to do the right thing. The incentive to do so may be less financial and regulatory and be more created by the desire to enhance social and economic welfare, and to avoid criticism of poor performance either on the part of the utility or on the part of the government.

Hurdles to expansion

There may be other starting points that represent hurdles needing to be overcome in designing the recipe for transmission expansion. The hurdles may exist in the legal framework or even in the constitution of a country that has a state-owned monopoly transmission owner. Transmission may be a strategic activity reserved to the state under the constitution, such as in Mexico. Another potential legal or constitutional restriction we have observed is that it may not be possible for expropriation rights to be granted to private sector companies. It is important to look for all potential hurdles to efficient transmission expansion that cannot be easily overcome. If it is not possible to design a recipe to overcome them, a change of law or even of the constitution may be necessary.

Regulated and Market-based Expansion Projects

Two approaches to transmission expansion

It is necessary for us to design our recipes for successful transmission expansion for cooking in different ovens. We must allow for expansion projects on a regulated basis and in a market environment. We need to ensure that the recipes for each environment have the best chance of success, because we may need to be able to utilize both approaches.

Regulated expansion projects

A common example of regulated expansion is where there is a single transmission system that is a regulated natural monopoly, where the owner has an obligation to provide transmission service to any user that applies for it. The regulatory regime would apply the appropriate rules and incentives to ensure that the regulated entity makes efficient investments and that it provides service on acceptable terms for connection to or interconnection with its system and for use of its network. The primary role of the regulator would be to decide what costs of the expansion would be recoverable and how they should be allocated among users. The transmission planning and investment approval process would probably be subject to scrutiny by the regulator both as to need and capital expenditure. These subjects are discussed in more detail in chapters 6 and 7. However, in many jurisdictions, there is no reason that a new entrant should not propose a regulated expansion project.

The types of projects that are typically regarded as projects suitable to be carried out on a regulated basis are those that would be difficult for market participants to implement, *i.e.*, where there has been a market failure. This might be because they carry a strong element of common good or public interest, where the benefit is shared by all users, *e.g.*, projects that strongly enhance reliability, reduce local or regional market power, contain significant economies of scale and benefits for future users, or where direct beneficiaries are extremely difficult to identify.

Expansion driven by the market

The other approach is to look toward market mechanisms to drive transmission expansion. Market participants would make transmission investments in response to price signals with respect to the costs of congestion and losses that they bear. The investment would be made on the basis that the developer would take the entire usage or revenue risk without the protection of the regulatory regime as to cost recovery. For example, generators may be motivated to build transmission to obtain access to buyers of energy in high-priced areas or where the payments they receive for being constrained off the system are inadequate to compensate them.[32] Load serving entities, particularly those without the automatic ability to pass through the full amount of their costs, could be willing to invest to protect themselves against increases in the costs of congestion and losses and in order to be able to import energy from low-cost generators outside their areas.

As we shall see in chapter 5, tradable transmission property rights are a key ingredient of market-based transmissions expansion.[33] A workable system of financial or physical rights that provides an equivalent to transmission property rights can be used or sold by market participants to guarantee the costs of energy and transmission[34] or to give them physical scheduling rights to utilize the capacity addition.[35] The sale by investors and the acquisition by market participants of these rights would help to remove many of the uncertainties associated with revenue-earning ability of the investment.[36]

It is important not to inhibit market-based expansion, because it may have the best chance of success in certain circumstances. It may be that the regulatory regime will fail to effectively incentivize transmission developers to build needed transmission capacity, or the necessary degree of coordination between regulators may be lacking to incentivize the building of interconnectors. Market participants are most likely to respond to opportunities to build transmission that will alleviate congestion or that will enable them to arbitrage price differences between regions. They may prefer to use new controllable power-flow technologies such as high voltage direct current (HVDC) cables and be attracted to projects that are not embedded in existing mesh systems.

Markets should have a good chance of ensuring that opportunism and innovation flourish both in terms of technology and finance. A market participant is likely to act quickly and to innovate in order to build needed transmission or interconnector capacity. As discussed in chapter 2, the building of DC interconnectors on a merchant basis might result in a quicker increase in the transfer capability between transmission systems than on a regulated basis.

However, Paul Joskow has written:

> *Transmission investment decisions do not immediately strike me as being ideally suited to relying entirely on the invisible hand. Transmission investments are lumpy, characterized by economies of scale and can have physical impacts throughout the network. The combination of imperfectly defined network property rights, economies of scale, long-lived sunk costs for transmission investments, and imperfect competition in the supply of generation services can lead to either under-investment or over-investment...*[37]

His optimism that the market would provide needed transmission has waned.

Joskow expresses skepticism with respect to market-based approach to network enhancements:

> *The primary initiatives for these types of network enhancements are best left to the owners of the existing network in the context of good RTO/ISO planning processes since the enhancements are often physically and economically inseparable from the existing network and its operation.*[38]

Therefore, we need to design recipes for a transmission expansion scheme that accommodates both market-based and regulated investment in transmission and to reconcile the issues that lie at the interface between them.

NOTES

1. Sanford V. Berg, and Jinook Jeong, "An Evaluation of Incentive Regulation for Electric Utilities," *Journal of Regulatory Economics*, vol. 3, no. 1 (March 1991), p. 45-55

2. Eric Hirst, "Expanding U.S. Transmission Capacity," *Edison Electric Institute* (August 2000). Eric Hirst, and Brendan Kirby, "Transmission Planning for a Restructuring U.S. Electricity Industry," Edison Electric Institute (June 2001)

3. Nick P. Winser, "Division of RTO Responsibilities: Slicing and Dicing," presentation to FERC Technical Conference (February 19, 2002), Docket No. RM01-12-000

4. Commission of the European Communities, Directive Concerning Common Rules for the Internal Market in Electricity, 96/92/EC, *Official Journal*, L 027, 30/01/1997, pps. 0020-0029

5. Proposal for a Directive amending Directive 96/92/EC, Concerning Common Rules for the Internal Market in Electricity (2001/0077 (COD), EU, Council Presidency text (March 14, 2002) published on the EU Council website

6. Sally Hunt, "Making Competition Work in Electricity," presentation to the CMS Cameron McKenna Third Electricity Conference, Washington, D.C., Aug. 8, 2002

7. James V. Barker, Jr., Bernard Tenenbaum, and Fiona Woolf, "Governance and Regulation Power Pools and System Operators: An International Comparison," *World Bank Technical Paper No. 382* (1997)

8. "Response of National Grid USA to Questions Posed by the Commission," FERC Docket No. EL01-80-000

9. James Barker, Jr., Bernard Tenenbaum, and Fiona Woolf, "Governance and Regulation of Power Pools and System Operators: An International Comparison," *World Bank Technical Paper No. 382* (1997)

10. Beatriz Arizu, William H. Dunn, Jr., and Bernard Tenenbaum, "Why System Operators Must Be Truly Independent," *The World Bank Group Private Sector and Infrastructure Network Note No. 226* (January 2001)

11. Regional Transmission Organizations, Notice of Proposed Rulemaking: Comments of Professor Paul L. Joskow, Docket No. RM99-2-000 (August 13, 1999), p. 18

12. Thomas-Olivier Nasser, "The Hidden Value of Transmission Assets," *The Electricity Journal* (1999), pp. 69-78

13. "Italy: Grid management, ownership to be merged?" *Power in Europe* (March 11, 2002), pp. 20-21. ENEL's agreement to divest its transmission subsidiary, Terna, may have been motivated by the low rate of return allowed to Terna by the regulator

14. Order No. 2000, Regional Transmission Organizations, 18 CFR Part 35, Docket No. RM99-2-000 (issued December 20,1999)

15. FERC Order Providing Guidance on Continued Processing of RTO Filings, Docket No. RM01-12-000 (issued November 7, 2001)

16. FERC Notice of Proposed Rulemaking, 18 CFR, Part 35, Docket No. RM01-12-000 (July 31, 2002) ¶125 and 126

17. John D. Chandley and William W. Hogan, "Independent Transmission Companies in a Regional Transmission Organization" (December 20, 2001), available on Professor Hogan's website, www.whogan.com

18. For a description of various hybrid models and the arguments surrounding them, see John D. Chandley and William W. Hogan, "Independent Transmission Companies in a Regional Transmission Organization" (December 20, 2001), available on Professor Hogan's website, www.whogan.com

19. Joint Petition for Declatory Order to Form the New England Regional Transmission Organization, Docket No. RT01-86-00 (January 16, 2001)

20. Appendix I of the Agreement of Transmission Facilities Owners to Organize the Midwest Independent Transmission System Operator, Inc. (the ITC-MISO Agreement), Docket No. ER01-3000-000 (August 31, 2001). Under the ITC-MISO Agreement, The International Transmission Company would share certain RTO functions with the Midwest ISO. The agreement was accepted for filing in a Commission Order dated December 20, 2001. International Transmission Company, 97 FERC P 61, 323 (2001)

21. Hogan makes the point that system operations and the markets for imbalances and congestion management use exactly the same energy and dispatch. System operations and market operations in very short time scales are must be coordinated by the system operator in order to maintain reliability. See William W. Hogan, "The Wholesale Pool Spot Market Must Be Administered by the Independent System Operator: Avoiding the Separation Fallacy," *The Electricity Journal* (December 1995), pp. 26-37

22. FERC approved a profit-driven ITC with a hybrid RTO under which certain functions were delegated to the ITC in its Translink order, Translink Transmission Company, LLC *et. al.*, 99 FERC ¶ 61, 106 (2002)

23. Alex Henney, "Transmission-Can it Transform from Cinderella to Princess," presentation to EUCI Transmission Conference, Miami (February 17, 2002)

24. Thomas-Olivier Léautier, "Regulation of an Electric Power Transmission Company," *The Energy Journal*, vol. 21, no. 4 (2000)

25. Nick Winser, untitled presentation to the Harvard Electricity Policy Group, San Diego, California (January 25, 2001)

26. William W. Hogan, "Electricity Market Design: Regional Transmission Organizations," presentation to Sixth Electricity Conference, International Bar Association, Section on Energy & Natural Resources Law, Berkeley, California (February 25, 2002)

27. Fiona Woolf, "The Unbundling and Rebundling of Transmission and Market Related Functions," *The Electricity Journal* (December 1996), pp. 44-51

28. Nick P. Winser, "Division of RTO Responsibilities: Slicing and Dicing," presentation to FERC Technical Conference (February 19, 2002), Docket No. RM01-12-000. Larry E. Ruff, "Defining and Allocating RTO Functions," Submission to Expand on Oral Statements at the FERC Technical Conference (February 19, 2002), Docket No. RM01-12-000. See also Response of National Grid USA to Questions Posed by the Commission, FERC, Docket No. EL01-80-000

29. Section 211 of the Federal Power, 1920, begins: "Any electric utility, Federal power marketing agency, or any other person generating electric energy for sale for resale, may apply to the Commission for an order under this subsection requiring a transmitting utility to provide transmission services (including any enlargement of transmission capacity necessary to provide such services) to the

applicant." Section 212 of the same Act begins: "An order under section 211 shall require the transmitting utility subject to the order to provide wholesale transmission services at rates, charges, terms, and conditions which permit the recovery by such utility of all the costs incurred in connection with the transmission services and necessary associated services, including, but not limited to...the costs of any enlargement of transmission facilities"

30. U.K. Department of Trade and Industry Electricity Transmission License: Standard Conditions, Condition D8B (September 27, 2001). A license is a regulatory instrument containing the detailed rules of the regulatory regime as they apply to the particular licensee

31. Carolina Power & Light Company, 95 FERC P61, 282 (2001). In this Order, the Commission directed the applicants to eliminate provisions that gave the transmission owners the right of first refusal to construct transmission facilities in their service areas. The Commission found that these provisions unduly limited the decisional authority of GridSouth over transmission planning, presenting the possibility of discrimination by self-interested transmission owners favoring their own generation and possibly precluding lower cost or superior transmission facilities or upgrades by third parties from being planned and constructed. ISO New England, Inc., 91 FERC P 61,311 (2000). In this Order, FERC directed ISO New England to revise its proposal to eliminate any decisional role transmission owners may have in the current expansion plan and concluding that only the ISO should identify expansions and upgrades required for reliability reasons

32. It is not the case that all transmission expansion will benefit all market participants, particularly those that are exercising local market power as a result of congestion or are able to benefit from being able to charge high prices because of a constraint

33. James B. Bushnell and Steven E. Stoft, "Grid Investment: Can the Market Do the Job?" *The Electricity Journal* (January-February, 1996), pp. 74-79. Eric C. Woychik, "Competition in Transmission: It's Coming Sooner or Later," *The Electricity Journal* (June 1996), pp. 46-58

34. William W. Hogan, "Market-Based Transmission Investments and Competitive Electricity Markets" (August 1999). This paper extends the earlier discussion in William W. Hogan, "Transmission Investment and Competitive Electricity Markets" (April 1998), available on Professor Hogan's website, www.whogan.com

35. Scott M. Harvey, William W. Hogan, and Susan L. Pope, "Transmission Capacity Reservations and Transmission Congestion Contracts" (June 6, 1996, revised March 8, 1997), available on Professor Hogan's website, www.whogan.com

36. Transmission prices have traditionally been based on sunk costs rather than short-run marginal costs. This creates a difficulty because the prices then bear no relation to the value that users might place on the use of the transmission system, particularly in a competitive environment

37. Paul L. Joskow, "Restructuring, Competition and Regulatory Reform in the U.S. Electricity Sector," *Journal of Economic Perspectives,* vol. 11, no. 3 (Summer, 1997), p. 132. See also, FERC, "The Transmission Task Force's Report to the Commission" (October 1989), p. 164

38. FERC, Regional Transmission Organizations, Notice of Proposed Rulemaking: Comments of Professor Paul L. Joskow, Docket No. RM99-2-000 (August 13, 1999), p. 16

CHAPTER 4
The Key Ingredients of Market Design and Transmission Pricing

Ingredients of this Chapter

Efficient market design and pricing are the key ingredients

If the transmission system is used efficiently it may not need to be expanded at all. If it is used inefficiently, it may result in expansion that could otherwise have been avoided. The key to efficient use lies in the principles of market design and transmission pricing. These principles will also create the signals and incentives that will both stimulate and facilitate investment in transmission.

This chapter begins by explaining the complexities of mesh transmission systems and some of the realities that need to be taken into account in market design and transmission pricing. We analyze the effects of market freedom and the need for central coordination of the electricity market—ingredients that appear to be incompatible. Then we look at congestion and the value of scarce transmission. The most important ingredient is to get the prices right. We will look at postage stamp and locational pricing, recognizing that access to and usage of the transmission system are closely interlinked. We explain why the right model is one that integrates the energy markets and the pricing of scarce transmission.

The important principles of market design are then discussed, because they are key ingredients of the creation of efficient price signals to investors in transmission. We will look at transmission access and usage in more detail, from the perspective of allocating access and scarce transmission capacity efficiently. Then there will be an explanation of transmission usage in practice, covering nodal pricing, zonal pricing, and other approaches. This will be followed by a discussion of the creation of transmission rights, of which there are a number of flavors or models. Transmission rights are an important ingredient of transmission expansion, particularly from the point of view of facilitating market-based expansion. Finally, in this chapter, we will discuss the principles of transmission pricing, and the recovery of fixed and sunk costs.

If Only Life Were Simple

No networks

If transmission systems consisted entirely of controllable, radial (single) lines, where power flowed from a generator to a consumer along a direct path consisting of lines and valves, life would be simple. It would be possible to charge directly for the power flows on each line, because they could in effect be labeled, directed, and tracked. The valves could be closed in the case of non-payment. There would be no network interactions or externalities. In a competitive market equilibrium, the price of usage would be equal to the difference in the prices of electricity at the point of production (source) and the point of consumption (sink). In other words, the difference is the cost of transportation. The owner of the line could buy at source and sell at sink, profiting from the difference in locational prices. Buyers and sellers could be given physical rights to use the system to reflect the volumes they contract to sell and buy along the "contract path" that lies between them.

No economies of scale

Life would also be simple if there were no economies of scale or lumpiness in transmission investments. If the investments came in small increments with constant or increasing marginal costs, it would be possible to expand transmission capacity by adding the cheapest available increments, with anyone allowed to make the investment. The investment would be profitable as long as the difference in locational prices between source and sink exceeded the cost of the increments. In equilibrium, the cost of the last increment added would just equal the resulting expected difference in prices. The profits collected from the purchase and sale of power would be just enough to pay for the market-based investment. In equilibrium, all of the earlier investors could expect a capital gain equal to the difference in the market clearing prices and the cost of their investments.

No barriers to entry

If the market were out of equilibrium (perhaps because of growing demand or changing cost structures), the price differential between two locations might be higher than the marginal cost of investment. The existing transmission owners would enjoy high profits resulting from scarcity rents. New investors would see this as an incentive and an opportunity to build. The early movers would capture some of the scarcity rents, but would also add to capacity and reduce the gap between price and marginal cost. If there were no barriers to entry, the extra profit incentive of the scarcity rent would encourage new investment until equilibrium returned, where further expansion would no longer be profitable.

Three equations

If the equilibrium prices were to differ at the source and the sink, there would be three important equations:

- The amount of power purchased and sold at source and sink would equal the capacity created by the transmission investment[1]

- The power flowing on the controllable transmission line would be equal to the same capacity
- The expected price difference between locations would be equal to the marginal cost of expansion[2]

Powerful network externalities

For the most part, transmission systems are a mesh or reticulated network, and power flows along the path of least impedance (the vector relationship of resistance and reactance). It cannot be directed to a particular consumer. This creates what are called network externalities, and they are significant. The reality is that power flows across every parallel path between source and sink. Even where there is a controllable line that is part of a larger system, the impact on the controllable line cannot be fully separated from the pattern of use of the rest of the system. There can be significant economies of scale, and an efficient level of investment might come in lumps that change equilibrium prices substantially.

This means that at least two of the three important equations can be destroyed. The amount of power purchased and sold at the source and sink could still equal the expanded transmission capacity.[3] However, the power actually flowing may not be the same. If there are substantial economies of scale, the important linkage between the expected price differences and the marginal cost of investment might be broken, because the expected price difference between locations may not be equal to the marginal cost of expansion. The power flows across individual lines would not be equal to increments in capacity, and the simple model of charging for the flows on a transmission line (and denying service to those who did not pay) would not support efficient investments. If there are significant economies of scale, and efficient investment eliminates the price difference between the locations, there might be no incentive to invest at all.

Allocation of costs and benefits becomes complicated

The allocation of the costs of short radial connections from a plant to a nearby transmission system is relatively straightforward, because the benefits of the connection accrue to the owner or user connected to the grid, and the associated transmission costs could be allocated to it. The close match between benefits and costs would allow for investment charges that would be consistent with operation of a market. It is for this reason that transmission pricing often treats these investments separately. However, investments on the main transmission system are much more complex, because the allocation of costs and benefits is no longer obvious.[4] Moreover, as Figure 4-1 shows, there are network interactions that make it difficult or impossible to define the total capacity of the network. It is simply not possible, in any sufficiently interconnected or reticulated network, to define the capacity of the network, because it is strongly affected by the pattern of usage moment by moment. Except for truly radial systems, transmission congestion can change the capacity of the system, depending on the configuration of the load and generation.

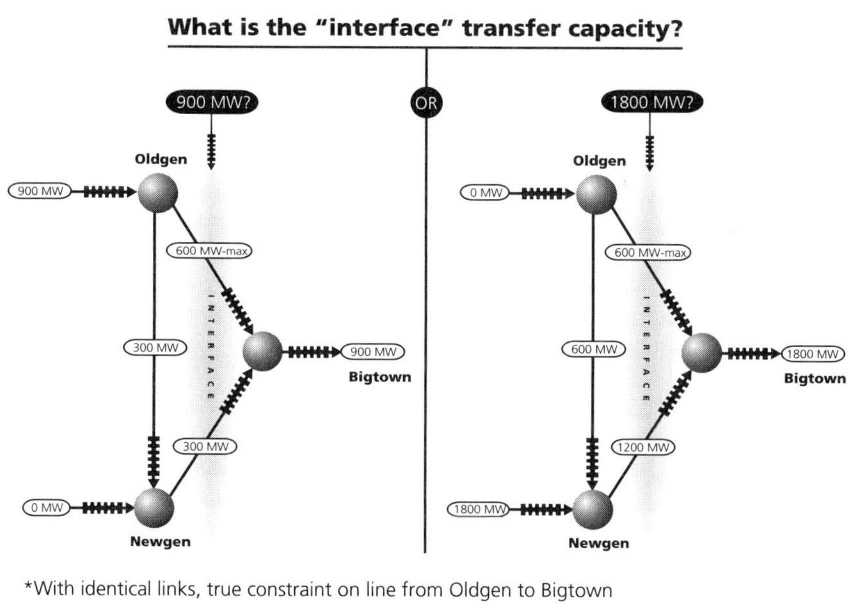

Fig. 4-1 Power transfer capacity varies with load (with identical links, true constraint on line from Oldgen to Bigtown)

The reality that contracts do not match demand and supply

To complicate things further, buyers cannot guarantee to limit the electricity that they consume, and sellers cannot guarantee to produce the exact amount they have contracted to sell. Buyers need a mechanism to enable them, at short notice, to sell any capacity or energy they have contracted for, but do not need. They also need to be able to purchase and sell quantities of energy that they need that are different from the amounts for which they have contracted. Generators need a similar mechanism to enable them, at short notice, to sell energy they are unable to produce from their uncontracted capacity and to purchase any shortfalls in the amounts they have contracted to supply, but are not able to produce.

For all of these reasons, the creation of wholesale markets in electricity has resulted in a paradigm shift to a coordinated spot market and locational marginal pricing, and transmission rights that are described in more detail in this chapter. This leaves the question of the recovery of fixed or sunk cost allocation, which we will discuss at the end of this chapter.

Central Coordination and Market Freedom

Ingredients that appear incompatible

Successful competitive markets depend upon the interaction of the private decentralized decisions with respect to trading and investment that market participants are free to make. The transmission system provides the marketplace for the wholesale electricity market.[5] However, it requires a degree of central coordination, particularly with respect to system operations. Central coordination and the commercial freedom that characterizes competitive markets often appear to be ingredients that do not go well together. However, this should not be the case in a well-designed electricity market.[6] The apparent conflict has been resolved in the successful electricity markets that allow the system operator to coordinate the necessary real-time balancing or spot markets with congestion management and the procurement of ancillary services in order to maintain reliability. Then market principles and commercial freedom can be consistently applied to achieve the flexibility and decentralized decision-making that effective electricity markets require and on which they thrive.[7]

Commercial freedom can create congestion that needs to be managed

In the same way that society has freedom to utilize the public road network in a way that can create bottlenecks and congestion, the commercial freedom accorded to market participants can create bottlenecks and congestion on the transmission system that are difficult to anticipate. It would be very inefficient if society were to build new roads or transmission lines to alleviate congestion without first exhausting all methods available to manage the use that is made of scarce transportation or transmission capacity, to ensure that it is used efficiently and that expansion is really needed.

As we saw in chapter 1, the early stages of most deregulated markets have been characterized by increased congestion, and it has not always been the case that the market design has incorporated the best mechanism to manage congestion.[8] If congestion is not managed, the costs to the consumer will increase, because the transmission lines have a finite capacity that cannot be exceeded if reliability and security of supply are to be maintained.

Because of the highly interactive nature of system balancing, system reliability, and congestion management, it is necessary to look for solutions in the design of the market or the trading arrangements with respect to signals and incentives that encourage efficient use of and investment in transmission. It must be remembered that investment in transmission will only be economic or justified in the consumer interest if the costs of construction are less than the short-run marginal costs of congestion and losses. Therefore, it is important to design a market that creates clear signals as to the costs of congestion and losses. It must also create a market-based mechanism in order to give market participants the commercial freedom and the opportunity to make decisions as to whether they are prepared to pay (and hedge) the costs of congestion in order to alleviate the congestion by building new transmission facilities or lines.

As we discussed in chapter 2, the congestion will cause the system operator to change the dispatch of (redispatch) the schedules of the plant or transactions, with the result that less economic generating plants will have to be dispatched in order to ensure that supply matches demand and reliability is maintained. Put another way, if there is a binding constraint on a transmission line between two areas, the generating plants on the "import" side of the constraint will have to increase production. Generating plants (which may have lower prices or be more economic) on the "export" side of the constraint will have to decrease production, by comparison to the schedules they had originally proposed or preferred. The cost risk associated with this uneconomic redispatch will ultimately be borne by consumers who, for the most part, are not in a position to manage it. Large customers and other market participants should be able to do so if the key ingredients described later are incorporated into the market design.

When transmission is scarce it becomes valuable

In the world of the vertically integrated utility, transmission congestion is seldom a serious problem, because the utility can control all of the generation on the system. In a competitive wholesale market environment, however, congestion is a critical issue, because as soon as the transmission capacity becomes scarce, it becomes valuable. Those who are able to obtain access to the scarce transmission capacity to serve their customers will capture the value. Those who are unable to obtain access will incur additional costs in purchasing energy for their customers from generators with plants on the customer side of the constraint. Generation and transmission become substitutes that depend on location and relative costs. The spot prices of energy will be higher in this area, because of the restriction of cheaper supplies from outside the area created by the constraint.

The question then arises of how the value of scarce transmission is quantified. Following the general principles of market economics and transportation, the value of a congested transmission line (the congestion rent) is equal to the price of energy in the area on the import side of the congested line minus the price in the area on the export side of the line. The reference to import and export areas assumes a transaction between a generator or supplier located in the area "behind" the constraint from which it had originally hoped to export to its customer on the other side of the constraint. That customer had, in turn, hoped to import energy from the generator. The customer had entered into the transaction on the basis that the agreed price of energy within the import area was cheaper than the price of energy in the area local to it. In short, the value of congested transmission is equal to the difference in energy market prices from one location to another, because the transmission system is, essentially, the transportation system of energy from one place to another.[9] This means that the methods for allocating or pricing scarce transmission are inextricably linked with the price of energy in the wholesale market.

Price Signals Are Important—Get the Prices Right

Prices motivate efficient production, consumption, and investment decisions

The most important recipe for success is to get the prices right. In a competitive market, prices provide the most important signal that will motivate production, consumption, efficiency, investment, and innovation. The key ingredient in electricity markets and trading systems is to define short-run marginal prices that accurately reflect what the system operator must do to:

- maintain reliability
- balance supply and demand on the system
- manage congestion efficiently and economically

This is critical, because short-run marginal costs drive investment decisions.

The system operator needs to manage all of these three elements consistently, effectively, and not separately, because it will use the same energy to carry out all three functions. For example, calling upon a particular generator that may be the most economic to balance the system may also create or relieve congestion. The same actions may or may not affect reliability of the system.

The importance of getting the prices right cannot be over-emphasized. Efficient price signals provide the platform for the marketplace to work effectively. Once they are in place, bilateral contracts and all kinds of decentralized physical and financial trading, including self-scheduling and self-commitment, can be accommodated and coordinated. Indeed, the degree of flexibility and market freedoms that can be created is remarkable. Market participants can be given complete freedom, whether they trade on a long-term or a short-term basis. It will facilitate their investment decisions.

Postage stamp pricing

Many different transmission pricing or rate design methods have been used over the years to recover fixed or sunk transmission costs. These methods are also necessary in the new era of open grid access. The traditional approach of many regulators has been to require transmission owners to provide "firm" transmission service in return for charges designed to yield annual revenues that are equal to the embedded cost of the transmission facilities, namely a reasonable rate of return on the original cost (less depreciation, but including operations and maintenance costs and taxes). The charges were not distance or location sensitive, and it did not matter how much of the interconnected transmission system was being "used." This gave rise to the name "postage stamp pricing." Users could send power anywhere on the system. The concept also arose (particularly in the U.S.), that if power flowed through several transmission systems, it was said to flow along a fictional "contract path" consisting of transmission facilities between the points of delivery and receipt. Postage stamp pricing is very common, even in deregulated markets such as Norway, Australia, and Ontario.

Pancaking

Another complication that has caused inefficient use of the grid as a result of disincentives to trading is termed "pancaking" in the U.S. As the name implies, this involves paying a stack of transmission charges to different transmission owners in order to trade across the multiple transmission boundaries. The disincentives result from the requirement to pay multiple transmission charges, but this is not the only problem.

The concept of paying fixed and variable transmission costs under postage stamp pricing (the same price per MW, irrespective of location) where there is no congestion, involves treating fixed costs as variable costs. If fixed transmission costs are treated as variable, it can become profitable to invest in expensive downstream generation that reduces measured (metered) use of the transmission system. In this way, an investor will avoid the variable charges that are now higher than they would otherwise have been if some of the fixed costs had not been treated as variable. As a result, the total costs of the system increase and have to be borne by other users.

This problem manifests itself as strange behavior, such as selecting a connection point for a generating station deliberately in order to be on the down side of the meter so that the output of the plant is not taken into account in allocating the transmission charges.

The single, network-wide price recommends itself on the grounds of simplicity, but it may create incentives on generators to bid to exploit any known constraints. Pancaking is inefficient and inhibits access to the system and to trade. Indeed, if the margins are thin, the disincentive to trade can be significant. Distortions, true-ups, approximations, and distributional effects can be created, for example, where local customers pay less than traders across the same system.

An obvious solution would be to merge the transmission systems together and create transmission charges for the resulting single system or market. However, those users located on a lower cost system will not want to have to pay higher prices as a result of the merger with higher-priced transmission systems. As a result, the concept of "license plate pricing" arose, under which those located in one area would pay the postage stamp prices of the local transmission owner, but they would be able to use the whole of the interconnected system (*i.e.*, all of the systems comprised in the region of the relevant market).[10]

Allocation of direct and indirect costs

Postage stamp pricing and pancaking are issues that relate to the allocation of transmission charges among users. Cost allocation is termed in the U.S. as "ratemaking," covering all cost elements, both direct and indirect. Fixed (sunk) costs as well as short-run transmission costs are allocated through postage stamp pricing. Modern market design and transmission pricing separates the short-run, indirect costs of congestion and losses from the direct fixed costs of the existing assets (and the maintenance and asset management costs).

At the end of this chapter, we shall discuss the design of efficient transmission prices based on the principle of marginal costing. The principle means that it is more efficient to charge only the marginal cost for actual use and to recover the investment costs some other way. However, most of the costs of transmission are investment costs, and if they were simply allocated to users by dividing them by a number representing total usage, the resulting charge per unit of use would be much higher than the marginal cost of usage. We know that for the purposes of transmission expansion, we need to focus on the short-run marginal costs, namely the costs of congestion and losses that form the basis of investment decisions. These costs are recovered through a transmission usage charge. The costs of connecting or interconnecting a generation or load to the system are usually recovered directly from that user through a connection charge. The remainder of the fixed or sunk investment costs is recovered through a transmission access charge. We shall start by looking at the short-run marginal costs of congestion and losses.

The logic of locational prices

An alternative to a single price for the entire system could be different prices in different zones or different prices at every node on the system.[11] A single price makes sense when there are no constraints and no losses on the system but, as we have seen, when there is congestion (and no system has infinite capacity), the normal principles of economic dispatch will mean that some expensive generators will be dispatched at some locations and other cheaper generators will not run at all in other locations. The reason for this is, simply, that it is not possible to transport all of the energy from the cheaper locations to the more expensive areas if transmission constraints exist. Therefore, it makes sense to adopt a locational pricing system, because no transmission system represents an infinite busbar or conductor where no congestion will ever arise (nor would it be cost-effective for it never to be constrained).

The locational prices give efficient signals not only for the production and consumption of energy, but also for the construction of new generation and transmission facilities. Locational pricing will encourage generators to locate in areas where energy is more expensive and discourage them from locating in lower-priced areas that already have sufficient generation. The same incentives are created for consumers that should be encouraged to locate in low-cost areas. The locational prices also provide important signals for investment in new transmission facilities. If there is a significant difference in prices between nodes or areas, a clear signal will be created that investment in new transmission capacity would be valuable. The additional transmission capacity would enable more power to flow from generators in lower-priced areas to higher-priced regions. The marginal value of the new transmission investment will be equal to the price difference between the nodes or areas. This will form the basis of the investment decision, *i.e.,* whether the building of new capacity will be efficient and the investment justified (applying the basic principles outlined in chapter 2). The signals are important for market-based investment, as we shall discuss in chapter 5. They are also important for regulated expansion, because they will help the regulator to determine the need for the investment.

Cost socialization

Cost socialization (a "rolled-in-rate," in U.S. parlance) spread across all users on a $/kWh basis will result in a very blunt locational incentive, because it does not create signals that define for users exactly what their costs of using the system really are. In the interests of fairness and efficiency, costs should be allocated according to some measure of the benefits and of cost causation. The only alternative to signals from locational prices is centralized command-and-control decision-making on where to site generation and load in order to achieve efficient use of the transmission system. If costs are socialized and rates are "rolled in," distributed generators may be forced to pay for a portion of the transmission assets being used by other generators to compete against them. Competitors could find themselves taking "cut-the-wires" action to avoid excess transmission charges. The action would directly affect reliability and, even if it does not create stranded costs (which it could), it would shift the cost burden to other users.[12]

The right model integrates the energy market and the pricing of scarce transmission

To quote Sally Hunt:

> *The right congestion model is the integrated model, which properly integrates the energy market with the allocation and pricing of scarce transmission. In this model, scarce transmission is allocated as part of the optimization process. It is allocated so that the value of transmission is maximized. Traders who schedule transactions across valuable transmission lines are charged a transmission usage charge (a bottleneck fee) being equal to the energy price dif-*

> *ference between the two ends of the transaction. That way, there is no difference for traders between (1) paying a bottleneck fee and shipping energy from the export zone to the import zone, and (2) selling their energy in the export zone and buying it backing the import zone. There is no reason to squabble over allocation rules...*[13]

Efficient transmission access and usage are interrelated

The mechanisms for access to and allocating the use of the transmission system are obviously interrelated. Transmission usage is highly interactive with market design and transmission expansion, as we have seen. Pricing both access and usage is another key ingredient of good market design and of good transmission expansion schemes. Indeed, it is not easy to separate the issues that need to be addressed with respect to transmission access and pricing, congestion pricing, and transmission expansion. They all depend upon clear pricing signals being given to market participants and transmission owners to ensure that the transmission system is used in the most efficient manner and that the investment decisions are correct. The difficulty is that there are many ingredients. There are also several different flavors and seasonings.

The Important Principles of Market Design

Combine market operations with system operations

It has been demonstrated that a well-designed electricity market benefits from a system operator that is also a market operator to some extent. The reason for this is that real-time operations should be driven by the voluntary offers and bids of market participants to buy and sell energy that the system operator can then use to balance the system, manage congestion, and maintain reliability of the system with its security limits. The alternatives would be inefficient. At one end of the spectrum would be a centrally administered arrangement with the costs socialized. At the other end of the spectrum, there would be a market-driven arrangement that market participants might be able to manipulate, which would be very difficult for the system operator to coordinate to maintain reliability and security of supply. It is impossible for all of the activities of the system operator to be contestable. Only one entity can operate the system in real time. This is the most workable compromise to create the maximum market freedom.

In this way, the system operator can define market-clearing prices to pay those who offer energy and charge those who bid to take energy from the system in the real-time and short-run markets. The coordination of real-time and short-run forward markets is a natural consequence of the integration of the needed central coordination with market forces and commercial freedoms of market participants. In effect, the system operator is using the offers and bids to balance supply and demand, manage congestion, and maintain reliability, using energy that is priced at market prices that will support the commercial objectives of market participants. All of these purposes use the same energy. The system operator does not need to run all of the markets. Robert wilson points out:

> *"The [I]SO's responsibilities include real-time operations that protect system reliability, but its authority to intervene in forward markets is limited to cases where prior commitments promote reliability, such as day-ahead scheduling of transmission."*[14]

Although some of the actions of the system operator are uncontestable, its actions are largely based on the competing bids of market participants. Coordination through the system operator is unavoidable, as reliability must be maintained and supply and demand must balance.[15]

Open access to the spot market is necessary

Without open access to the imbalance or spot market operated by the system operator, parties to contracts would have to arrange themselves in a manner that ensured that the generation output followed the load moment-by-moment, and the load followed the output exactly. This is expensive, inefficient, and can only be achieved with a considerable degree of imprecision. The system only needs to balance on an aggregate basis. Mandatory balanced schedules are not needed and are, in any event, not commercially desirable. If the contracting parties choose not to match their schedules, the differences can be easily absorbed in the real-time market or in the short-term forward markets. Last minute purchases and sales can be settled at market-based, short-term, or spot prices. In other words, generators have a market for their uncontracted output. Consumers and retailers have a dependable market in which they can obtain energy to meet their needs for which they may not have sufficient contract coverage to meet their retail contract obligations. It is often said that this facilitates entry by new suppliers and retailers. In short, once the system operator has defined the market clearing prices to settle purchases and sales of energy to balance supply and demand, manage congestion, and maintain reliability, market participants are free to pursue a wide range of commercial transactions and investment choices, with price signals consistent with the system operator's wider reliability and other obligations.

Apply market-clearing prices and locational marginal pricing

Whatever services the system operator provides, it should price them efficiently, using market-clearing prices derived from market offers and bids, and principles of marginal cost pricing. As explained earlier, the prices must be consistent with security-constrained economic dispatch. They should reflect the marginal cost of redispatch required to relieve congestion and accommodate each transaction. These are the concepts on which all spot pricing and locational marginal pricing are most efficiently based.

Nodes are grid locations where the system operator defines spot prices using nodal pricing. Nodal prices differ by location as a result of congestion and losses. The differences may be positive or negative. Not all locational pricing systems reflect marginal losses as yet, as we shall discuss later.

For this reason, it is helpful to apply the principles of locational marginal pricing to define market prices for spot energy and transmission. The basic principle, and a key ingredient from the point of view of transmission expansion, is that the price for imbalance or spot energy at each location on the grid should be defined as the incremental cost of dispatch of meeting the last increment of load at each location, given the price offers and bids the system operator receives, the actual dispatch, and the constraints on the transmission system affecting the dispatch.

The settlement process becomes quite straightforward. Generators would be paid for their injections at the nodal prices for their respective locations, and loads would be required to pay for their withdrawals at the nodal prices of their respective locations. Bilateral contracts would be fully accommodated by the system operator, and deviations or imbalances from the bilateral schedules would be settled at the applicable nodal prices, so that generators injecting less or more than their contracted amounts would be charged or paid the nodal price at their locations for their deviations. Similarly, loads withdrawing more or less than their contracted amounts would be charged or paid the nodal price at their locations for their deviations.

Therefore, it follows that the price for transmission usage for any transmission schedule for a bilateral contract transaction would be the marginal cost of energy purchased or sold in the spot market that was required to accommodate the contract or schedule. The price for the transmission usage would be equal to the difference in the nodal price at the point of injection and the nodal price at the point of withdrawal. The same usage charges that apply to contract transactions could also be applied to transactions with parallel transfers (loop flows) across the transmission system resulting from transactions derived from external systems. This avoids absorbing redispatch costs or imposing curtailments.

If price signals are not right, inefficient administrative action will be needed

The objective of any market design is to provide all transmission system users with efficient price signals that reflect congestion and expansion costs and give market participants efficient incentives regarding the consequences of their transmission usage.[16] These efficient price signals are a key ingredient for transmission investment decisions as well as short-term usage decisions that will dictate whether it is economic to build new transmission facilities. They will give the market participants the commercial freedom to respond to the price signals without any significant concerns that their decisions will undermine reliability.

If the prices are not right, in order to maintain reliability and avoid overloading the circuits, the system operator will have to undertake administrative action to put in place restrictions that will curtail transactions, impose potentially arbitrary, non-market solutions, and provide penalties that may well interfere with market forces. They result in inefficient usage of the transmission system.

These restrictions typically involve *pro rata* physical curtailments, such as the NERC's transmission loading relief (TLR) procedures. Experience in the Pennsylvania New Jersey Maryland (PJM) power pool during 1997 provided a lesson as to what can happen if pricing is inefficient. Restrictions were imposed not only on access to the transmission system, but also on access to the spot market. Market participants in PJM were allowed access to the spot market in implementing TLR transactions to support internal transactions, while import and export transactions enjoyed no such support in order to avoid having the redispatch costs imposed on the PJM members. The PJM rules did not allow open access to the spot market until the independent system operator (ISO) implemented nodal pricing (as we shall discuss later).

Minimize opportunities for gaming

There may be short-run strategic behavior or price manipulation (gaming) by market participants that gives rise to limits on market freedom and flexibility. These limits in the form of mitigation may be necessary as the market matures, because the system operator cannot allow actions that jeopardize system reliability and may artificially increase energy prices for which market participants and consumers will have to pay. When energy and transmission prices are consistent and accurately reflect the marginal costs of congestion in the economic dispatch process, the prices will encourage generators to follow the dispatch and discipline those that participate improperly. This should also minimize the need for constrained-on or constrained-off payments and opportunities for gaming. The market-clearing prices will assist in maintaining reliability. Both real-time and short-run market prices create the basis for forward price signals, siting new generation, and siting of loads in locations that will help to relieve congestion rather than exacerbate it and provide the signals for market-based transmission investment.

Combine marginal costing, real-time pricing and security-constrained economic dispatch

An important step is to combine the principles of marginal costing and of real-time market pricing with the security-constrained economic dispatch used to maintain reliability. If the imbalance or spot market prices are not derived from marginal costs, incentives may be created to use the imbalance or spot energy market in ways that could shift costs to others. A key ingredient is that transmission usage charges should reflect the full opportunity costs of any usage (*i.e.,* the marginal costs of any redispatch required to accommodate a transmission schedule), otherwise market participants will have incentives to over-schedule transmission usage that will be inefficient and increase congestion. Accommodating these schedules through the imbalance or spot market would then shift redispatch costs on to others, encouraging them to over-schedule, creating further congestion.

The best framework for an electricity market is elegantly contained in Hogan's well-known diagram contained in Figure 4-2.

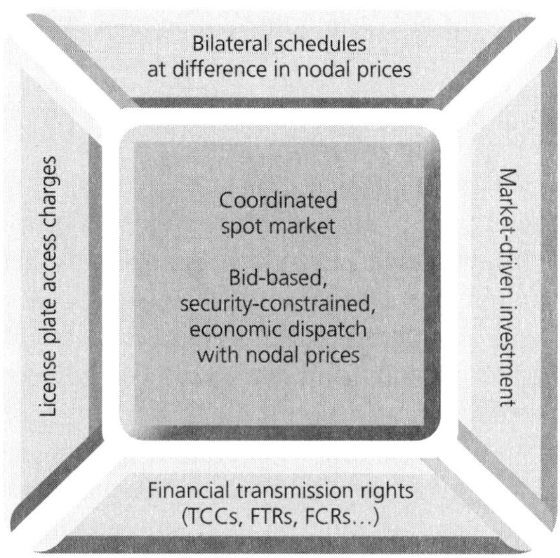

Fig. 4-2 Electricity market framework

Transmission Access and Usage

Transmission usage in practice

The way transmission usage would work in practice is that any transmission user would submit its final contract schedule to the system operator, which the system operator would accommodate through a bid-based, economic redispatch arranged by the system operator, provided that the redispatch bids were available and that the user agreed to pay the marginal cost of the dispatch to accommodate the transaction. The user would, in effect, agree to pay a bottleneck fee, which would be equal to the nodal price difference between the points of injection and the point of offtake. However, if the expected fee were more than the price the user was willing to pay, it could choose to curtail its transaction. In effect, the opportunity cost of transmission usage equals the marginal cost of redispatching the system to accommodate the usage that is equal to the difference in nodal prices at the point of injection and the point of off-take.

Counterflows

In order to allow the parties to make decisions more effectively, the system operator will continually publish the nodal prices as near to real time as possible. The transmission user can also implement a schedule that creates a counterflow, alleviating the marginal cost of redispatch for which it would be compensated. The compensation would be defined by the savings in opportunity cost and marginal redispatch costs as defined by the nodal price differences between the points of receipt and delivery for the counterflow schedule. This is another method of providing market-based, efficient congestion management for which market participants can be paid the market value of the congested interface.

In short, this security-constrained economic dispatch process would produce the corresponding prices at each location, incorporating the combined effect of generation, congestion, and also losses. In terms of its own supply and demand, each market participant could see a single price, which is the short-run marginal-cost price of power at its location. If an explicit transmission price is necessary, the natural definition of transmission is supplying power at one location and using it at another. The corresponding transmission price would be the difference between the prices at the two locations.

Trading hubs

Another ingredient can be added to the design that facilitates trading. Financial trading hubs, based on any aggregation of nodes, can be created. The hub prices would be defined by reference to the fixed weighted average of the prices at the nodes included in each hub. Market participants may then trade to, from, or at these hubs and have their energy trades and transmission rights settled by the system operator, using the hub prices. In the same way, retailers can aggregate nodal prices in any way that is convenient to facilitate settlements with loads at multiple locations. Figure 4-3 shows trading through nodes and hubs on a contract network.

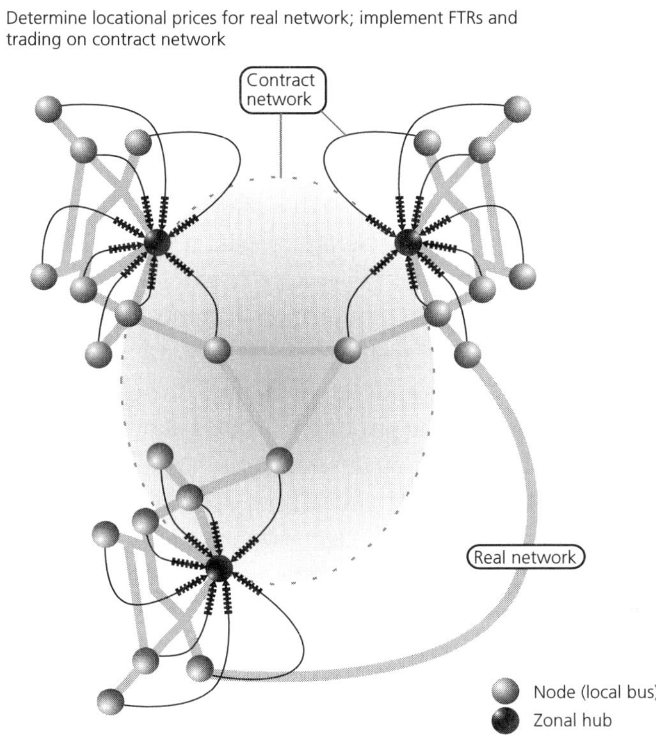

Fig. 4-3 Contract network connects with real network (determine locational prices for real network; implement FTRs and trading on contract network)

Zonal pricing

Some jurisdictions have managed without locational prices, preferring to utilize zonal pricing to give the locational incentives to generators and loads. England and Wales provide an example of this, where it was originally thought that the cost of implementing a nodal pricing system would not be outweighed by the benefits. In other cases, zonal pricing, notably in PJM, California, and Australia, has created difficulties.

Zonal pricing suffers from the defect that it does not necessarily allocate costs efficiently or fairly among transmission system users. It is attractive because it appears to be simple and to deliver the revenue requirements of the transmission owners. However, it is based on an assumption that all prices within a zone are the same, and it is seldom the case that the differences between all nodal prices within a zone remain negligible. In fact, pure nodal pricing can prove simpler to implement than an artificially constructed zone. The zone may be artificial in that its boundaries may reflect physical constraints, or it may have been designed to minimize the impact of congestion pricing on particular generators or loads.

Shi-jie Deng and Shmuel Oren proposed a priority-pricing scheme for zonal access to transmission systems that is uniform across all buses in a zone. They describe the scheme as "a hybrid of priority insurance and postage stamp approach." The priority insurance allows for network users' self-selection, which in turn provides economic signals for the allocation of scarce transmission resources. They find that efficiency gains can be improved if the network is divided into more zones that allow different insurance premium schedules to be offered in different functional forms. There is mathematical complexity in attempting the market equilibrium. Their calculations for discrete priority levels resulted in efficiency losses ranging from 3-5%, but they were not expected to be persistent. They regarded this to be consistent with the result of Richard Green[17] who calculated efficiency losses due to zonal aggregation (with no intrazonal priority pricing) in England and Wales and estimated these losses at less than 1% of social welfare. They recognized that further research was needed to explore how the proposal might be implemented and how the details concerning the management and allocation of risk and surpluses will persuade the ISO to "do the right thing."[18]

The PJM experience

PJM originally introduced a zonal pricing system in March 1997 with a spot market that computed a single price for the entire PJM system, ignoring transmission constraints. Generators were not paid if they were constrained off—in other words they did not have firm transmission service. Market participants could also schedule bilateral contracts and were charged a share of the total costs of congestion on an average basis. In June 1997, it was possible for a customer to buy from the spot market dispatch at $29, or it could arrange a bilateral transaction with a constrained-off generator in the West at a price of approximately $12. The small average congestion costs would not affect the choice. This created the incentive for constrained-off generators to enter into bilateral contracts, giving them priority that would result in the ISO being forced to constrain off some other competing generator. The ISO rapidly came to the point where it had no more controllable generating units with which to manage transmission constraints. Where demand exceeded supply on one side of the constraint, there was no available generation to meet the shortfall. The ISO was forced to resort to administrative mechanisms to prohibit bilateral transactions or declare a "minimum" generation emergency during times of system peak. The gaming came to a head in June 1997, and the ISO intervened by banning certain types of non-firm bilateral trades.

In April 1998, this system was replaced by a full nodal pricing system. The PJM system has some 2,000 nodes between which prices can differ. If it were true that there had indeed been a few zones, there would have been a few clusters of locations where the average prices were the same and the standard deviations would have been the same. William Hogan showed that there was substantial dispersion even in the first six months and that there were between 689 and 825 locations within the PJM area where the price points did not overlap. Between October 1998 and March 1999, there were 842 different locations with distinct prices. This meant that over the first six months, between 52 and 94 zones would have been necessary, and in the second six months, 61 zones would have been necessary (on the basis of a $1 per MWh threshold for the maximum deviation of prices across a zone).[19]

In the nodal pricing system, the perverse incentives did not arise. In the same circumstances that prevailed between April and September 1997, customers purchasing power from the spot market in the East would have seen a price of $89. Although they could have arranged a bilateral transaction with the generator in the West at $12, they would have had to pay a transmission charge of $77 ($89-$12), making them indifferent at the margin, just as intended. Customers in the West would have paid $12 and would have had no incentive to change. Every generator would be producing at its short-run, profit-maximizing output, given the prices. The market equilibrium would have supported the necessary dispatch given the presence of transmission constraints.

Difficulty of defining zones

The principle by which zonal boundaries should be defined is by differences in marginal costs during constrained periods.[20] They should not, according to Steven Stoft, be based on accurate congestion boundaries (California proposal in 1996) or the physical congestion of lines or paths.[21] This approach was adopted in California, where transmission constraints were thought to be relatively small within each zone, but large between zones. The reality proved to be very different. A usage charge was imposed on all users transferring energy across the zonal interfaces, calculated by reference to bids submitted by a scheduling coordinator to increase or decrease output in a particular zone. Stoft showed that this idea of a boundary or zonal interface was not well defined in the presence of loop flows.[22]

Zonal approaches are not simple

Both the theoretical work and also the experience in practice indicate that zonal approaches are not so simple and that nodal pricing is not so complex as originally thought. Stoft demonstrated that, in its complete form in a network with N constraints, N+1 prices may be enough to characterize the economic information completely.[23] All other prices would simply be weighted scalings of these prices, and the weights would depend only on the impedances of the network lines. However, it is not easy to define zones. It is possible to have zones with different prices, but without constraints between them. There can be zones within zones. Much depends upon the way in which the definition process is started by grouping nodes or buses together.

With strictly radial connections, locations within and between unconstrained zones would have a common price. Therefore, it is attractive to think that things could be simplified by aggregating these locations to achieve a few distinct zones on the basis that the locations would all have the same prices. This may result in major averaging of locational prices, which then masks and blunts locational signals. This can make life simpler for market participants and reduce the locational costs to certain participants at the expense of others. Certainly, if multiple locations actually do have the same prices, then there is no point to averaging them into zones. The purpose of aggregation is to reduce the number of prices, not the number of nodes or locations.

Defining zones on mesh systems

Another problem is that the definition of zones inevitably becomes more problematic in the case of a mesh system with loop flows. As usual, the radial line examples are overly simplistic and do not represent the real world. It had been assumed that differences in prices between zones would be caused only by transmission constraints that would be defined by reference to lines that connect the zones. It was also assumed that the differences in prices within the zones could only be caused by congestion on lines within the zone. These simplifying assumptions resulted in a further assumption that zones could be well defined and that what happened within a zone could be treated independently of what happened between zones or independently of what happened in other zones.

However, on any mesh system, we know that the interaction of the various parts of the system must always be taken into account, and in the more typical case of loops and parallel paths in a network, prices could differ within and between "unconstrained" zones as a result of the indirect effects of remote constraints. The prices within a single zone can differ even though there is no binding constraint within the zone. The difference in prices between the buses within the zone arise not as a result of constraints within the zone, but because of the loop flow effects interacting with the binding constraint between the zones. The determination of prices within a zone cannot be made independent of the effects of constraints outside the zone. In a strongly interconnected network, a wide variety of pricing patterns would be possible. In fact, with loop flows, it is possible for a single binding constraint to result in different prices at every location on the system, reflecting the fact that every location has a different impact on the constraint.[24]

Aggregation into zones may add to complexity and distort efficient price incentives and signals. There is a danger in adopting a zonal pricing approach, because it will distort the nodal prices by assigning uniform prices to regions in which the price changes gradually from one interzonal node to another. In conclusion, the assertion that conversion into zones will simplify the pricing problem is not supported by analysis of the conditions that exist on a looped network. This is because different nodal prices arise, because every location has a different effect on the constraint.

We now examine some examples of the difficulties and effects of zonal pricing.

The zonal approach in Australia

The Australian National Electricity Market (NEM) is divided into four zones represented by the four states that participate in the NEM (Queensland, New South Wales, Victoria, and South Australia). There is a reference node within each of these four states. In spite of the fact that Queensland is a large state with a long transmission system and several large intraregional constraints, it represents a single zone. There is a major constraint between the Tarong Power Station and the Brisbane regional reference node. Tarong is close to the point of interconnection of the Queensland-New South Wales interconnector (QNI). QNI was built to help overcome the capacity shortfall in Queensland, in order to provide cheaper energy and ancillary services from New South Wales, which has had a capacity surplus. There are also some very cheap coal resources in this particular area that have created incentives for the construction of several new power stations on the Queensland side of the border.

Because of the adoption of the zonal, as opposed to nodal approach to pricing, any one of these new power stations will be able to capture the

Queensland price that will stay at moderately high levels, because the Tarong constraint and the intraregional constraints will prevent the importation of cheaper energy. These new plants will not need to worry about being constrained off. They could, and doubtless will, bid low in order to be scheduled and dispatched and will receive the high Queensland market-clearing prices. The power will be sent across the border into New South Wales across QNI, which was not the object of the project when it was built. Perhaps these new plants would not have been built if they could only have captured the local price rather than the higher Queensland zonal price.

The zonal approach in England and Wales

The English solution that is still in place was to divide the transmission system into zones and to add discounts or premiums designed to incentivize the location of generation near demand to avoid the cost of building transmission. However, the locational prices that resulted were insufficiently differentiated, and the costs of gas supply made it economic for new entrants to locate gas-fired generating plants in areas where there were already significant transmission constraints. The original generation zonal charges were set at levels between £3.13/kW in the Northeast and zero in London, but by 1997-1998 the differential was between £7.98 and minus £10.11 in the far Southwest. The differential in the charges in effect on April 1, 2002 was £8.38/kW in the North and minus £9.86/kW in the far Southwest.[25] Losses had increased from 1.6% in 1990 to 2.4% in 1999, because the location of generation was a long distance from the load centers.

David Newbery is not convinced that nodal pricing would have made things better in England, given the underlying problems of market power in the Pool. A substantial amount of generation was committed before the zonal price signals were put into effect. The compensation payments to a constrained-off plant provided perverse locational incentives. The payments would encourage generators to locate in an area where exports were constrained and to bid zero for periods when they could expect to receive constrained-off payments. He suggests that NGC should, perhaps, not have offered firm transmission to generators who chose to locate in these zones, because their capacity was less valuable. He points out that the locational charges were revised by averaging the extra costs incurred by entry (or exit) over the relevant zone, rather than charging entrants the increase in total cost resulting from the entry.[26]

Richard Green suggests that moving to optimal nodal pricing in England and Wales could increase welfare by approximately 0.6% of the cost of generation. Gains from optimal prices will be greater when their effect on investment signals to generators is taken into account.[27]

Costs of inefficient location

It is clear that inefficient pricing results in inefficient location and use of the transmission system. Inevitably, this will have a cost to consumers. Newbery quotes another statistic from NGC, that a new power station in the North generating 100 MW only meets 93 MW national demand average across the system at the predicted 2003/2004 winter peak, while a 100 MW located in the peninsula Southwest meets 110 MW demand by alleviating power losses. It meets 19% more of the demand than the Northern station. Since all the capital costs should ideally be collected at the system peak, this differential implies that lifetime generating costs might be 12% too high for an incorrectly located station (on the assumption that half the cost is capital and operating losses add another 4%).[28]

Newbery and Pollit[29] demonstrated that the costs of inefficient location are large, and they can be in the order of 6.5% of the wholesale electricity price annually.

California is moving away from zones

California adopted a design that relied on a few large pricing zones that ignored critical features of the grid and the many locations that had important impacts on transmission constraints. The design adopted interface rights between zones and relied on decentralized trading between market participants in separated markets for ancillary services, balancing, and congestion management, instead of a security-constrained economic dispatch coordinated by the system operator. The original design established a separate power exchange and balanced scheduling requirements. The unfortunate outcome of this design is well known, and the California ISO (CA ISO) published a comprehensive redesign proposal in January 2002. The zonal system is criticized by the CA ISO because it has had to manage the system through real-time adjustments and by making frequent modifications to the rules in order to mitigate innovative strategic behavior and gaming by market participants. The report announcing the proposals for a new design involving locational marginal pricing and financial transmission rights observes:

> *...it will be far simpler, and more transparent, to design forward [congestion management] procedures to be as consistent as possible with the real-time operating needs of the grid.*[30]

In its comments on FERC's operational audit of the CA ISO, it was more specific:

> *Specifically, the CA ISO proposes to develop and implement a congestion management system that utilizes a full (detailed) network model and security-constrained economic dispatch with locational marginal pricing.*[31]

Zonal pricing is inefficient

In short, the simplification that is involved in a zonal approach will inevitably involve some loss of efficiency and result in cost shifting. The experience to date indicates that nodal pricing is not as complicated as originally thought, and the inefficiency that accompanies zonal pricing is not a price worth paying for simplicity. Even though views differ on the flavoring of financial transmission rights (discussed later), there is increasing consensus that locational marginal pricing (LMP) should form the basis of congestion management.[32]

Issues Where Locational Marginal Prices Are Not Adopted

Average out-of-merit dispatch costs approach

The New England ISO originally calculated congestion charges on the basis of the costs that it incurred in dispatching plants out of merit as a result of congested paths. These costs were averaged and charged to each load in proportion to the amount that it took relative to the total load on the system in the relevant settlement period. This is the approach that was initially adopted in England and Wales. Although it is simple to understand, it does not produce price signals that help to manage or alleviate congestion or that indicate when or where it would be economic to invest to expand the system to reduce congestion. This may be for the understandable reason that at the outset, the designers of the market rules believed there was no significant transmission congestion problem to manage. The New England approach was subsequently dropped in favor of LMP and financial transmission rights (FTR).

Type of service to be provided and compensation for lack of service

In a LMP system, all service is of a uniform type or standard and is regarded as firm, subject to the concept that scarce transmission capacity is allocated so that those who value it most highly must pay a transmission usage charge. There are obviously several different ways not to implement locational pricing. None of them are recommended, because they do not facilitate efficient use of the transmission system.

Where LMP has not been adopted, prices need to take into account the type of transmission service that is being offered. In many countries, the only type of service that is contemplated is what is called "firm" service, although the U.S. has extensive experience with "non-firm" service. Firmness is, in effect, a reflection of the degree to which the service can be interrupted and the periods of notice to be given by the transmission owner before the interruption takes place. Firm service can permit the transmission customer to change its transactions without incurring a charge in real time. There can be different prices for different degrees of firmness or, put another way, there can be it different levels of compensation for failing to provide service in different situations. However, Sally Hunt and Graham Shuttleworth argue that it is not desirable to allow utilities to offer service without payment of compensation if service is not provided, as it will act as a disincentive to operate, maintain, and expand the system efficiently. [33]

There is fairly widespread confusion regarding access rights around the world and also confusion regarding compensation for lack of transmission service. A number of systems do not impose penalties or compensation for lack of service at all. Argentina and Australia impose penalties, and Norway has introduced a system of penalties based on unserved energy. California transmission owners offer a number of degrees of firmness. PG&E made a substantial payment in respect of a transformer failure that prevented imports into the San Francisco area. Compensation to generators for being constrained off the system, which is typically included in market rules, could be regarded as part of the payment for transmission access.

In theory, users could choose a particular service to obtain access to the transmission system. They could enter into contracts for short-term or long-term transmission service with predefined conditions for access by reference to the prevailing market prices or system demand at particular times or seasons. Obviously, the compensation terms for lack of service could be specified and calculated by reference to opportunity costs or other financial losses suffered.

The U.S. gas industry has thrived on offering both firm and non-firm (interruptible) service and has been able to use the ability to do so to increase its revenues. The concept of non-firm or interruptible service has enabled gas pipeline owners to sell capacity that in aggregate was greater than the capacity of the pipelines.

However, there are regulatory issues that need to be addressed because of the unequal bargaining strength of the parties in negotiating transmission service contracts. Regulators might also need to monitor the degree to which non-firm service was provided or interrupted and the compensation paid, to give them a sense of whether the service contracts were operating fairly. It would also make the life of the system operator more difficult, because it would have to take into account all the different degrees of firmness and notice periods in order to establish which users had priority at which times. The protocols that had to be put in place in California to deal with different degrees of firmness and existing transmission rights were extremely complex and somewhat artificial. Like all administered, rather than market solutions, it creates a burden for the system operator and will tend to lead to inefficiencies.

Transmission Rights

Transmission rights are a key ingredient

Transmission rights may be financial or physical. In other words, they may be represented by financial instruments only, or the rights may allow for further physical scheduling of transfers across a congested interface. They are tradable property rights that can be used in a variety of ways to raise finance, provide security, or hedge risk. In the same way as other property rights in other industries, they are a key ingredient of the recipe for successful market-based (and also regulated) expansion.

Financial risks associated with LMP

From a transmission expansion perspective, there are two main advantages to the nodal pricing system:

- Building generation that increases congestion will tend to be discouraged
- Siting plant at locations to relieve congestion should be encouraged

While the risks associated with physical delivery are essentially eliminated under a nodal pricing system (firm service is provided), there are financial risks associated with locational marginal prices and associated usage charges that will involve some degree of uncertainty. The market participants will not know the effect of congestion on spot energy prices and transmission usage charges until after the fact, when the system operator has dispatched the system and calculated the nodal prices at each location.

Financial transmission rights (FTR) have been devised to provide a hedge against this uncertainty. FTRs are also known as "transmission congestion contracts" (TCC) in New York and "fixed transmission rights" in PJM and are identical except with respect to their method of allocation. FTR is a more generic term that could include TCCs that give rights to congestion rents only or rights to both congestion and loss payments. FTRs can provide a method to lock in the transmission usage charge in advance and to allow market participants to obtain access from one location to another, even though there may be congestion between them. They can also offset the impact of congestion on prices at any location.[34] A location for settlement purposes is a specific bus or node or the weighted average of several nodes.

In the presence of transmission congestion, a generation contract is necessary but not sufficient to provide the necessary long-term price hedge. A bilateral arrangement (such as a contract for differences) between the customer and a generator can capture the effect of aggregate movements in the market, when the single market price is up or down. However, transmission congestion can result in significant price volatility and movements that are different depending on location. Congestion might affect the customer and the generator differently, giving the customer a higher locational price and the generator a low locational price. The generator would be unable to provide the natural back-to-back hedge on fluctuations of the short-run market price. It would need an additional mechanism to hedge the price movements created by congestion.

Some jurisdictions such as Argentina and New Zealand have attempted to put in place nodal pricing without a system for FTRs that is clearly unsatisfactory, as it provides no mechanism for market participants to hedge against the uncertainty described earlier. The Australian National Electricity Code Administrator, in its draft report released in October 2000, suggested increasing the number of price regions or zones in the National Electricity Market to between 9 and 13, but did not recommend a move to full nodal pricing, on the basis that it might involve approximately 340 nodes with separate prices. The report does, however, make the point that without a firm hedging mechanism through fully developed FTRs, nodal pricing would expose participants to largely illiquid markets and unacceptable risks.[35]

Advantages of FTRs

The rights provide a number of advantages:

- They can provide a hedge against uncertain transmission or energy costs
- They can provide a degree of price insurance for load serving entities
- They can assist in meeting installed capacity obligations in conjunction with parallel property rights (similar to FTRs)[36]
- They can reward investment with rights that have a value and are tradable, so that investment can be realized
- They can encourage the retirement of an older plant by allowing the generator to sell its transmission rights to a new generator wishing to build a new plant at the same or an adjacent site, freeing up transmission capacity that could be utilized effectively by another generator or load without the need to expand the system[37]
- They may reduce incentives to make investments in inefficient transmission
- They may help to deal with market power issues

Spot-market locational prices define the opportunity costs of transmission that would determine the market value of the transmission rights without requiring physical trading and without restricting the actual use of the system. These transmission rights are equivalent to tradable physical rights. The market design creates the possibility of relying more on market forces to drive transmission expansion.

Market power

Cardell, Hitt, and Hogan[38] and Nasser[39] showed that in the presence of constraints, the behavior of oligopolistic players could be very different from their behavior in simple models of competition in the wholesale energy market. In particular, they found that a dominant firm might increase rather than decrease output if it were behind a constraint, and lower prices at other locations as a means of exploiting the constraints on the system in order to foreclose competition. The exercise of market power depends upon the strategic location of the generating plants on the system as much, if not more so, as market share. Moreover, market power may be exercised even if prices at specific locations are low.

Nasser showed that certain players possess local market power and, unless a maximum bid were specified, would be able to extract potentially infinite market power rents. He showed that the ability of some generators to profit from local market power depends crucially on the presence of limited transmission capacity.[40] Vernon Smith, in a laboratory experiment, suggested that generators could exploit their market power at the expense of transmission owners when uncontracted.[41]

There was a debate as to whether, in the absence of FTRs or TCCs, congestion rents would be captured by the active buyers and sellers, rather than the holders of the FTRs, giving little or no economic value to the holders of FTRs. This would mean that they would not be such a useful market power mitigation mechanism.[42]

Jurgen Weiss conducted a number of experiments with industry subjects, examining the impact of seller concentration and demand-side bidding on the existence or nonexistence of congestion rents. He found that in almost all cases, sellers with market power and active players were not able to preempt or eliminate congestion rents and, therefore, FTRs would have substantial economic value to their holders. He concluded that they were an efficient way of mitigating market power when LMP is adopted.[43]

In summary, market power can be mitigated in a market by contract through a combination of contracts for differences (CfD) and FTRs. In this way, volatility in market-determined nodal prices can be hedged. The incentive for non-competitive behavior should be eliminated if market participants are fully contracted.[44] If market participants are fully contracted, *i.e.*, they have contract coverage in the form of TCCs and CfDs corresponding to a feasible dispatch, no one will have the incentive to make disadvantageous grid investments. In other words, an investor would, in effect, have to compensate all other participants for externalities. Newbery has expressed uncertainty that sellers would choose to be fully contracted and forego their market power.[45]

Mechanics of FTRs

FTRs are designed to be directional. They are defined from one node or point to another and denominated in MW. In some contexts, they are referred to as "point-to-point" rights to distinguish them from flowgate or other transmission rights that will be discussed later. However, we shall refer to them as FTRs.

The holder of an FTR is entitled to the difference in the nodal prices between its choice of two points, which is equal to the congestion rent (transmission usage charge) multiplied by the number of MW corresponding to the number of rights that it holds. The funds to pay the rights holders come from the settlement surplus collected by the system operator from its market operations. The surplus arises naturally, because the usage charges are represented by differences in nodal prices paid by transmission users and because the aggregate revenue from spot energy prices paid by loads will be greater than the payments made to generators. The system operator keeps nothing. The FTR holder receives compensation for out-of-merit dispatch costs, and the transmission user pays the true opportunity costs. The transmission owner will receive the regulated fixed charges for the existing system.

In its standard market design, FERC proposes that if a revenue deficit occurs, it would be made up by transmission owners whose facilities are out of service, so that the holders of FTRs (called CRRs) would receive full protection against congestion costs. Revenue surpluses (after the rights holders have been paid) would be rebated back to the transmission owners and used to offset their other revenue requirements.[46]

FTRs create commercial freedom

The rights can be used to hedge (in whole or in part) the usage charge for any transaction, and it would be unnecessary for the rights to match the transaction. This will give market participants complete commercial freedom as to how they acquire or trade the rights for risk management purposes. Sometimes they may look for a perfect hedge to match a particular transaction, and at other times they may make the investment in transmission rights to manage only certain trading risks. The FTRs do not confer an exclusive right to use the line in question, unlike physical rights. The payments made under them would accrue as if the holder were the holder of a physical right that had sold it to the highest bidder. The result is as if the rights have been automatically traded and assigned to other users who place a higher value on them.

FTRs should not interfere with dispatch

The transmission rights must not undermine the system operator's security-constrained economic dispatch, nor should transmission rights be held to restrict the access of others. For this reason, the quantity of rights that are allocated or auctioned should be limited by the simultaneous feasibility test (discussed in more detail later) of requested rights for a set period of time. The requests for quantities of energy injected and withdrawn at particular nodes, together with the price offers or bids for the rights, are entered into a dispatch model, which includes all potential constraints on the transmission system. Rights auctions must be undertaken on a centralized basis by the system operator and repeated at specified intervals (*e.g.,* weekly, monthly, yearly).

Initial allocation of rights to existing transmission capacity

An important and often controversial issue is how the initial rights relating to the existing transmission capacity are auctioned or allocated initially. The incumbent transmission owners that made the investment may feel that the rights should be allocated to them or, if the rights are auctioned, that the owners should receive the auction revenues because the owners made the initial investment. New entrants will prefer to have the opportunity to acquire the rights they need in a primary auction rather than in a secondary market, because they do not want to be dependent upon the decision of other market participants as to whether they will sell or hold on to their rights.

Load-serving entities, existing transmission contract holders, and users that are regulated and risk averse may feel that the rights should be allocated to them, in proportion to the current use that each of them (or its load) makes of the transmission system, because in a broad economic sense, the existing users and their loads have paid the sunk costs of the existing transmission assets or will do so through their existing transmission contracts. They need the FTRs in order to hedge the risk of price fluctuations and to help them meet any installed capacity requirements (in conjunction with a parallel set of property rights). If they were unable to hedge the risk in this way, they would seek to pass it on to their customers (through regulated prices if retail competition does not exist). In theory, if their customers move, the rights should move with them.

It has been argued, however, that load-serving entities are unlikely to be prepared to trade their rights in secondary markets. These entities will tend to feel that if they sell the rights and the rights prove to have been valuable, they will be criticized. If load-serving entities hold on to them and the rights have no value, nobody will notice. If they hold on to their FTRs, it will be very difficult for new entrants to acquire rights, and the FTR markets will be illiquid.

Auction revenue receipts

The marketers and resource owners advocated a "business solution" to this problem. Auction revenue receipts (ARR), *i.e.,* the right to receive the revenues from the initial FTR auctions rather than FTRs themselves, would be allocated to transmission customers or users in order to convert their existing transmission service contracts or entitlements into FTRs. They would then participate in the FTR auctions and receive the proceeds under their ARRs. The system operator would auction the FTRs in the usual way, and, therefore, no user would be in a position to hoard the initial allocation of FTRs. The participation by the existing transmission customers and users in the initial FTR auction would, essentially, force them to value the rights they would like to receive and to set a floor or reserve price. If someone else were to bid a higher price for the same FTR and only one of the FTRs were feasible, the auction receipts would still provide funding to hedge future congestion costs.

However, it is not easy to place a value on FTRs at the outset of a new trading arrangement or market. Existing transmission users are concerned about the resources they must commit in order to participate in a mandatory auction intelligently. They have been worried that market participants might bid for rights in order to enhance their transmission rights positions, or even to monopolize them. This is particularly true of transmission users that do not own transmission facilities, but rely on their existing transmission service contracts under which they may well have paid for their transmission service in advance. They will require that their existing contracts should be honored in exactly the same way as they have been in the past, and their rights may be very long term.

The difficulty that the existing transmission customers and users have faced has resulted in a compromise solution to phase in FTR auctions over an initial period. Under this arrangement, FTRs would be allocated to the holders of existing transmission contracts and load-serving entities on an interim period, after which they would be required to participate in a mandatory auction for all rights. This compromise was designed to balance the need for liquidity in the FTR market with the concerns of existing transmission users, uncertain of their ability to value the rights that they are allocated appropriately. The rights would be allocated to existing users based upon their current use of the transmission system. Under this scheme, the existing users would have a trial period to become comfortable with determining the value of the rights that have been allocated in order to prepare themselves for the mandatory auctions.

The U.S. displays a range of these options:

- PJM and New York— allocated FTRs as financial rights to existing transmission users and had a mandatory auction for the remaining residual rights
- New England—proposed that auction revenue rights be allocated to existing transmission users and that all FTRs should be subject to a mandatory auction
- California—held a mandatory auction of FTRs that conferred physical rights and allocated auction revenues to existing transmission users

FTRs for sunk costs and new investment

The general principle is that transmission owners receive payments reflecting embedded or sunk costs for the existing transmission assets and investment costs for new additions. These payments can be cost-based in accordance with traditional regulatory principles, or market-based. Actual construction and ownership of the assets could follow various arrangements with the ultimate economic value defined in the FTRs. Those paying the regulated fixed charges for the existing transmission assets receive the associated FTRs. The FTRs could be allocated to users paying access charges to the transmission system or sold in an auction with the revenue used to reduce the access charges. New investment, either by existing transmission owners or third parties, would create new FTRs. If the investment is included in mandatory access charges (payable by all users), the investment would be regulated in a traditional way and the incremental FTRs would be allocated to those paying the access charges. If the investment is market-based and not included in mandatory access charges, the pricing could also be market-based and the investors would receive the incremental FTRs.[47]

Clearly, most of the FTRs will be based on the existing transmission capacity and locational marginal prices. FTRs act efficiently as a mechanism to price the use of those assets. However, the question of the allocation of sunk costs must be dealt with. The value of the FTRs to market participants is unlikely to be the same as the regulated cost. It is not easy to efficiently deal with the allocation of the difference between the value of the FTRs and the regulated cost.

One idea is to treat sunk transmission costs in the same manner as sunk generation costs. In other words:

- Use market pricing for FTRs associated with existing grid by carrying out an auction
- Include the delta (whether positive or negative) in a non-bypassable grid access charge
- Allow the owners of the existing transmission assets to receive regulated prices or costs
- Require the captive customers or ratepayers to pay or receive the market delta (depending on whether it is positive or negative)[48]

Auction and application mechanics

An auction is one of the mechanisms for allocating FTRs. It is desirable as an element of the market even if FTRs are also allocated using other mechanisms. There are three primary reasons:

- It ensures that feasible FTRs are made available to market participants
- It allows FTR holders to reconfigure FTRs acquired through other mechanisms
- It allows market participants to sell counterflow forward, making additional congestion hedges available in forward markets

The frequency of the auctions can vary. For example, PJM holds monthly auctions for FTRs corresponding to uncommitted grid transfer capability. The New York ISO holds periodic auctions for long-term FTRs (corresponding to uncommitted grid transfer capability and transfer capability initially allocated to native load), as well as monthly reconfiguration auctions.

In both PJM and New York, buyers can bid for an FTR between any combination of receipt and delivery nodes, including hubs. In New York, bids must be between locations (buses, zones, or external proxy buses) at which the New York ISO calculates prices in the day-ahead market. In PJM, bids must be between receipt and delivery buses, including the PJM hubs. Each bid to buy an FTR must state the maximum price per MW that the buyer is willing to pay for the FTR. It will pay no more than its bid, but may pay less. Each bid to sell an FTR must state the minimum price per MW the seller is willing to accept for the FTR. It will sell its FTR for no less than its bid, but may receive more.

The awards will be subject to the constraint that the awarded FTRs be simultaneously feasible in a security-constrained dispatch in conjunction with all outstanding FTRs. This will enable the system operator to collect enough congestion charges to pay congestion rents to the FTR holders.

Each buyer in the auction will pay the market-clearing prices for the FTRs that it purchases, and each FTR seller receives the market-clearing prices for the FTRs that it sells. The prices of all FTRs sold in the auction are related through the impact of the transmission flows associated with those FTRs on binding transmission constraints. Therefore, the auction values all FTRs, whether or not they are bought or sold in the auction and the system operator will publish the prices of all FTRs relative to a reference bus.[49]

This centralized FTR auction provides useful means of reconfiguring FTRs. Because the FTRs are determined by the bids of market participants and not the decisions of the transmission owners, FTRs purchased in the auction, may be different from those sold and may even be different from any of the outstanding FTRs. Reconfigurations are necessary to facilitate hedging risks with respect to changing trading patterns.

FERC Order 2000

FERC's Order 2000 requires each system operator to provide tradable transmission rights that can hedge locational price differences resulting from congestion and requires the transmission rights system to promote an efficient dispatch.[50] Order 2000 also requires efficient pricing signals with respect to short-run operations and long-run investments. To quote Order 2000:

> *[E]very RTO must establish a system of congestion management that establishes clear rights to transmission facilities and provides market participants with price signals that reflect congestion and expansion costs (p. 489)*
>
> *[W]e will require the RTO to implement a market mechanism that provides all transmission customers with efficient price signals regarding the consequences of their transmission use decisions. We are convinced that efficient congestion management requires that transmission customers be made aware of the cost consequences of their actions in an accurate and timely manner, and we believe that this is best accomplished through such a market mechanism (p. 382).*

In its proposed standard market design, FERC mandates both LMP and point-to-point FTRs (which it calls "congestion revenue rights" or CRRs)[51]

PJM and New York have created effective systems of FTRs that entitle the holders to the difference in nodal prices (or the congestion component, where the nodal prices also include marginal losses). These rights provide the means to acquire a perfect hedge, because the usage charges, which cannot be known with certainty in advance (since they are calculated ex-post), are calculated by reference to the same nodal price differences. As we have observed, an important advantage of FTRs is that they are purely financial and do not interfere with the system operator's security-constrained dispatch. The rights are quite separate from the physical use of the system in a number of ways:

- FTRs are not required to permit physical access to the transmission system
- There is no requirement to match the rights to the scheduled transaction or vice versa
- The market participants can choose whether they participate or not in the markets operated by the system operator, whether or not they hold FTRs, and the dispatch will not be affected by the rights that they hold in any way
- The rights holders will receive the market value of their rights (the nodal price differences), whether or not they carry out a matching or non-matching physical transaction

In short, they do not interfere with the physical dispatch of generation and the ability of the system operator to maintain reliability.

Physical Transmission Rights

Need to match rights to use of the transmission system

It is possible to establish a system of physical transmission rights. The physical approach, in effect, entails market participants acquiring rights that match their expected transactions in order to utilize the transmission system and obtain physical access. On a complex network, the burden on the system operator would be considerable in ensuring that each physical transaction that was scheduled was accompanied by exactly matching physical rights to use the transmission system. The main difficulty for the market participants is that if they make last-minute changes to the quantities of energy they trade or the points of input or off-take of their transactions, they would have to acquire new rights (or exchange the rights they held for other rights) for their last-minute changes or transactions. If they were unable to do so, perhaps because the secondary marketing physical rights was insufficiently liquid, they would not be able to obtain access to the transmission system and might lose the value of any physical rights they held that no longer matched their proposed trades.

Because of the laws of physics and the fact that power will flow in a manner that cannot be physically directed by the system operator, physical property rights on a complex network would be difficult to define, and, in practice, all users will be sharing each others' transmission rights. The most efficient answer is for the system operator to ignore who owns what rights and instead dispatch a plant in a manner that ensures that the entire transmission system is used most efficiently. This is the reason that financial rights are preferred by system operators and market designers.

Market power concerns with physical rights

Another problem with physical transmission rights is that they can be hoarded by a market participant that could exclude its competitors from using the transmission system, simply by withholding the physical rights it had acquired. It might have no intention to use the rights until the physical scheduling deadlines had expired. If the holder of physical rights can withhold any of those rights from other users when they are not being used, it is evidence of abuse of market power.[52] For this reason, "use-it-or-lose-it" rules have been implemented to prevent hoarding of the rights and to encourage efficient use of the transmission system. To the extent that any holder does not confirm its intention to use the rights at the specified time to the system operator, it will lose the right to use them. The market rules will set the specified time to enable the system operator to conduct a final auction of rights that are not being used.

However, these rules essentially require the market participants to trade their physical rights in the secondary market before they are certain of whether they will need them nearer to real time. The idea of trading on a recallable basis would leave the market participants in a difficult position if the rights they had acquired in this way were to be recalled at the last minute, because there would be little time to trade again to hedge the risks created for their schedules. Therefore, physical rights on a network can inhibit the commercial freedom that market participants need.

Market participants would be discouraged from entering into more profitable transactions or more efficient dispatch, if it meant having to change their physical rights at short notice. From the perspective of the system operator, the allocation of physical rights would define the dispatch, which might not be the most efficient that the market participants and the system operator could arrange at short notice. Efficient dispatch to relieve unexpected congestion would be much more difficult and, in some circumstances, impossible to achieve in real time, if the system operator were to respect the physical rights.

Interference with efficient dispatch to maintain reliability

FTRs are clearly superior to physical rights, because they can be acquired and traded independently from the physical dispatch of energy. By contrast, physical rights may interfere with efficient dispatch and congestion management by the system operator and create additional transaction costs and risks for market participants. Efficient dispatch and efficient congestion management coordinated by system operator are important ingredients in ensuring system reliability when the grid is constrained.

Physical rights on European and Scandinavian interconnectors

Physical rights can be used in suitable circumstances. They have been successfully used for access to interconnectors such as the England-France direct current (DC) interconnector and the interconnectors across the Baltic Sea. The Cross Sound Interconnector in New England was financed by the sale of physical rights. It was not possible to use financial rights, because, at the time, neither market at either end of the interconnector (New England and New York) had yet fully defined their locational pricing nor transmission rights regimes. However, they are convertible into FTRs. They are suitable in these cases, because a DC interconnector is not an integral part of a mesh network, and transfers across interconnectors tend to be scheduled separately as if they were generators or load at the boundaries of the system, because they are radial and controllable lines. There has been a significant regulatory oversight of the rights auctions to ensure that there was open access and that the "use-it-or-lose-it" principle was adopted.

The issue of long-term capacity reservation, particularly in favor of affiliates, is a difficult one for regulators. The European Commission (EC) has disallowed it in the case of three Baltic interconnectors, as we shall discuss in chapter 5. It raises concerns as to the potential for abuse of market power with respect to scheduling transfers across the interconnector or transmission line and within the markets at either end. Michael Albers of the Competition Directorate of the EC has issued a paper indicating that capacity reservation is only justified if it is necessary to secure a revenue stream to enable the financing to build a new line or interconnector to be raised.[53]

There are transmission rights that represent a hybrid between financial and physical rights, where the holders of financial rights are given priority over other users of the transmission system in the event that physical curtailments of transactions are necessary. In effect, this is equivalent to a physical right that cannot be withheld.[54] A physical right that cannot be withheld can be economically equivalent to a financial right if the spot market can clear. If it cannot clear, administrative rules will have to be used to curtail transactions and allocate priorities.

Effects of rights allocation mechanisms on market power

Paul Joskow and Jean Tirole, in their comparison of the welfare properties of financial and physical transmission rights, demonstrated that both types of rights can enhance electricity seller or buyer market power in essentially the same ways. Without use-it-or-lose-it rules, the physical rights would have potentially worse welfare properties than financial properties, because they can reduce utilizable transmission capacity by being withheld from the market. Therefore, it is important to take into account the potential adverse effects on market power in the design of rights allocation mechanisms and regulatory rules regarding the use of the rights and the potential for hoarding them. The effects, however, depend upon the detail of the transmission rights market and the nature of the market power, including the location of the buyers and sellers. They describe their conclusion as simple and intuitive:

> *The possession of financial rights by a producer in the importing region by a consumer in the exporting region aggravates their market power, since financial rights give them an extra incentive to curtail their output or demand to make the rights more valuable. In contrast, the possession of financial rights by a monopsony in the importing region mitigates its market power by giving it an incentive to raise price in the importing region.*[55]

Rights Are a Key Ingredient for Market-Based Expansion

FTRs to facilitate transmission investment

In effect, FTRs create tradable property rights that can be sold to raise finance to build new lines and interconnectors. They can be used to secure the financing or to create a certain revenue stream that will assist in raising the necessary finance. A key ingredient in the market design is to ensure that the incremental FTRs that become possible because of the expansion are awarded to those who fund or invest in the expansion.

The investors can allocate or auction FTRs in advance, and the resulting revenues can be used to finance the investment. The rights can be traded in forward secondary markets, and, in the same way, the initial allocation of the rights will reflect the market value of using the grid. Forward trading will reflect the market's forward valuation of the use of the transmission system. Put another way, the prices will reflect the expected values that the market places on avoiding congestion charges. In this way, FTRs provide a measure of the market value of upgrading the transmission system to relieve the same congestion. Typically, traders will want to acquire rights between points where they regularly buy and sell energy so that they receive back exactly what they pay in congestion charges. In this way, they are paid in advance for the right to the short-term use of the transmission path in question, free of charge.

The effectiveness of FTRs as tradable property rights that can be used to raise finance to expand transmission systems depends upon the rights being valuable and the predictability of that value over time. The rights can be valuable, because they will give the rights holder the ability to move power from one end of the line that it has funded to the other without having to worry about paying congestion charges. The rights will be valuable when there is congestion that will create incentives to build when and where additional transmission capacity is needed. The incentives will match the need, because the alternative to building will be to pay the cost of congestion.

FTRs are also useful for regulated transmission projects. The FTRs that relate to the additional capacity that is created can be auctioned, and the proceeds can be applied to reduce the fixed costs.

Long-term FTRs

Most FTRs that have been auctioned with respect to the capacity on the existing transmission system have had a relatively short term—one week, one month, three months, one year, or five years. To encourage investment in transmission, the FTRs allocated to the investors should ideally have a term that is sufficiently long to assist in repaying the debt and providing a reasonable period for a return on the equity. Although the experience to date is mainly with short-term FTRs, it is clear that market participants are becoming increasingly interested in longer-term FTRs. They might like to be able to conclude power purchase contracts of similar length in some cases.

Determining incremental FTRs following expansion

An important issue is the measurement of the investment in order to decide what FTRs should be granted to the investor. This means that another key ingredient of the market design is a predefined method for determining the incremental FTRs made possible by an expansion. In this way, the award of the FTRs enables investors to receive the market value of their expansion and to remove some of the market or usage risk associated with the investment. A policy on measurement of incremental investment and property rights will need to be put in place in each market. PJM awards incremental rights to investors in facilities associated with new generation interconnections and also grants related capacity rights. PJM and New York are both developing policies for measuring rights associated with pure transmission investments. Measuring what is incremental inevitably involves establishing the baseline use, but this can change as a result of either participant bids for TCCs (as in New York) or designated network-based FTRs (as in PJM).[56]

It should be possible to look at what is incrementally feasible given the existing uses. It should also be possible to compare the same bid on the existing and the expanded system, and calculate what is incremental. Although this sounds simple, there are complications because of the numerous feasible incremental uses, but an auction device should be able to deal with it. Under this auction device, the investor is treated as a third party and gives pseudo-bids of its preferences (weights) for the contents of a portfolio of FTRs. The system operator will solve for the most valuable combination. The existing FTRs are held constant, and the value of choice is maximized. The system operator will solve for any destruction of existing FTRs by the new investment through counterflow FTRs, and the investor would buy those FTRs that were negative.

Simultaneous feasibility test

The simultaneous feasibility test is another key ingredient that is necessary to ensure that the system operator is always able to collect enough revenues from nodal prices and usage charges to be able to make payment in full to the holders of the FTRs, even if the dispatch is different from that assumed in the simultaneous feasibility test. The test involves ensuring that the quantity of rights issued between all locations never exceeds the quantity of physical transactions that can be accommodated on the transmission system under expected system conditions.[57] In essence, the simultaneous feasibility test defines how many FTRs can be issued. If there is a danger that the system operator is unable to realize sufficient revenues to fund the FTRs, their hedging value will be diminished.

In an interconnected network, it is not possible to define the capacity of the network without also defining the pattern of usage. Except in the case of truly radial systems, transmission congestion can change the capacity of the grid, depending on the configuration of load and generation.

This difficulty in defining the capacity of the grid without specifying the pattern of use does not mean that it is impossible to define capacity or property rights. For example, it is always possible to determine if a particular pattern of usage of the transmission grid would be within the security-constrained dispatch limits required for reliability. The system operator makes these judgments all the time, with an affirmative conclusion implicit in every dispatch. Hence, it is legitimate to define any feasible pattern of transmission system use as being within the limits of the system, and the associated input levels can be defined as transmission rights.

The restriction on the allocation of new FTRs after a grid expansion is that both the old and the new FTRs must be simultaneously feasible after the expansion. If the set of FTRs corresponded to a feasible dispatch, the payments required by FTR holders would never exceed the aggregate of the congestion rents collected by the system operator. This is important from the point of view of system operator solvency and incentives to invest. Bushnell and Stoft showed that if the original set of allocated FTRs initially matched the actual dispatch, and the new FTRs were allocated to investors under the feasibility rule, then detrimental grid modifications would be prevented.[58]

Figure 4-4 gives an example of the concurrent auction of FTRs.

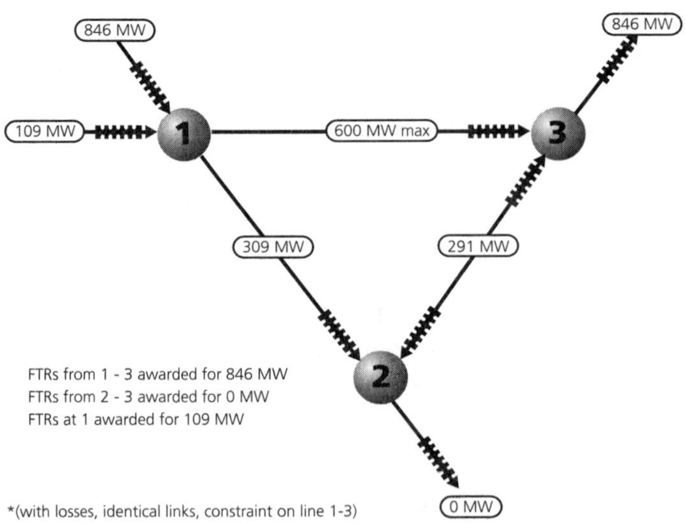

Fig. 4-4 Concurrent auction of financial transmission rights (with losses, identical links, constraint on line 1-3)

Effect of expansion on congestion rents

The question is often asked as to what will happen to the value of the FTRs if further upgrading is carried out at a later date in order to alleviate the remaining congestion or new congestion that may have arisen as a result of increased usage of the transmission facilities in question.[59] The particular concern is that the difference in the nodal price at either end of the line in question (*i.e.*, the congestion rent that an FTR holder will receive) may well reduce or even disappear as a result of the expansion. The building of new generation on the import side of the expanded facility may also reduce the price differential and, hence, the congestion rent.

This argument does not negate the value of FTRs in facilitating expansion. At the time that FTRs are allocated, they can be used to hedge existing and future costs of energy sales and purchases and future congestion costs. The rights may have value in both the current and forward markets that could be realized at the time the rights are allocated and at any time afterwards. The holding of FTRs by a generator in a low-priced area will have value as hedge against the location of new, low-cost generation in a high-priced area, hedging both price reductions and loss of market share. The holding of FTRs by loads can provide a hedge against risk management costs, generator market power, and their shares of the costs of ancillary services and reserves. FTRs can help to deal with the free rider problem and difficulties with forming coalitions to carry out expansion projects. The application of the simultaneous feasibility test protects the definition of the FTRs, not the value. In competitive markets, the value of investment can never be precisely guaranteed, but in this case, the property rights will be protected.

It is useful to analyze the issue in the following way. The cost of expansion is equal to the cost of congestion at the margin. In theory, it would be possible to expand the system (if it could be done relatively cheaply) in small increments so that, at any moment, the transmission capacity would equal the exact amount the market needed for equilibrium. However, it is more efficient to make a single larger-capacity addition at one time. The investment could be made on a speculative basis, the FTRs could be sold or auctioned afterwards, and the revenues could be utilized to defray the costs of the investment. The investment would have been carried out on the basis of an expectation that the revenues would be greater than the cost of the expansion, but the expansion may give the opportunity to free riders to use the system too. This strategy is risky, so the more likely approach is to carry out the investment after an auction or sale of the future FTRs, in which case the investors will know what the auction or sale proceeds are and can estimate more exactly whether the revenues exceed the expected costs of the expansion.

In short, if the transmission investment eliminates congestion entirely and the FTRs have no value, it may not be a problem as such. The investment might still be cost-effective, because it allows more cheap power to substitute for expensive alternatives. In other words, cheaper, but previously isolated generators will be able to sell their power into the same market at the same prices as everyone else. Loads will be able to purchase energy at lower prices from previously isolated generators.

If it is impossible to auction the rights, either in advance or after the investment has taken place because the beneficiaries are difficult to identify or there are too many of them to negotiate contracts with or to attract to an auction, there could be a market failure, in which case a regulated approach to the expansion will be needed. Even in this situation, it may not be necessary to adopt the immediate solution of socializing the costs.

Regulatory concerns over secondary markets

The establishment of mechanisms for secondary trading of both financial and physical rights has been the concern of both market participants and regulators, even though regulators do not (or should not) regulate the prices. The market may be centrally administered as in Ontario, which will be adopting a market-based system of FTRs in relation to its interconnections with neighboring jurisdictions, administered by the independent market operator (IMO). In other cases, such as the interconnector between England and France, its owners will auction physical rights and facilitate secondary trading. The concerns of the regulator are the obvious issues of open access, market foreclosure, and the abuse of monopoly power. They will be concerned about the market power effects of long-term capacity reservations and the use of reserve or minimum prices to force up the bidding.

Obligations and options

There are two flavors of FTRs—"obligations" and "options." The markets in the northeast of the U.S. have implemented obligations. An obligation requires the holder of the FTR to pay the difference in nodal prices if the difference is negative. This could arise if the nodal price at the point of offtake defined by the FTR were higher than the nodal price at the point of injection defined by the FTR.

Obligations may sound unattractive to market participants, because they run the risk of having to pay for a negative difference as well as receive the difference, if it is positive, but they can provide hedges for proposed counterflows that help to relieve congestion. Counterflows will often reduce the marginal cost of any congestion or dispatch. When the price difference is negative, the holder of an obligation FTR would either provide or schedule the counterflow and be paid the difference in nodal prices to offset its obligation to make a payment. It could also pay the marginal redispatch costs for the FTR's location, which, once again, will be defined as the difference in nodal prices between those locations. This might be more of a problem for those who trade on an entirely speculative basis than for those who trade in order to hedge the uncertainty of their congestion costs.

In essence, an obligation is a method of selling congestion management services on a forward basis. The idea is that the holder is hopeful that the amount paid in advance to take on the obligation will turn out to be greater than the cost of either providing the counterflows in real time or paying for the redispatch. Because counterflows help to alleviate congestion and expand the use of the grid, it is argued that obligations are helpful in accommodating more transactions on the transmission system and, therefore, allow for more rights to be issued under the simultaneous feasibility test.

By contrast, "options" provide a hedge against the congestion or usage charges when the nodal price difference is positive, but there is no obligation to pay the difference when it is negative. Clearly, options will be preferred by market participants that are not in a position to schedule counterflows, such as retailers with no generation. Therefore, it follows that if all the FTRs are options, no holder will be obligated to provide or pay for counterflows and they are unlikely to be forthcoming. This may mean that counterflows, cannot be used to the fullest extent to manage congestion and maximize the use of the transmission system, so that fewer transactions can be permitted and the system operator will have to allocate fewer FTR options. Also, options are not decomposable relative to a hub.

Problems with the simultaneous feasibility test with respect to options

The simultaneous feasibility test is straightforward if the FTRs are obligations, but, hitherto, there has not been a simple method for applying the simultaneous feasibility test to options. If this problem can be solved, FTRs could be defined as either obligations or options or both, and the system operator could issue some of each according to the needs of the market.[60] It looks as if a solution can be found by the application of what might be described as "brute force." It is a question of load flow analysis and the number of calculations that would have to be made under it. It appears that the solution would involve considerable computing power to carry out a large number of calculations. In practical terms, this means that the solution many turn upon the cost of the computer time rather than the software.

The appeal of options and forward obligations

Hogan has drawn attention to the attractions of options and longer term obligations and has suggested a way forward:

> *Transmission rights defined as forward obligations for the existing capacity are easier to analyze for revenue adequacy, but options and longer term obligations have great appeal. Transmission rights defined as obligations could be offered over longer periods and without perfect connection to the physical capac-*

> *ity of the grid. The tradeoff between partial analysis of the feasibility of options and the market value of transmission rights defined as options could be evaluated by a Gridco [transmission owner] that was prepared to take the business risk in exchange for an appropriate return. The Gridco could make a business judgement about the risks and rewards.*[61]

Managing the risk of revenue deficits

There is no guarantee of revenue adequacy for FTRs if conditions on the transmission system change. For example, forced outages can prevent certain transactions from taking place that might reduce congestion and, as a result, reduce the congestion revenues collected by the system operator. If the system operator has insufficient revenues to fund all of the outstanding FTRs, a rule will be needed to deal with this shortfall.

The most obvious answer is to place the risk in the hands of the party best able to manage it, namely the transmission owner, which, under contract with the system operator, could be incentivized to maintain the transmission system in a manner that minimizes the risk of revenue inadequacy caused by outages. It could be required to make up the revenue shortfalls attributed to transmission outages and other capacity reductions within its control. It is not clear what mechanism could be put in place to reward the transmission owner with a higher return if it met superior maintenance targets, however, if as a result there were no congestion. In other words, there is no point in creating incentives for the transmission owner to create additional capacity that is not needed, because it will have no value. If, however, the transmission owner were able to increase capacity by superior maintenance and asset management in a manner that was predictable and sustainable for the requisite period, it should be entitled to FTRs representing the additional capacity for the relevant period and be able to sell them in a daily or weekly auction, for example.

The system operator may be able to use congestion revenue surpluses from earlier periods to fund revenue shortfalls in later periods. Shortfalls could be shared by the holders of the FTRs on a *pro rata* basis. Equally, a deficit could be funded from a levy on the transmission tariff. However, none of these solutions is likely to place the major risk of transmission maintenance and repair in the right hands. The advantage of placing this risk with the transmission owner is that it is in the best place to manage it and doing so will encourage efficient maintenance.

Designing an incentive scheme will not be as straightforward as it appears. A transmission owner does not have full control over all circumstances that result in the reduction of transmission capacity or outages. Clearly, it has no control over the weather, earthquakes, or hurricanes. In a region or control area with multiple transmission owners, it might be difficult to establish which transmission owner is responsible for what element of the system that suffered a reduction in capacity and to what extent the reduction in capacity or outage was either the direct result or increased by the actions of another transmission owner. Joint cost allocation schemes could be difficult to establish.

In any event, it will be important to ensure that the incentives have the effect of making the critical parts of the transmission system available at times of peak usage. The incentives should allow for flexibility to move maintenance outages and to ensure that maintenance is not postponed indefinitely where it is needed. Focusing on key performance indicators may be of little value to customers in terms of quality of service. An overall incentive scheme designed in terms of responsibility for revenue shortfall with a negotiated floor would probably be the most appropriate approach.

Revenue surpluses

Revenue surpluses arise from market participants that do not hedge, *i.e.*, they did not make the relevant constraints binding in the FTR auction, and, by default, the surpluses are held by the transmission owners on behalf of those paying the access charges.

If no one bought FTRs, all the congestion rents would be surplus and would flow to reduce transmission access charges. In the end, those who pay the access charges collect all the congestion rents. In the auction, they receive the expected value of the rents for the FTRs that market participants want. In the hourly dispatch, they receive the residual congestion charges for the FTRs that market participants did not want.

Loss hedging FTRs

Work on the development of loss hedging financial transmission rights (FTR) is proceeding. By definition, in accounting for losses, the aggregate of the FTR awards cannot be perfectly balanced. The net difference is the energy that would be required to meet the loss requirements of the FTR dispatch. In principle, an individual FTR could be defined for inputs and outputs at any node, or in any combination. In the aggregate, the simultaneous feasibility requirement is only that the net inputs must equal the net outputs plus losses. There is no need for individual FTRs to balance with losses, only for the aggregate to balance in order to preserve feasibility in the system. Under the usual congestion-only FTR formulation, market participants are fully exposed to variations in the cost of losses. The purpose of the development work is to include the effect of losses in FTRs and provide a perfect hedge for both congestion and loss costs for future use of the transmission system.

A number of designs for defining FTRs that hedge losses as well as congestion have been evaluated. There is the potential for some formulations of loss-hedging FTRs to expose the holder to energy price risks and uncertainty as to the ability and willingness of market participants to bid for FTRs that expose them to these energy price risks. Some designs could expose the system operator to an unhedged energy price risk, and a preferred approach has emerged.[62]

Under this approach, auction participants can bid for and offer balanced FTRs between locations, or unbalanced FTRs at a location. Unbalanced FTRs are individual injections of the specified quantity of megawatts (MW) or withdrawals of the specified quantity of MW at a particular bus. The unbalanced FTRs to inject MW are the equivalent of selling energy forward, and those to withdraw MW would be the equivalent of buying energy forward. The balanced FTRs would identify the quantity of energy that would be input at one location and withdrawn at another location. The overall award of both types of FTRs would be constrained so that a simultaneous feasibility test could be passed that would include losses. In aggregate, the FTRs would reflect injection quantities greater than the aggregate withdrawal quantities. The FTRs in total would reflect the actual losses for an equivalent level of loads and pattern of injections.

Flowgates

Another flavor of transmission rights grew out of concerns over the potential complexity of the point-to-point FTR scheme on a system with several hundreds, if not thousands of nodes. It was felt to be computationally cumbersome.[63] The main concern was that secondary markets in individual FTRs could be illiquid, because there could be very many different directional rights.[64] This resulted the development of "flowgate rights" (FGR), originally proposed as an alternative form of financial transmission right to FTRs. FGRs have not yet been implemented, whereas FTRs have been implemented in PJM and New York and are proposed for New England, Argentina, and California.

An FGR involves a major simplification to the effect that only flows over certain "commercially significant" congested physical transmission elements or flowgates would have to be tracked, rather than the flows between all nodes. These flows could be continually calculated on large systems using linear power transfer distribution factors (PTDF). PTDFs are typically calculated as the linearized fractional flows induced over particular transmission lines (*i.e.*, over the elements of the transmission system being evaluated for congestion) that have been called "flowgates" (hence, the name flowgate right, or FGR) by an injection of 1 MW at one node and the withdrawal of 1 MW at another node.

Since there are fewer congested flowgates than nodes, the transmission rights market should be more liquid. The rights would be issued for the flowgates identified and auctioned by the system operator, and the market participants could obtain a hedge against the LMP-based congestion charges calculated in the real-time energy market. The advocates of the FGR model also argue that flow-based rights are insensitive to the system topologies or changes in load levels[65] or power flows in the same way as FTRs. As a result, the quantity of rights should be more stable than the quantities of FTRs, which need constantly to be reevaluated and reconfigured as system conditions and transactions change. PTDFs also change because they are based on physical flows. On an alternating current (AC) network, the ability of power to flow depends upon the pattern of use of all parts of the network. The PTDFs will vary over time, and market participants could require many FGRs.

Figure 4-5 shows the differences between the definition of transmission capacity for the purposes of the contract path fiction (under FERC Orders 888 and 889),[66] FGRs, and FTRs.

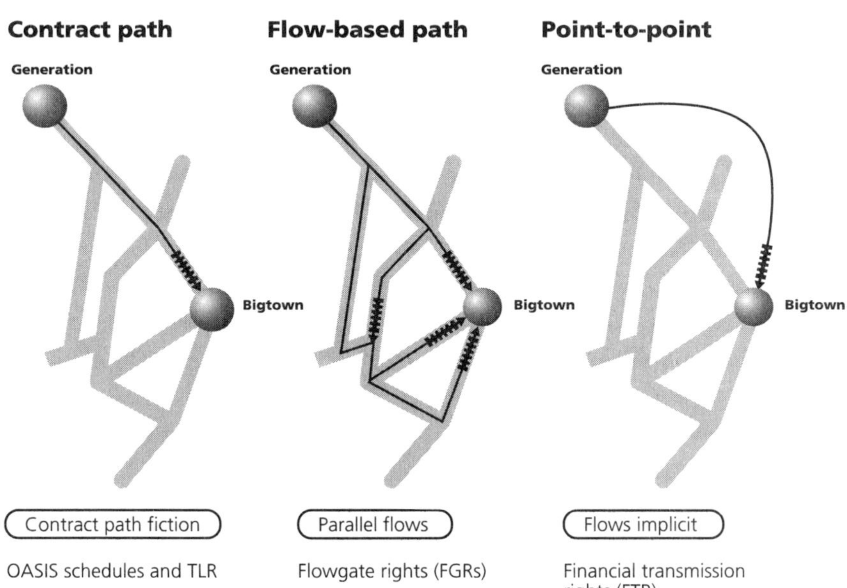

Fig. 4-5 Transmission capacity definitions

Other issues

FGRs have to be cast as obligations. The same considerations as to options and obligations apply to both FGRs and to FTRs. It is not the case that FTRs are naturally obligations and FGRs are naturally options. The question of how to deal with counterflows also arises.

Although a full system of flow-based rights has never been implemented, the concept is that they would be auctioned annually and traded in a secondary market. The holders of the rights would recalculate the amount of the flowgates that they needed, based on PTDFs provided by the system operator, and enter into the secondary market to purchase the appropriate quantities, if available. Because it is likely the congested flowgates would change over time, the idea was developed that two sets of rights could be issued and bundled together—"commercially significant flowgates" rights and other "commercially insignificant" rights.[67]

Andrew Ott of PJM has explained that in PJM, the Eastern interface limits can vary by 4,000 MW over time and by up to 1,000 MW within 30 minutes.[68] In other words, in PJM, FGRs would not provide the simplification over FTRs originally sought. Because FGR holders may not hold rights to flowgates congested by their transactions, they could "insure" themselves against marginal congestion costs for which they are actually uncovered. A solution is to use the set of PTDFs in the original auction for settling the FGRs in real time. This is intended to provide transmission price certainty in advance to rights holders.

Unlike FTRs, where the surplus of the congestion rents might reflect unhedged transactions, market participants could buy FGRs representing any unsold capacity on specific elements of the transmission system.

Attractiveness of FGRs

The chief attraction to a transmission investor is that it can acquire FGRs in the assets it actually constructs rather than having to select a set of FTRs from one point to another, or from a point to a trading hub. Simultaneous feasibility tests are not required, and secondary trading should be straightforward. Although FGRs do represent a simplification and therefore do not provide perfect hedges, not all market participants will be looking for a perfect degree of insurance.

The ISO uses PTDFs to calculate power flows across a constrained path. The FGRs can be purchased in the same proportions as the PTDFs. If an impending outage of an element of the transmission system is published, market participants can trade the FGRs in a secondary market and adjust the mix accordingly.

FGRs capture the main effects of FTRs, which are more exact in tracking the power flows. The disadvantage of the less perfect hedge the FGRs provide could be compensated for by the additional liquidity in the secondary markets they would create. FTRs could also provide a good transition from physical rights to FTRs.

Criticisms of the flowgate model

The main criticisms of this model have focused on the assumptions of long-term stability in the amount of flowgates capacity and the number and location of significant flowgates.[69] The model depends upon simplifications and approximations of the DC load flow along with the assumptions about stable topology and static PTDFs over time. The set of potentially-congested flowgates that forms the basis of the rights may not be a good prediction of the actual patterns of congestion in deregulated markets, particularly in view of the fact that experience in most deregulated markets is that the number of congested paths both increases and changes significantly.[70] The criticism has also been made that contingency constraints need to be taken into account, because they will increase the number of potentially-binding constraints.[71]

A further difficulty with flow-based rights is that the model assumes that the amount of transmission capacity is relatively fixed with respect to each flowgate, and is insensitive to changing power flows. Available capacity is essentially dependent upon reliability considerations and changing patterns of supply and demand. For example, system operators deal with reactive power limits to avoid voltage collapse by imposing limits on real power flows. Many system operators revise these limits regularly to reflect changing patterns of load.

There are a number of other technical difficulties that reflect the complexity of determining power flow limitations. Moreover, the system topology used to calculate PTDFs is not particularly stable, as some may have assumed. It would change when lines go out of service or in response to the use of phase-angle regulators. Nonlinearities in network flows and constraints will result in changing PTDFs. These factors will cause deviations in the amount of FGRs traded and actual physical flows. These deviations, however, would not occur as frequently in the FTR approach that requires simultaneous feasibility of point-to-point transactions.

The FGR, flow-based rights model has not yet been implemented. There is no operational experience to demonstrate that it would provide all the desired properties that market participants and investors in transmission need.[72] The model could create, through the inherent simplification, increased uplift costs and a lack of the necessary locational pricing signals for new generation and transmission expansion. However, according to Shmuel Oren, it should not be necessary to socialize the uplift costs.[73]

Choosing the right flavor of transmission rights

There has been a debate as to which of FTRs and FGRs was the superior model. It is claimed that FTRs and FGRs are not mutually exclusive, and a system to allow both types of rights to be allocated, auctioned, and traded simultaneously has been proposed.[74] The Midwest ISO proposal submitted to the Federal Energy Regulatory Commission (FERC) proposes both types of rights. The reason for this is that there is a major physical constraint in the region, and market participants have utilized a similar concept in the past. Market participants may feel more comfortable with FGRs with respect to this constraint, given their knowledge of a set of physical assets.[75]

Another scheme was proposed by Seabron Adamson, involving a portfolio of transmission rights and an unequal quantity swap mechanism. It represents continuous-time price formation and price discovery at a hub, while minimizing the transactional and information costs of trading FTRs.[76] The regulator in England proposed yet another model through an auction of physical transmission access rights.[77]

Auctions

The transmission access rights auction was conceived primarily as a market-driven mechanism to assess demand for expansion and to overcome the motivation of transmission owners to limit capacity, increase rents, over-state capital expenditure ex-ante, and under-spend ex-post. However, as Dieter Helm points out:

> *Since the network is, in economic theory terms, in part a public good, with externalities from one part of the network to others, and complementary with the rest of the economy, the market is likely to under-provide.*[78]

He does not think that auctions are an alternative to regulation. The difficulty lies in combining them.

An auction is, in theory, a good way to forecast demand and to overcome the transmission owner's near monopoly of information, but it only gives an indication of what the market is willing to pay at one moment in time. The response in an auction will reflect market participants' short-term horizons and also market power. In a dynamic marketplace, arrangements would have to be put in place for multiple auctions over various time horizons as well as secondary trading arrangements to meet the needs of market participants, which would be complex and expensive to implement.

Defining the capacity to be auctioned on a highly reticulated system is also complex if the rights are to be physical and not financial. The idea was abandoned initially on the grounds that the costs would outweigh the benefits (mainly the cost of half-hourly locational settlement). In any event, NGC has been successful in bringing the costs of constraints down to a fairly low level under TSS incentives.

An auction can be used for allocating existing capacity and new capacity, the size of which is finite (e.g. a new DC interconnector), but to use an auction to define the amount of new capacity on an AC network is much more difficult. Given the long lead-in times and the economies of scale involved, the auction response would have to be very precise and committed in order to be reliable. If the market is to be relied upon in this way, it would have to be very liquid and reflect time horizons sufficient to cover the period of the investment. This mechanism would require a liquid and transparent futures market, which usually becomes much thinner over longer periods.

An auction can supply only one piece of evidence of demand for expansion. The regulator could use it as market testing of the capital expenditure budget that it is asked to approve. It would be possible for the regulator to put in place some sort of incentive scheme on the lines of a performance target scheme discussed in chapter 7, with a floor and a ceiling, for the transmission owner to invest in line with auction signals. This could deal with divergences in expectations formed at the time of the approval of the capital expenditure budget. However, it would not obviate the need for regulation of investment plans and other regulatory incentives to encourage long-term investment, particularly to maintain reliability, reduce system and market operations costs, or to reduce market power. Regulation would be required to deal with the very long lead-in times involved and the need to take a regional or national view, which would not be reflected in the bids made by market participants in their narrow self-interests in the auction. The regulator would also have to deal with the questions of auction revenues that greatly exceed the cost of the investment, shortfalls between the cost of the investment and the auction proceeds, and what proportion of the capital expenditure should be funded through an auction mechanism.

None of these alternative models undermines the strength and integrity of the nodal pricing approach. Moreover, the increasing experience indicates more widespread acceptance, particularly in the U.S., that FTRs are a good model. Indeed, as we have discussed, FTRs have been included in FERC's proposed standard market design.[79] It is difficult to sustain an argument that an approach accommodating both FTRs and FGRs should be implemented, when the early indications are that market participants can manage with FTRs on their own. The dual system involving both approaches would, on the face of it, be complex to implement, and it seems unlikely that traders would find it really useful to be able to switch between one type of right and the other on a regular basis.[80] It is also not entirely clear that the concerns over liquidity of the secondary markets in rights would be overcome by going down this path.

Criticisms of FTRs

The main criticism of FTRs is that the large numbers of nodes (and therefore combinations of nodes on which FTRs will be based) means that the secondary markets in a particular set of FTRs could be illiquid. Liquidity of markets is not usually a policy issue, but it has caused concern, particularly to retailers and marketers. The growing practical experience of FTRs indicates that the point-to-point right does provide a perfect hedge from congestion charges (but not for losses) if the holder does not deviate from the covered transaction. Although it is said that there is a large number of potential node-to-node combinations and that any change in the configuration of a particular right requires establishing simultaneous feasibility with all other outstanding rights, this does not appear to have limited the development of the market. Although the secondary markets may not be fully developed as yet, the liquidity of the primary auctions has increased considerably, reflecting the learning by the market participants. In order to deal with an apparent lack of flexibility for market participants in forward markets, the design was adapted in PJM and New York to increase the frequency of primary reconfigurations auctions and to develop hub nodes or zonal aggregations of nodes. Day-ahead congestion bidding was implemented in PJM, which allows market participants to lock in the upper level of the congestion charges they are willing to pay, which is characterized as a short-term, point-to-point option right.

Those advocating flowgates argue that because point-to-point rights are defined independently of the transmission system topology, they include "PTDF insurance."[81] In other words, the auction will become revenue inadequate, and the cost of fulfilling the obligations associated with the set of outstanding rights will be allocated to market participants by the system operator through an uplift. However, this does not appear to be correct. The issue is essentially one of infeasibility of point-to-point rights that would result in revenue inadequacy, and it is a problem for both FTRs and FGRs.

There may be circumstances where there is revenue inadequacy because of transmission outages that are not within the control of the system operator, but are within the control of the transmission owner. As we have seen, it would be possible to put in place rules or incentives for transmission owners to maintain their lines and to be liable for any revenue inadequacy for failing to do so. FERC has proposed this in its standard market design (allowing for outages due to *force majuere*).[82] If there are any long-term changes in PTDFs due to investment in the transmission system, the resulting set of transmission rights is re-auctioned to the users who value them most highly. As a result, any PTDF insurance provided by the system operator would not increase over time.

In short, the theory of point-to-point FTRs is borne out by practical experience. FTRs are increasingly utilized, better understood, and liked by market participants. Given the complexities of the transmission system, the shifting patterns of constraints and power flows, and the realities of system operations, the point-to-point FTR approach gives the best answer and a key ingredient of good market design.

FGRs are attractive because of their relationship with the physical assets that an investor might construct and could, in theory, be a stepping-stone to the FTR ideal. They will appeal particularly to generators that build radial lines in order to obtain access to high-priced markets, because physical delivery will be important to them. Over time, they may be more concerned with hedging a variety of differently located transactions as market prices converge and the markets merge. An investor is likely to take a long-term view, however, and regard FTRs as having wider appeal in secondary markets, because they can provide perfect hedges that would be more attractive to a wider range of market participants, including those that wish to conclude transactions, but own no physical assets with which to hedge their risks.

The Design of Efficient Transmission Prices

The concept of marginal pricing

Most economists agree with the proposition articulated by Harold Hotelling that charging all investment costs to usage of a facility would be both excessive and inefficient, and that it is more appropriate to charge only the marginal cost for actual use and recover the investment costs in some other way.[83] In the context of a transmission system, most of the costs represent fixed investments, and if the transmission system were priced by taking all of the investment costs and dividing them by a number representing total usage, the resulting charge per unit of use would be much higher than the marginal costs of usage. This would deter buyers and sellers from using the transmission system and drive a cost wedge between them. The concept of marginal pricing had significant impact in many sectors and on many policy issues. It was reflected in the common two-part tariff that takes the form of a fixed fee and a lower usage payment, which tends to make more transactions efficient.

Objectives of an efficient pricing scheme

The objectives of an efficient transmission-pricing scheme are straightforward:

- The efficiently incurred operating, maintenance, and investment costs should be covered to reward transmission owners adequately
- Long-term incentives to maintain, upgrade, and expand the transmission system must be efficiently created
- The pricing system should create good locational signals as to when and where to use the system to encourage its efficient use by market participants
- There should be no barriers to access the transmission system on a non-discriminatory basis
- The payment of multiple transmission charges or tariffs to different transmission owners within an interconnected system (pancaking) should be avoided
- The charges should enable new generation and loads to be connected quickly and efficiently
- Prices should not distort energy market investment decisions and create perverse incentives for inefficient bypass of the system (or part of it)

The design of an efficient pricing scheme should reflect the following principles:[84]

- Where the relevant beneficiaries or market participants can be identified, the costs that are directly attributable to them should be paid by them
- The costs recovered through the transmission prices should include the short-run costs where the short-term marginal costs of transmission usage (congestion and losses) can be identified
- The costs should also include long-term costs such as the costs of connecting a generator or load to the transmission system
- Other long-run costs should be recovered from market participants in a manner that minimizes inefficiencies in the running of the electricity markets and in a manner that is seen to be fair

For this reason, many transmission-pricing schemes consist of three parts—a transmission connection (interconnection) charge directly attributable to a generator or load, a transmission usage (or congestion) charge, and a transmission access charge. This contrasts with the traditional two-part pricing approach in which there is a fixed charge and a variable price based upon actual usage. Consumption decisions will be based on the variable price that should, accordingly, be set as close as possible to marginal cost.

Connection charges

The concept of a connection or interconnection charge is that it should enable the transmission owner to recover long-run costs that are directly attributable to a particular user, primarily the costs of connecting the generator or load to the transmission system. In this way, the transmission owner can recover more of the total costs of the transmission investment than would be the case if the principles of marginal costing were applied and were recovered solely in transmission access charges.

Generators and directly-connected loads usually enter into transmission connection agreements (called interconnection agreements in the U.S.) that ensure that connections or interconnections are provided, maintained, and modified on terms that are efficient and fair to the parties. There will be separate tariffs or use of system agreements that define the terms and charges for the use of the system. Generators, in particular, are concerned about access to the system on stable and predictable terms, given the long lives of their assets. The same is true of large directly-connected consumers. The consumers may be able to rely on a good measure of regulatory protection, whereas generators will look to the market rules for assurance of long-term access.

The main debate regarding connection and interconnection charges in systems that have not adopted locational-marginal pricing relates to whether they should be calculated on a "deep" basis or on a "shallow" basis. The difference between the two approaches turns upon the certainty with which it is possible to identify whether the cost of reinforcing the system at points that are remote from the point of connection can really be attributed to the new connection or interconnection. It also depends upon whether consequential reinforcement is also necessary for system reasons related to anticipated future generation developments or load growth, where other users are likely to be the beneficiaries of the reinforcement as well as the generator or load.

The advantage of charging on a "deep" basis is that fewer costs have to be socialized and recovered in access charges. However, the more remote the reinforcement is from the point of connection, the more contentious the allocation of the cost to the new generator or load, or, indeed, to a third party transmission developer connecting to the system becomes. Obviously, the nearer the reinforcement is to the generator or load, the less the scope for argument as to which user is the true beneficiary. The short point is that it is very difficult to create a bright line and a simple and transparent rule for deciding which are deep and which are shallow connection charges.

However, while charging only the shallow connection or interconnection charges to a generator or load is less contentious, it suffers from the disadvantage that it results in more remote reinforcement costs being socialized in access charges. This cost socialization may also result in an inefficient side effect. It may create incentives for the generator or load to locate at a point on the system that is inefficient, both from the point of view of system expansion and also minimizing the costs other consumers may have to bear. The reason for this is that the investing facility owner will not be required to pay the full costs of connecting to the system, if it only pays the shallow costs. In effect, it is being subsidized. This may overstimulate generation solutions where transmission solutions would be more efficient. This would not be the case in a system that had adopted locational marginal pricing (LMP).

Transmission usage charges

Whereas connection charges relate to direct, fixed-cost recovery specific to a user connecting to the system, usage charges relate to short-run pricing of the indirect costs of transmission. Charging for transmission usage is simple when the energy is priced on a locational marginal cost basis, because it includes the effects of congestion. The locational marginal cost-based prices can also include the cost of losses at a particular location. A distant part of the import area might have higher losses and, therefore, a higher locational marginal price than a different part. Scheduled contracts are charged the price difference between the origin and the destination of the contract, and spot transactions are charged implicitly.

Locational marginal or opportunity cost pricing will not, however, pay for the existing transmission assets. The fixed charges of the transmission system reflect investments that, in principle, have been made to avoid paying the opportunity costs of losses, congestion, and more expensive generation. Largely as a result of economies of scale, these fixed charges would be greater than the congestion opportunity costs after the investment was made. FTRs could be described as a long-term guarantee that would be obtained in exchange for these fixed-charge payments. They would convert the short-run opportunity cost prices into long-term transmission arrangements.

Transmission access charges

It is left to transmission access charges to recover the rest of the costs that are not recovered by usage and connection charges. There is always considerable debate as to the basis on which access charges should be allocated. The issue is all about who pays how much of the residual costs that the transmission owner is entitled to recover.

It is hardly surprising that market participants seldom agree on the allocation of the costs of the existing transmission system. The costs are sunk, and typically, they exceed the transmission congestion opportunity costs of using the system. A large part of the reason for this, as we have observed, is the effect of the economies of scale. Most market participants would prefer to avoid paying the sunk costs while using the system and to pay the continuing cost of congestion, rather than the other way round. If the sunk costs were to be recovered in prospective payments, there must be a requirement to pay these costs as a precondition to using the grid.

The need for access charges does not give any indication as to the form the charges should take. A simple principle is that they should be priced to minimize distortions in energy market investment decisions, but even this does not indicate any clear, best way to allocate these charges. In practice, the access charges might be collected in proportion to usage, but, in this case, it should be at the last point of connection, in order to minimize perverse incentives for inefficient bypass (*e.g.*, by locating on a particular side of a meter) to avoid paying for sunk costs.

Who should pay?

The final consumers are the ultimate source of revenue, and it is appealing to charge them directly. This is the typical arrangement in many countries, but in the United Kingdom, Argentina, Sweden, and Norway, sunk costs are charged to the generators and retailers in various proportions. As Figure 4-6 indicates, these charges look somewhat arbitrary on paper.

	Generators	Consumers
England & Wales	27%	73%
Argentina	80%	20%
Sweden	30%	70%
Norway	54%	46%
Others	0%	100%

Fig. 4-6 Split of charges between generators and consumers (Source: Alex Henney, "Transmission Pricing In Competitive Markets–Lessons From The World," EUCI Transmission Conference, Miami, Florida, February 5, 2002)

There is an argument that if the generators have to pay some proportion of the costs—even if they are passed on to their wholesale buyers—it will create some awareness that generators should utilize the transmission system efficiently.[85] In other words, some locational signals will be created as well. It should also create incentives for generators to invest in transmission. This approach could avoid the danger of cross-subsidizing generators distant from load centers by comparison to those located nearer to the load centers.

LRMC methodology and tariffs

The correct signal to the developer of the new generating plant is that it should bear the marginal costs it imposes on the system with respect to congestion, losses, and reinforcement. If there is insufficient capacity in the system, then indirect, short-run marginal cost (SRMC) gives the correct signal. However, some jurisdictions use long-run marginal cost (LRMC) methodology and tariffs. In the United Kingdom, NGC utilizes an LRMC substitute called investment cost-related pricing. The disadvantage of using an LRMC approach for a signaling location is that it involves the regulator or the transmission company in making assumptions about the future with which there may be disagreement. If the assessment is made on a probabilistic basis, the issue arises as to whether charges should be calculated by reference to anticipated future outcomes, or whether it is better to wait and find out what the operational outturn actually demonstrates to be the case. It could result in perverse signals and delay investment.

Methodologies for the allocation of access charges

There are many different methodologies used to allocate charges for those parts of the network that are shared. For example, California has used the concept of the contribution that load makes to peak demand. Sweden based the charge on the latitude of the location of the user. Australia, New Zealand, and Argentina have used power flow models to determine the use of the system. Norway adopted a measure of winter peak-rated capacity for generators and the hour of regional coincident maximum demand for loads. Finland has utilized a time of use energy measurement, and Spain has used a mix of capacity and energy.

In the U.K. there is an annual, locationally-differentiated charge related to the costs of expansion in an area, levied on the three peak hours (the "triad"), according to the annual usage in those hours. In PJM, all customers (not just transmission users) are charged according to the units they use or consume. There remains the debate as to whether the addition of a regional component in transmission access charges would capture the effect of some long-term, location-specific transmission costs that LMP might not capture.

The use of a methodology involving regional peak demand makes sense, because most transmission systems have been sized to meet peak demand. However, if the necessary interval metering is not in place, there will be little point in charging by reference to peak demand, as consumers will be unable to see the price signals. If the charges are made on a capacity (kW) basis, the result may be an energy market price with higher peaks and lower demand than if the charges were based on energy (kWh).

Many jurisdictions do not have any rules relating to generators embedded in distribution systems that serve load connected to their local distribution systems "behind the meter." They may spill only a small amount of energy on to the transmission system, relatively infrequently. Should they pay transmission charges on a "gross" basis with respect to all the energy they are capable of producing that might spill on to the transmission system at some time, or should they pay on a "net" basis with respect to the infrequently spilled energy? Should they be treated simply as negative loads? Whatever the answer, the embedded generator will benefit from the reliability of the transmission system. Norway, Finland, and Ontario have adopted "net" and "gross" charging principles that are set politically. The issue was greatly debated in California with respect to municipal utilities and large industrial customers with their own generation. The principle that makes the greatest sense is that load beyond the transmission system should pay for reliability. The savings in the cost to the main transmission system provided by the embedded generators could be calculated, and the probability of the need to call on the reliability or support of the main transmission system could be computed. The embedded generators could be charged on a net basis together with an expected reliability charge.[86]

New Zealand methodology

New Zealand does not have a specific regulatory regime for the electricity industry. In the absence of a regulatory regime, the sunk costs cannot, in practice, be recovered from parties that do not benefit from them, because they simply will not agree to pay. Transpower made a carefully reasoned proposal that, in all fairness, they should be allocated to users that enjoy the benefits:

- South Island generators—who, through the HVDC link, enjoyed both higher clearing prices, usually set by North Island power stations, and increased production
- North Island consumers—who obtained clearing prices lower than would be the case if the North Island transmission system were not connected to the South Island grid. The largest load centers—Auckland and Wellington—are in the North Island

Transpower believed that the sunk cost charges should be fixed in order to minimize the new opportunities for the charge to be treated as variable (*i.e.*, passed to consumers at a cents per kWh rate), which, if it occurred, would create incentives for consumers to change behavior on account of sunk costs, creating the economic distortions we discussed earlier.

Transpower maintained that charging the sunk costs to South Island generators prevented fixed costs from being treated as variable, because it did not change the incentives to offer generation at SRMC, and so acted as a fixed charge. The only alternative, charging North Island consumers, would be likely to lead to the treatment of fixed costs as variable costs through retailers, who could be expected to recover their costs as a cents per kWh charge to consumers, or through a similar treatment of fixed line charges. This was regarded as economically inefficient.

Transpower consultation paper

In January 2002, Transpower published an excellent consultation paper setting out draft design principles that it would use in developing its pricing methodologies. Applying the same principle to sunk and fixed costs so that the allocation should not distort investment decisions and be non-bypassable, Transpower also made proposals with respect to variable costs and the costs of new investment. The principle that would apply to variable costs was that they should be allocated in accordance with the principles of marginal cost pricing. New investment costs should be allocated in full to the expected beneficiaries. This would create consistent signals for the trade-off between transmission, generation, and demand-side solutions and indicate how the resultant sunk costs should be allocated. Figure 4-7 demonstrates this thinking.

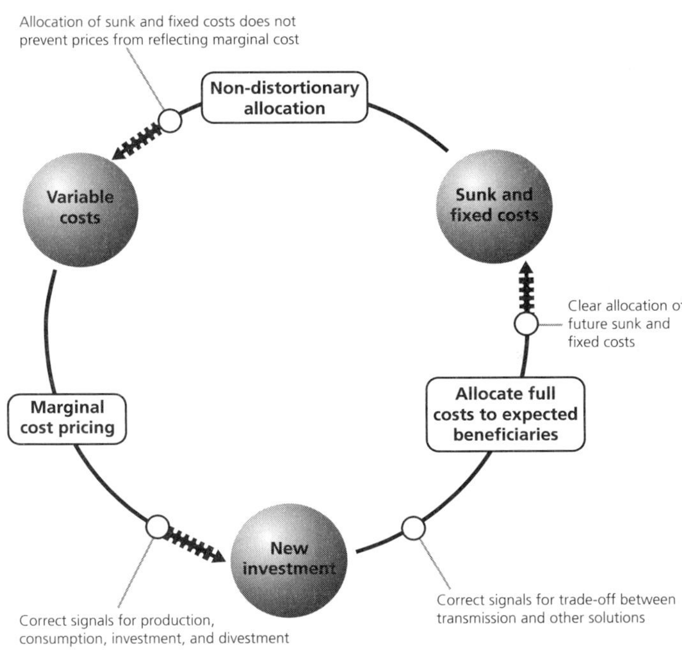

Fig. 4-7 Pricing signals in an efficient transmission market (Source: Transpower New Zealand Limited)

Both mandatory and maximizing criteria (*i.e.*, principles that can only be maximized because they involve trade-offs) are applied, as Figure 4-8 demonstrates.

	Criterion	Efficient means of charging
Mandatory criteria	Integrated approach to transmission pricing	The pricing signals for sunk, fixed, and variable costs, grid use, and new investment must be as consistent as possible
	Marginal cost pricing	Variable costs should be charged to the causer
	New investments charged to beneficiaries	New investment full costs allocated to expected beneficiaries
	Not exceed economic value	Charges must not exceed the economic value of the service to the customer charged for it
	Preclude cross-subsidies	Charges must be between the stand-alone cost and incremental cost of providing the service to the customer
Maximizing criteria	Least distortionary sunk and fixed cost allocation	Sunk and fixed costs should be allocated in a way that minimizes distortions to production/consumption and investment decisions made by grid users
	Sunk costs should be recovered by fixed and unavoidable charges	Sunk costs should be recovered by fixed and unavoidable charges
	Fixed costs should be recovered by fixed charges that are unavoidable in the short term	Fixed costs should be recovered by fixed charges that are unavoidable in the short term
	Sunk and fixed costs allocated	The transmission pricing methodology for the recovery of sunk and fixed costs of transmission is an allocation
	Resolve conflicts consistently with energy policy	Where any principles conflict, those conflicts should be resolved in a manner that is most consistent with the delivery of electricity in an efficient, fair, reliable, and envionmentally sustainable manner to all classes of consumer

Fig. 4-8 Criteria for an efficient cost allocation (Source: Transpower New Zealand Limited)

Transpower recommended that the sunk and fixed costs should be allocated in a manner that did not prevent variable prices from reflecting marginal costs. The pricing signals arising from the allocation of the sunk, fixed, and variable costs, the use of the transmission system, and new investment should be as consistent as possible. The pricing methodology for the recovery of each category of cost should contribute to the overall signal in order to support and not distort the signals created by each category.[87] All of these principles make good sense and are the ingredients of efficient transmission pricing.

Link award of FTRs with obligation to pay access charges

Collection of the access charges could be simplified by the simultaneous allocation of the benefits of the existing transmission system through the award of FTRs. Hogan suggests that one approach might be to award the initial ownership of the FTRs along with the long-term obligation to pay the access charges. Another possibility would be to auction the FTRs and apply the revenues to reduce the required payment for the transmission system. The remaining costs would be collected as access charges for all users. These access charges would pay for the existing transmission system assets that allow open access for use of the system. All users would pay the transmission usage charge for congestion, *i.e.*, the second part of a two-part tariff, and the holders of the transmission FTRs would have the transmission hedge.[88]

The Key Ingredients From Market Design for Transmission Expansion

The best ingredients

Every recipe begins with a list of the ingredients. The ingredients from market design and transmission pricing are critical from the point of view of efficient use of the grid, minimizing the need for expansion, and creating the necessary price signals for expansion. Our list of the best ingredients includes:

- LMP in order to integrate the energy market and the pricing of scarce transmission (covering both congestion and losses)
- combined marginal costing, real-time pricing, and security-constrained economic dispatch
- nodal rather than zonal pricing, with the use of trading hubs if simplification makes sense
- a system of point-to-point financial transmission rights (FTR) that are sufficiently long-term to meet the needs of investors (unless the new facility is a controllable, radial line, in which case physical rights may be sufficient, although they will need to be subject to use-or-lose-it rules and closer regulatory scrutiny)

- a fair and efficient system for dealing with the allocation of short-, medium-, and long-term transmission rights to the existing transmission capacity and new capacity additions that will facilitate new entry in both generation and transmission
- a transparent auction system for allocating and trading FTRs
- FTRs that are both obligations and options and cover losses (when the systems are fully developed to allow this)
- an efficient transmission pricing scheme, consisting of three parts—a transmission connection or interconnection charge, a transmission usage (congestion) charge, and a transmission access charge
- an efficient pricing methodology for the recovery of sunk and fixed costs that allocates them to transmission users in a manner that does not distort new investment decisions and prevents fixed costs from being treated as a variable

From a Transmission Expansion Perspective—Do We Need All These Ingredients?

We have looked at a broad range of market design and transmission pricing issues in the context of a number of social welfare enhancing objectives. For the narrower purpose of transmission expansion, we could take the view that not all these ingredients are entirely central to our needs. Indeed, we could become quite cynical as to how good these ingredients really are to create the signals and incentives that we need for investment and efficient use of the transmission system. For example:

The nodal price signals are too volatile for a decision to invest a large sum of money in assets with lives of more than 30 years

Price volatility is a feature of the energy markets as well as the transmission rights market. There are hedging instruments such as contracts for differences and FTRs to help investors to absorb risks. Notwithstanding the volatility, significant investment decisions have been made in both generation and transmission based on nodal price signals using these hedging instruments.

No regulator would base its decision to approve a project on nodal price differences

FERC and the regulators in New England (notably Connecticut) have based their decisions to approve the CrossSound Interconnector, the development of which is based on nodal price differences. FERC has also approved Project Neptune (described in chapter 5) on this basis. Nodal prices provide a signal of need to the regulator.

Investors and regulators will respond better to a well-prepared transmission plan than to nodal prices

The existence of nodal prices does not preclude a well-prepared transmission plan. Investors and regulators still have to go through an analysis of whether the project is needed and who are the beneficiaries, which can be difficult. Nodal prices can help to signal the timing and need for investment more accurately. They can also be used to identify the beneficiaries (as in Argentina, or through the use of the reference network described in chapter 8).

LMP is all about the cost of congestion and losses. Transmission expansion is not simply about reducing these costs. Its benefits are wider, because expansion can reduce market power, increase competition, facilitate imports of low-cost power, decrease price volatility and the cost of hedging, and reduce ancillary services cost

There is no reason that both merchant and regulated developers should not make the case that a transmission project can achieve these wider benefits. Indeed, merchant transmission links could be used to provide ancillary services and reserves at lower cost. It may be difficult to remunerate the developer for creating a common good such as the reduction in market power or the benefits of increased competition, except through regulated expansion. However, loads could find the benefits valuable and may be prepared to pay for them. Other market participants may be prepared to buy transmission rights to enhance their competitive positions.

Nodal prices are the product of the optimization of engineering power transfer distribution factors (PTDF) that are converted into nodal prices. For example, there could be more than 1,000 nodal prices when there are only five congested lines. The information the investor wants is where the five congested lines are. The prices on specific lines may be hidden

PJM publishes this information for every five-minute interval. The publication of nodal prices does not preclude the publication of other information. The constraints are identified in real time. Price signals, whether they come from nodal prices or some other source, will never identify exactly what the best project is. This involves a variety of transmission planning skills, including integration of system studies, evaluation of substation space, knowledge of rights-of-way expansion potential, as well as permitting and siting process expertise, among other things.

Security-constrained economic dispatch does not simply involve applying constraints defined by normal operating and planning standards. The system operator applies local reliability constraints and other calculations that are hidden from view. It is something of a black box, which cannot be ideal

The system operator will need to apply every constraint that it knows about, whether driven by normal operating and planning standards or local conditions. The constraints are not hidden, and the system operator should characterize all system limitations, to the best of its ability, in a transparent manner.

The system operator can also change equations to calculate the limits to define constraints in the security-constrained dispatch. This can create additional transmission capacity and act as a substitute for transmission expansion. It is also something of a "black box," because it is usually entirely within the discretion of the system operator

System operations do involve making judgments, given the nature of the system, the laws of physics, and the fact that operations take place in real time with energy that cannot be stored and where supply and demand must be kept in balance. It is simply a fact of life. Voltage and stability limits can vary, and it may appear that the system operator is using a "black box" to create additional transmission capacity, but the chances are that planned transmission projects will not be significantly affected. In any event, if the system operator can change the constraint equations to increase capacity without physical expansion of the system, it would be beneficial. Indeed, nodal prices would give it signals when to do so.

The system operator's decisions are not contestable, and it may be impossible in some market designs for market participants to offer adjustment bids near to real time for congestion management and balancing and to self-provide ancillary services and losses

The system operator's decisions are being made in real time or very close to real time, and there is a limit to how much contestability can be achieved. In any event, contestability may not produce efficient outcomes for the consumer. It is true that some degree of coordination by the system operator in real time is necessary, because the same energy is being used to balance the system, manage congestion, and maintain reliability. The point is not particularly relevant to LMP and transmission expansion.

Market participants will never know how to value long-term FTRs, and therefore, how to bid in auctions. Secondary trading in them will be virtually impossible

It is clear from the experience in PJM and New York that market participants are learning how to value FTRs and to trade in them. Indeed, they have learned how to value and to trade in five-year FTRs in New York. There is no reason to believe the same will not be true of longer-term FTRs when they become available.

The lack of liquidity in the market for trading FTRs between hundreds, if not thousands, of different nodes is the major concern

The problem of the lack of liquidity in the market for trading FTRs can be mitigated by the creation of trading hubs. This has been successfully implemented in PJM. There is an argument, however, that there will be an incumbency advantage for those that hold FTRs at the central hubs, where the liquidity will be highest and the transaction costs lowest. This will mean that the hubs could become difficult to change over time, which will put those wishing to conclude transactions remotely from the hub at a disadvantage. One solution to this problem would be to allow the relevant transmission owner to be counterparty to the hedges and allow it to utilize the surplus rents it receives to sell or give hedges to market participants. In this way, it could exchange anticipated rents for a fixed fee and remain revenue neutral. The transmission owner could play a role in equalizing the prices to overcome the lack of liquidity created by the inherent incumbency advantage of existing FTR hubs. It would also create incentives for efficient investment by transmission owners. In effect, it would create a fixed differential from the hub price. The disadvantage is that consumers may become indifferent as to whether they use a congested path because of this fixed fee, and the transmission owner should not be exposed to risk if a consumer reduces its consumption.

Investors would prefer FTRs that are specific to the facilities they build. FGRs are more specific to assets than FTRs and form a better base for long-term rights. FTRs need to take into account all of the assets of the transmission system used to create the FTR, the long-term status of which is difficult to guarantee

Investors will be hoping to sell transmission rights to market participants that are looking for hedges. Some of them may be looking for a significant degree of insurance that only FTRs can provide. FGRs do not provide perfect hedges. For the most part, market participants concluding transactions on a mesh system will not be particularly concerned that their FTRs are not specific to certain specific facilities, provided they have the right hedging properties.

On the other hand, FGRs do have their basis in the physical properties of the system and may be a more attractive hedging instrument for generators dependent upon a radial line for access to high price markets. FTRs are more attractive for hedging transactions. All trading in the wholesale market can be characterized as transactional. The whole of the transmission system will be available throughout the life of the asset to provide the necessary hedging properties, by definition (even if elements of it may change). Although it may be difficult, initially, for investors to know which set of FTRs to choose, they will be able to reconfigure them and will soon learn how to utilize and trade with them in secondary markets.

If we are finding it difficult to make a choice between FTRs and FGRs, will it not be easier to allow for both in a joint transmission rights model?

Point-to-point FTRs and FGRs are very similar and are functionally the same. The way in which they are calculated is not, in concept, very different. In effect, FGRs involve a quasi-nodal calculation using power transfer distribution factors (PTDF). It is another way of calculating the same values, and both sets of rights can clear in the same format. Other similarities include casting the rights as options rather than obligations, which is a problem for both FTRs and FGRs. The only advantage in allowing for a joint transmission rights system involving FTRs and FGRs appears to be that an investor that owns generation and is seeking access to a high priced market might like to start with FGRs on the line or flowgate it has constructed, and move to FTRs as the markets merge in order to obtain more perfect hedges for its transactions. However, it is not clear why it would want to revert to FGRs at a later stage. It is also not clear why energy marketers and retailers would prefer FGRs, because they would primarily be interested in hedging transactions.

The software to calculate the feasible number of FTRs uses a computational method involving multiple computer runs to check whether constraints are violated. Typically, relatively few constraints (100 or so) need to be included. In a joint model, all constraints will be limiting, and implementation of this model will depend upon solving the constraints selection problem out of potentially thousands of constraints.

At the moment, it looks as though it would be a computational headache to offer both types of rights on a joint basis. Ultimately, the possibility of implementing a joint model may turn upon the cost of software development, the cost of computing time, and whether the calculations can be done near to and within real time.

Conclusion: We need all these ingredients

The arguments in response to the cynical points of view lead us to the conclusion that our list of key ingredients can be justified. Ideally, they should all be in place, but any of them will provide benefits that facilitate our recipes for success in transmission expansion.

NOTES

1. This assumes that the grid is built to the level where total customer value of service would equal the total cost of service, which means that the respective marginal values would also be equal, *i.e.,* that marginal value of service equals marginal cost of service or MVOS = MCOS. See Mohan Munasinghe, *Electric Power Economics* (London: Butterworths, 1990), pp. 230-244. Even in fully regulated electric system expansion the economic reliability criteria should be the same. See Eric Woychik, "Regulatory View of Capacity Valuation in California," *The Energy Journal,* vol. 9 (1988), pp. 39-42, and Mohan Munasinghe and Arun Sangvi, "Reliability of Electricity Supply, Outage Costs and Value of Service," *The Energy Journal,* vol. 9 (1988), pp. 1-18

2. William W. Hogan, "Market-Based Transmission Investments and Competitive Electricity Markets," Center for Business and Government, John F. Kennedy School of Government, Harvard University (August 1999). This paper gives a comprehensive and very readable account of the key ingredients of market design that will bring about efficient market-based investment in transmission facilities. For further explanations see William W. Hogan, "Transmission Investment and Competitive Electricity Markets," Center for Business and Government, John F. Kennedy School of Government, Harvard University (April 1998). See also, William W. Hogan, "Competitive Electricity Markets: A Wholesale Primer," Center for Business and Government, John F. Kennedy School of Government, Harvard University (December 1998). All these papers are available on Professor Hogan's website, www.whogan.com

3, This assumes that consumers of electricity can register their preferences by responding to price, which is an important long-term objective. In this way consumers state their respective marginal value-of-service (MVOS). This is equivalent to a customer self-selecting non-linear pricing, or fluid "demand-response." See Stephen Brown and David Sibley, *The Theory of Public Utility Pricing* (Cambridge, England: Cambridge University Press, 1986), pp. 61-127. Hung-Po Chao and Robert Wilson, "Priority Service Pricing, Investment, and Market Organization," *American Economic Review* (December 1987). Hung-Po Chao, Shmuel Oren, Stephen Smith, and Robert Wilson, "Priority Service: Market Structure and Competition," *The Energy Journal,* vol. 9 (1988) pp. 77-95

4. The so-called "common-cost" problem arises and demands some form of traditional ratemaking for fixed cost recovery. The usual issue is "rolled-in" (socialized) treatment versus "incremental" costing for major transmission additions. Rolled-in ratemaking is preferred by those that do not want to face full incremental costs. But

any deviation from responsibility to pay full incremental costs is by definition a subsidy. See Stephen Brown and David Sibley, *The Theory of Monopoly Pricing* (Cambridge, England: Cambridge University Press, 1986), pp. 53-55. "To be subsidy-free, revenue from each possible grouping of services must at least equal the incremental cost of that grouping." Ibid. at p. 55, fn 11

5. As Paul Joskow explained: "Inadequate transmission investment in a decentralized system is a stubborn problem that undermines wholesale market performance." Paul L. Joskow, "Regional Transmission Organizations: Don't Settle For Nth Best (N>>1)," presentation to the Harvard Electricity Policy Group (September 2001)

6. The key ingredients of good market design are becoming clearer in the light of operational experience in deregulated markets around the world, but the principles have long been enunciated by a line of eminent economists. See Paul L. Joskow and Richard M. Schmalensee, *Markets for Power: An Analysis of Electric Utility Deregulation* (Cambridge, Massachusetts and London, England: The MIT Press, 1983). Fred C. Schweppe, Michael C. Caramanis, Richard D. Tabors, and Roger E. Bohn, *Spot Pricing of Electricity* (Norwell, Massachusetts: Kluwer Academic Publisher, 1988). Larry Ruff, "Stop Wheeling And Start Dealing: Resolving the Transmission Dilemma," *The Electricity Journal* (June 1994). William W. Hogan, "Coordination for Competition in an Electricity Market," Center for Business and Government, Harvard University (March 2, 1995) available on Professor Hogan's website, www.whogan.com. William W. Hogan, "A Wholesale Pool Spot Market Must Be Administered by the Independent System Operator: Avoiding the Separation Fallacy," *The Electricity Journal* (December 1995)

7. John D.Chandley, "A Standard Market Design for Regional Transmission Organizations" (September 17, 2001) published in *The Electricity Journal* (December 2001), pp. 27-54. See also Sally Hunt *Making Competition Work in Electricity* (London: John Wiley & Sons, 2002)

8. In England and Wales, there was originally no mechanism and uplift costs that included congestion costs, spiraled to over £500m p.a. until the uplift management (TSS) incentive arrangement was put in place to manage it. The same phenomenon was encountered in those jurisdictions that either had no mechanism or that adopted administrative solutions (*i.e.*, curtailments) and socialized the costs. See chapter 2

9. Fred C. Schweppe, Michael C. Caramanis, Richard D. Tabors, and Roger E. Bohn, *Spot Pricing of Electricity* (Norwell, Massachusetts: Kluwer Academic Publisher, 1988)

10. The concept was created by Jose Delgado of The American Transmission Company

11. Professor William W. Hogan "Contract Network for Electric Power Transmission," *Journal of Regulatory Economics*, 4 (1992), pp. 211-242. Fred C. Schweppe, Michael C. Caramanis, Richard D. Tabors, and Roger E.Bohn, *Spot Pricing of Electricity* (Norwell, Massachusetts: Kluwer Academic Publisher, 1988)

12. In its standard market design, FERC proposes to roll in, on a region-wide basis, all high-voltage network upgrades (*i.e.,* new capacity) of 138 kV or above. Consistent with its proposal for interregional service pricing, costs would be allocated to the region that benefits from the expansion, which may or may not be the region where the facilities are located. FERC Notice of Proposed Rulemaking, 18 CFR Part 35, Docket No. RM01-12-00 (July 31, 2002). ¶ 191-201

13. Sally Hunt, *Making Competition Work In Electricity* (London: John Wiley and Sons, 2002)

14. Robert Wilson, "Architecture of Power Markets", *Econometrica,* MS #3582 (July 2002) p. 30 in electronic version. The article provides a very readable account of many features of good market design. See also by the author, "Ingredients of Liberalized Power Markets", EPRI (November 2001)

15. The countervailing view (expressed by Alex Henney) is that the system operator does not have to operate a balancing market. While it may receive and pay for accepted offers and bids for services to help it to balance the system and maintain the stability of the system, and may use an optimizing algorithm, that is the end of its essential role as system operator per se. Another entity—a balancing market operator (or real-time system administrator) can use the information resulting from the system operator's dispatch to derive prices from the marginal units that can be used to measure the price of congestion across a system (either nodally or zonally) and to settle deviations between schedules and outturn using those prices. Alex Henney and Tim Russell, "Lessons from the Institutional Framework of Transmission, System Operation, and Energy Markets in Most West European Countries and Some Other Countries: The Case for TRANSCOs," filed with FERC under Docket No. RM01-12-000.

16. FERC Order 2000 emphasizes these objectives at p. 300 and p. 489. FERC 18CFR Part 35, Docket No. RM99-2-999, Order 2000 (issued December 20, 1999)

17. Richard Green, "Electricity Transmission Pricing: How Much Does it Cost to Get it Wrong," Power Working Paper PWP-058, University of California Energy Institute (May 1998)

18. Shi-Jie Deng and Shmuel S. Oren, "Priority Network Access Pricing For Electric Power," *Journal of Regulatory Economics*, 19:3 (2001), pp. 239-270

19. William W. Hogan, "Getting the Prices Right in PJM: Analysis and Summary: April 1998 through March 1999. The First Anniversary of Full Locational Pricing," Center for Business and Government, John F. Kennedy School of Government, Harvard University (April 2, 1999)

20. Richard D. Tabors, "A Market Based Proposal for Transmission Pricing," *The Electricity Journal* (June 1996), p. 61

21. Steven Stoft, "Transmission Pricing Zones: Simple or Complex?" *The Electricity Journal* (January/February 1997), pp. 25-31

22. Ibid.

23. Ibid.

24. William W. Hogan, "Competitive Electricity Markets: A Wholesale Primer," Center for Business and Government, John F. Kennedy School of Government, Harvard University (December 1998) available on Professor Hogan's website, www.whogan.com

25. The National Grid Company plc, "The Statement of Use of System Charges, Effective from 1 April 2002," available on NGC's website, www.nationalgrid.com

26. David M. Newbery, *Privatization, Restructuring and Regulation of Network Utilities* (Cambridge, Massachusetts and London, England: The MIT Press, 2000), pp. 264-269

27. Richard Green, "Electricity Transmission Pricing: How Much Does It Cost To Get It Wrong?" University of California Energy Institute Power Working Paper PWP-058 (1998)

28. David M. Newbery, *Privatization, Restructuring and Regulation of Network Utilities* (Cambridge, Massachusetts and London, England: The MIT Press, 2000), pp. 264-269

29. David M. Newbery and Michael G. Pollitt, "The Restructuring and Privatization of the CEGB-Was It Worth It?" *Journal of Industrial Economics* 45 (1997), pp. 369-303

30 California Independent Market System Operator, "Market Design 2002 Project: Preliminary Draft Comprehensive Design Proposals" (January 8, 2002), p. 14

31 Comments of the California Independent System Operator Corporation on the Commission's Operational Audit of the CA ISO, Docket No. PA02-1-000 (February 15, 2002), p. 25

32. William W. Hogan "Getting The Prices Right in PJM: Analysis and Summary: April 1998 through March 1999. The First Anniversary of Full Locational Pricing," available on the Professor Hogan's website www.whogan.com. Hung-Po Chao, "Flow-Based Rights for Congestion Management," presentation to MEET 2000 Workshop, Stanford University (August 17-19 2000).

33. Richard D.Tabors, "Forward Markets For Transmission that Clear at LMP," presented at Hawaiian International Conference on System Studies, Hawaii (January 2001)

34. Sally Hunt and Graham Shuttleworth, "Operating a Transmission Company under Open Access-The Basic Requirements," *The Electricity Journal* (March 1993). In PJM, a market participant that paid for a reinforcement was given what was described as "firm access." If transmission service was not provided, *e.g.*, because of constraints, it received no compensation

35. Scott M. Harvey, Susan Pope, and William W. Hogan, "Transmission Capacity Reservations and Transmission Congestion Contracts" (June 6, 1996), available on the Professor Hogan's website, www.whogan.com

36. John Tamblyn, "Managing Electricity Transmission Network Congestion," paper presented to the APEC Energy Regulators Forum 9, Kuala Lumpur, Malaysia (May 17-18, 2001)

37. Roy Shankar has explained to the author that a parallel set of property rights (locational injection rights) will be needed that are traceable and can be reconfigured to allow for delivery of capacity contracted to meet installed capacity obligations. He has developed the concept of a system of a generalized property right that can be used in various markets and also the implementation methodology

38. Contracts would match costs to payments that would be beneficial in systems where cost of building has already been incorporated into a tariff and the transmission owner is paid no extra when the line goes into operation. Sally Hunt and Graham Shuttleworth, *Competition and Choice in Electricity* (London: John Wiley & Sons, 1996)

39. Judith Cardell, Carrie Cullen Hitt, and William W. Hogan, "Market Power And Strategic Interaction In Electricity Networks," *Resource and Energy Economics* (October 1997), pp. 109-137

40. Thomas-Olivier Nasser, "Imperfect Markets for Power: Competition and Residual Regulation in the Electricity Industry," PhD dissertation, Massachusetts Institute of Technology (June 1997)

41. Ibid.

42. Vernon Smith, "Deregulation of Natural Monopoly Industries," paper presented to the International Conference in Industrial Economics, Madrid (July 3-5, 1996)

43. Shmuel S. Oren, "Economic Inefficiency of Passive Transmission Rights in Congested Electricity Systems with Competitive Generation," *The Energy Journal*, vol. 18(1) (1997), pp. 63-83. Steven Stoft, "How Financial Transmission Rights Curb Market Power," University of California Energy Institute Working Paper, PWP-049 (June 27, 1997)

44. Jurgen Weiss, "Congestion Rents and Oligopolistic Competition in Electricity Networks: An Experimental Investigation," Harvard Business School, Cambridge, Massachusetts, n.d. See also, by the same author "Behavior Questions in the Restructuring of the Electricity Industry: An Experimental Investigation," Ph. D. dissertation, Harvard University (1997)

45. Richard J. Green, "The Electricity Contract Market In England and Wales," *Journal of Industrial of Economics,* 47 (1) (1999), pp. 107-123. David M. Newbery, "The Power Markets and Market Power," *Energy Journal* 16 (1995), pp. 41-66. James B. Bushnell and Steven Stoft, "Electric Grid Investments Under a Contract Network Regime," *Journal of Regulatory Economics*, vol. 10 (1996) pp. 61-79

46. David M Newbery, *Privatisation, Restructuring and Regulation of Network Utilities* (Cambridge Massachusetts and London, England: The MIT Press, 2000), p. 267, and David M Newbery "Competition, Contracts and Entry in the Electricity Spot Market," *RAND Journal of Economics*, 29(4) (Winter 1998), p. 726-749

47. FERC Notice of Proposed Rulemaking, 18 CFR part 35, Docket No. RM01-12-00 (July 31, 2002), ¶ 251

48. William W. Hogan, "Market-Based Transmission Investments and Competitive Electricity Markets," Center for Business and Government, John F. Kennedy School of Government, Harvard University (August 1999), p. 25, available on Professor Hogan's website, www.whogan.com

49. Raymond Coxe, "Grid Planning and Expansion: Who? Where? When?" presentation to Harvard Electricity Policy Group (May 22, 2000)

50. Susan L. Pope, "An Empirical Assessment of the Success of Financial Rights Auctions in PJM and New York," presentation to EUCI Electric Power Market Performance Conference (September 20, 2001)

51. FERC Order 2000, 18 CFR Part 35, Docket No. RM99-2-000 (issued December 20, 1999), p. 333

52. FERC Notice of Proposed Rulemaking, 18 CFR part 35, Docket No. RM01-12-000 (July 31, 2002). FERC also reserves the power to mandate FTRs as options and flowgate rights at a later date.

53. Paul L. Joskow and Jean Tirole, "Transmission Rights and Market Power on Electric Power Networks," *RAND Journal of Economics*, vol. 31, no. 3 (Autumn 2000), pp. 450-487

54. Michael Albers, "Compatibility of Long Term Capacity Reservations with EC Competition Law," discussion paper prepared for the 6th Regulatory Forum for Electricity in Florence (November 2000)

55. Hung-Po Chao, Stephen Peck, Shmuel Oren, and Robert Wilson, "Flow-based Transmission Rights for Market Trading and Reliability Management," draft (April 2000). By the same authors "Flow-Based Transmission Rights and Congestion Management," *The Electricity Journal* (October, 2000), pp. 38-58

56. Paul L. Joskow and Jean Tirole, "Transmission Rights and Market Power on Electric Power Networks," *RAND Journal of Economics,* vol. 31, no. 3 (Autumn 2000), pp. 450-487

57. Roy J. Shanker, "New Transmission Investment Policies-Implementation Issues," presentation to Harvard Electricity Policy Group (May 22, 2000)

58. Scott M. Harvey, Susan L. Pope, and William W. Hogan, "Transmission Capacity Reservations and Transmission Congestion Contracts" (June 6, 1996), available on Professor Hogan's website, www.whogan.com. The test involves entering the requests for energy injected and withdrawn at particular nodes and the price bids for the resulting transmission rights into a dispatch model that includes all potential constraints that could affect reliability

59. James Bushnell and Steven Stoft, "Electric Grid Investment Under a Contract Network Regime," *Journal of Regulatory Economics,* 10 (1) (July 1996), p. 61-79. See also by the same authors "Grid Investment: Can a Market Do the Job?" *The Electricity Journal* (January/February 1996), pp. 74-79

60. The building of the Fourth Line from Comahue in Argentina attracted more generation to locate in the Comahue area and, as a result, congestion on the line has increased

61. Substantial progress is being made, however. See Susan L. Pope, "TCC Auction Feasibility Test with Pre-Existing TCC Options," prepared for the NYISO Market Structures Working Group (November 16, 2001)

62. Scott M. Harvey and William W. Hogan, "Loss Hedging Financial Transmission Rights" (January 15, 2002), available on Professor Hogan's website, www.whogan.com

63. I am indebted to Andrew Claxton of APX for explaining flowgates to me and making available to me number of papers that he cowrote with Ralph Samuelson on the flowgates model including a paper presented to the EU Florence Forum, "Making Markets Out Of Transmission," n.d.

64. There were also concerns over the degree of centralization of the primary FTR market run by the system operator by those who felt that the market should be left to provide financial markets. At the time the debate began, it was not the case that there was universal acceptance of the system of locational marginal pricing on which the FTRs are based. As time goes by, the degree of acceptance continues to increase

65. Hung-Po Chao, Stephen Peck, Shmuel Oren, and Robert Wilson, "Flow-based Transmission Rights for Market to Trading and Reliability Management," draft (April 2000). See also an article by the same authors "Flow-based Transmission Rights and Congestion Management," *The Electricity Journal* (October 2000), pp. 38-58

66. FERC Order No. 888, promoting Wholesale Competition Through Open Access Non-discriminatory Transmission Services by Public Utilities and Recovery of Stranded Costs by Public Utilities and Transmitting Utilities, 61 FR 21,540 (May 10, 1996), FERC Stats. & Regs. ¶31,037 (1996). FERC Order No. 889, Open Access Same-Time Information System and Standards of Conduct, Final Rule, Order No. 889, 61 FR 21, 737, FERC Stats. & Regs. ¶31,037 (1996)

67. Richard D. Tabors, "Forward Markets for Transmission That Clear at LMP: A Hybrid Proposal," presentation to Hawaiian International Conference on System Studies, Hawaii (January 2001)

68. Andrew Ott, "Can Flowgates Really Work? An Analysis of Transmission Congestion in the PJM Market from April 1, 1998, to April 30, 2000," PJM Interconnection (September 15, 2000)

69. William W. Hogan, "Flowgates Rights and Wrongs," Center for Business and Government, John F. Kennedy School of Government, Harvard University, Cambridge, Massachusetts (August 20, 2000), available on Professor Hogan's website, www.whogan.com

70. Both Professor Hogan and Andrew Ott point out that in PJM the set of actual congested lines was different from those forecasted over the first 27 months of opera-

tion and included 161 unique constraints. See William W. Hogan "Flowgates Rights and Wrongs," Center for Business and Government, John F. Kennedy School of Government, Harvard University (August 20, 2000), and Andrew Ott, "Can Flowgates Really Work? An Analysis of Transmission Congestion in the PJM Market from April 1, 1998, to April 30, 2000," PJM Interconnection (September 15, 2000). The early Texas experience supports their views

71. Larry E. Ruff, "Flowgates, Contingency-Constrained Dispatch, and Transmission Rights," *The Electricity Journal* (January/ February 2001), pp. 34-55

72. Larry E. Ruff, "Flow-Based Transmission Rights and Congestion Management" (July 22, 2000)

73. Shmuel Oren, Comments at FERC RTO Week (October 16, 2001), Docket No. RM01-12-000

74. Richard P O'Neill, Udi Helman, Benjamin F. Hobbs, William R. Stewart, Jr., and Michael H. Rothkopf, "The Joint Energy and Transmission Rights Auction: A General Framework for RTO Market Designs" (July 31, 2001)

75. Roberto F. Paliza, "MISO Market Design: Market Design and Congestion Management Overview," presentation to the Harvard Electricity Policy Group (November 30, 2001). See also Appendix 1 of the Agreement of Transmission Facilities Owners to Organize the Midwest Independent Transmission Systems Operator, Inc. (the ITC-MISO Agreement), FERC Docket No. ER01-3000-000 (August 31, 2001). Under the ITC-MISO Agreement, the International Transmission Company would share certain RTO functions with the Midwest ISO. The agreement was accepted for filing in a FERC Order dated December 20, 2001. International Transmission Company, 97 FERC p. 61, 323 (2001)

76. Seabron Adamson, "Design of a Transmission Rights Exchange" (December 20, 2000)

77. Alex Henney, " Transmission Access-A Case Study in Public Maladministration?" *Power UK*, 92 (October 2001). The auction proposal was heavily criticized and was abandoned. See also Tim Russell, "The Pantomme Horse Cannot Be Split in Two," *Power UK*, 88 (June 2001)

78. Helm Dieter, *Investment in Energy Networks: Auctions, Regulation, and Planning*, New College, Oxford, (April 26, 2002) p.2

79. FERC Notice of Proposed Rulemaking, 18 CFR Part 35, Docket No. RM01-12-000 (July 31, 2002), ¶ 241-254

80. John Chandley points out that "A point-to-point FTR is a complete set of flowgate rights for all of the possible constraints that could be binding between the two grid points defined by the FTR. In theory, therefore, there is a set of flowgate rights that corresponds to each point-to-point FTR. However, this correspondence does not hold as grid conditions change, thus complicating efforts by an RTO to use FTRs and flowgate rights interchangeably." John D. Chandley, "A Standard Market Design for Regional Transmission Organizations" (September 17, 2001). See also the article under the same title in *The Electricity Journal* (December 2001), pp. 27-54

81. Hung-Po Chao, Stephen Peck, Shmuel Oren, and Robert Wilson, "Flow-Based Transmission Rights and Congestion Management," *The Electricity Journal* (October, 2000), pp. 38-58

82. FERC Notice of Proposed Rulemaking, 18 CFR Part 35, Docket No. RM01-12-000 (July 31, 2002), ¶ 251

83. Harold Hotelling, "The General Welfare in Relation to Problems of Taxation and Railway Utility Rates," *Econometrica* 6 (1938), p. 242-269

84. See also Sally Hunt and Graham Shuttleworth, *"Competition and Choice in Electricity"* (London: John Wiley & Sons, 1996), p. 228 where four basic principles are set forth: (1) locational differentials should reflect true cost differentials; (2) those responsible for the indirect costs of transmission should be allowed to determine investment; (3) transmission service should be provided under tradable, long-term contracts that reflect all the system (deep) costs; and (4) technical standards should be replaced with economic standards

85. Sally Hunt, "Making Competition Work in Electricity," Presentation to the CMS Cameron McKenna Third Electricity Conference, Washington, D.C., Aug. 8, 2002.

86. Alex Henney, "Transmission Pricing in Competitive Markets-Lessons from the World," EUCI Transmission Conference, Miami (February 5, 2002)

87. Transpower New Zealand Limited, "Confirmed Pricing Methodology: Draft Design Principles" (January 25, 2002)

88. William W. Hogan, "Market-Based Transmission Investments and Competitive Electricity Markets," Center for Business and Government, John F. Kennedy School of Government, Harvard University (August 1999), p. 19, available on Professor Hogan's website, www.whogan.com

CHAPTER 5
Raising Agents: Capturing the Benefits of Market Forces

Ingredients of this Chapter

The conventional wisdom is that transmission expansion needs to be cooked in the regulated oven rather than in the open air of a market-based barbecue. Market-based transmission expansion is, however, a reality and is gathering momentum around the world. In this chapter, we shall look at the capture of market forces as a raising agent that is efficient and innovative in facilitating expansion. Market forces have a particular role to play in expansion using controllable lines and interconnecting regional markets. We shall try to avoid conventional thinking about natural monopolies and discuss whether markets in transmission, even though they may not be perfect, could produce a better answer in some circumstances than regulated outcomes. We shall look at the advantages and disadvantages of market-based expansion.

Two types of market-based expansion have been implemented, and we consider them both. The first type is what we shall describe as entrepreneurial market-based expansion. After looking at the experience, we shall discuss the key ingredients and the conditions that need to be put in place in order to facilitate market-based expansion by entrepreneurs. Then we shall evaluate the arguments of skeptics of this method of expansion and create some recipes for different kinds of merchant projects.

We shall then turn to the other form of market-based expansion that might be described as competitive bidding on an institutionalized basis. This can be carried out both as a backstop in the event of market failure and as the primary method of transmission expansion. We shall look at the experience in Argentina, Peru, and Colombia that provides useful case studies and some examples of lessons that can be learned. The supplement to this chapter describes the difficulties of implementing market-based expansion in the European Union.

Market Forces Are a Key Ingredient

A key raising agent

It is not necessary to be an economist to understand the benefits that society derives from market forces and that markets are not always perfect. Market forces are what might be described in culinary terms as an excellent "raising agent" (like yeast in bread). Markets raise the game of the firms participating in them. They create incentives for efficiency and innovation that benefit the consumer in the form of lower prices. In the context of transmission, market forces can, if they are captured effectively, make the role of the regulator both easier and more effective. Markets are also noted for their ability to encourage efficient investment, innovative financing, and new technological solutions.

Transmission does not easily lend itself to markets

Because of its natural monopoly character, transmission was not traditionally regarded as a function or service where markets had a significant role to play. The natural monopoly character is said to exist when technology, capital costs, and economies of scale are such that one firm or utility can provide a service cheaper than two or more entities. However, the presence of natural monopoly conditions does not mean that there is no place for markets or competition. If the cost advantages of a single firm are not particularly great, multiple competitors can exist in a market. There may also be pressure from potential entrants that want to limit the scope or scale of the natural monopoly, or even replace a monopoly in a particular segment of a potential market. Moreover, the incumbent monopolist may face competition from the suppliers of other services that can be substituted for it. In the case of transmission, distributed generation and demand-side responses that alleviate congestion are clearly competitors.

However, there seems to be very little appetite around the world to treat transmission as a sector in which it is impossible to capture the benefits of market forces or where new entrants are not permitted. Even though the market may be imperfect, it will still be worth incorporating some ingredients into the market and structural designs that introduce market forces into the transmission sector. This will be particularly true in jurisdictions where the incumbent transmission owner faces difficulty in raising finance for new transmission facilities. It is useful to avoid adding ingredients that act as a barrier to third-party investment and new entrants. If the market fails to respond, at least it will have had an opportunity to do so.

As William Hogan explains:

> *Market-based transmission investments confront a number of challenges. In practice, the difficulties can lead to a conundrum, a "Catch 22". The typical example of a transmission investment invokes the image of a large new transmission line, which might look to be too difficult to base solely on market decisions. As a result, the implicit assumption often is that only a regulated monopoly could manage the intended investment,*

> *and the discussion of market institutions defaults to the design of monopoly mechanisms. Market-based investments may never get a hearing, and would be foreclosed in the market design. However, with some innovation it is possible to envision at least a partial escape from this trap by designing market institutions that could meet some or all of the challenges and support market-based transmission investments.*[1]

The key ingredients of structural and market design were discussed in chapters 3 and 4.

The alternative to market forces is regulation

We know that existing transmission systems represent natural monopolies that require regulation. There is agreement among academics, regulators, and regulated utilities that regulation is not a perfect answer to the problem of the lack of market forces in the case of natural monopolies. This is one area where both theory and practice are consistent.

Paul Joskow and Richard Schmalensee have pointed out that:

> *Research over the fast past few decades makes it quite clear that regulation is neither perfect nor costless. ... In comparing regulated natural monopoly with any alternatives, we must recognize that we are comparing imperfect regulation with some alternative. Even if the alternative is not perfect, it could be better than the imperfect status quo. Similarly in looking for alternatives to the status quo,*

> *we should consider regulatory reforms that could improve performance of the regulatory process.*[2]

There is no criticism of regulators. As we shall discuss in chapters 6 and 7, they work under the enormous disadvantage of asymmetry of information. Their information is never as good as that of the utilities they regulate. It is never possible for them to create rules and incentives that are dynamic and capable of achieving the kind of innovative solutions that benefit the consumer as effectively as competitive market outcomes. Regulators can never achieve the perfect balance between flexibility and commitment in order to adapt their solutions to meet the changing needs of the consumer over time.

Is an imperfect market better than imperfect regulation?

Drawing on this idea, we should conclude that while the capture of market forces may be difficult, an attempt to utilize market-based approaches in transmission is worthwhile. But Joskow and Schmalensee warn us:

> *Clearly the alternative to regulated franchised monopoly is not likely to be anything close to a perfectly competitive market. In considering the natural monopoly question, it is important not to think of there being too simple extremes: natural monopoly and competition.*[3]

The starting point is to look at the overall objectives and the specific outcomes that will be beneficial for the consumer and for society. We need to choose the solutions that are likely to produce the most efficient results in terms of cost and speed. We also will look for innovation in financing and the choice of technologies. There may not be a perfect market, but there is also no perfect regulatory regime. An imperfect market may be better in some circumstances, for example, to implement the smaller expansion projects as well as controllable and radial lines, and may be worse for larger network investments.

Difficult questions

There are a number of difficult questions to which there are no black and white answers. For example, would a market with two or three potential transmission developers (even if one or more were regulated) be better than a single transmission owner under traditional regulation? Is there a way of determining costs and deriving prices by reference to responses in the market so that the difficulties of designing regulated prices and the lack of information on the part of the regulator can be overcome or, at least, minimized? There is evidence in England and Wales that regulation is capable of being adapted and improved to create incentives to induce the kind of welfare-enhancing outcomes that market forces provide. How easy is it to do this effectively? Can the consumer receive an adequate share of the benefits if the transmission owner has a monopoly of information about its system? In short, is there anything that can be done or any changes made that will help to simulate or facilitate the attainment of competitive outcomes that are, through no fault of the regulators, difficult to achieve under the regulated monopoly paradigm?

These questions arise whether or not the industry has been restructured. In a deregulated environment, the questions need to be considered not only in the context of creating a market in transmission, however imperfect, but also in the context of creating a marketplace for the wholesale trading of energy (with which transmission is so closely linked) in order to provide market participants with the commercial freedoms they require.

Potential advantages of a market-based approach

Merchant transmission is still in its infancy, but if the key ingredients discussed in this chapter are in place, the potential advantages of a market-based approach to transmission expansion are that:

- It should lead to transactional, financing, and technical innovation
- There will be a greater focus on the evaluation of options that may compete with regulated transmission and compete in the whole of the electricity value chain (such as changing system operations procedures or algorithms, alternative routes and technologies, distributed generation, and demand-side responses at peak periods), resulting in lower costs
- There will be a reduced need for regulatory oversight
- It may only be necessary to find a regulated pricing mechanism to incentivize the owners of existing transmission facilities to invest as a backstop, in the event of market failure
- There may be lower direct and indirect transmission costs, depending on the technology used and the location of the project on the system, because a market-based project is likely, for example, to improve reliability and reduce losses even though it was primarily designed to reduce congestion
- The risk of delay will be minimized to the extent possible (and there may be increased resourcefulness in obtaining permits and rights-of-way)

Potential disadvantages of a market-based approach

If there is to be complete freedom for an investor to respond to market forces by implementing an entrepreneurial transmission project that is not subject to the full rigors of regulation, a number of issues will need to be resolved. Some of these lie at the interface between the regulated and merchant approaches. The actual and perceived disadvantages of facilitating a market-based approach include the following:

- The merchant investor may be accused of "cherry picking" the easiest or most lucrative projects. The regulated transmission owners will have to handle the more complex and arduous projects, the costs of which will be borne by all consumers, but which will command only a lower, regulated return for a higher degree of risk

- There may be a significant usage risk because it is difficult for merchant investors to be remunerated by market participants except through a system of physical or financial rights to the new transmission facilities or by energy trading across the facilities, where they are permitted to do so
- Merchant investors will primarily be motivated by so-called "economic" projects involving the reduction of congestion and much less in achieving other benefits such as cost savings in ancillary services, reduced energy price volatility and reduced market power
- A project may give rise to "pancaking" of transmission charges payable to more than one owner unless locational marginal pricing is in place
- A merchant project may create stranded capacity on existing systems. It may also interfere with the plans of regulated transmission owners and developers, giving rise to stranded development and construction costs if the regulator decides that a project planned on a regulated basis should not proceed
- A merchant project may increase congestion, adversely affect reliability or increase loop flows on other systems, and there may be no machinery for compensation for these effects
- A large merchant transmission facility that increases the import potential into a small system could increase the amount of reserve the small system has to carry (because the worst contingency becomes the loss of a much larger infeed—the new facility)
- Merchant transmission may not capture and pass on to the consumer the benefits of inherent economies of scale
- The development, transaction, and negotiation costs may be excessive
- A market-based approach to transmission may result in the construction of a number of small projects that may be less efficient than a larger project
- The transmission expansion market is likely to be imperfect or not sufficiently liquid to put a downward pressure on transmission

costs—paradoxically, regulatory incentives may be more effective at achieving cost-savings for the benefit of the consumers
- There is a perception the market cannot be relied upon to meet public policy needs for new infrastructure
- The idea that the infrastructure upon which society relies should be provided by profit-motivated entrepreneurs with little or no regulation to extract excess rents and transfer them to consumers may be politically unacceptable

Arguments based on these issues are mounted to question whether market-based transmission expansion should be permitted or actively promoted by regulators. In addition to this, skeptics argue that the sheer difficulty of developing and implementing major transmission expansion projects, particularly the larger projects on a mesh network that capture economies of scale, makes it unlikely that they can be efficiently undertaken by market participants rather than regulated transmission owners. The regulator can force the beneficiaries to pay for new transmission facilities. Merchant developers do not have this protection.

Then there are the not-inconsiderable difficulties of obtaining the necessary permits, certificates, consents, land, and rights-of-way, where market participants may be at a disadvantage to incumbent public utility transmission owners. They may not have powers of expropriation or eminent domain. Landowners may demand higher levels of compensation from merchant developers that are perceived to be profit-motivated entrepreneurs with deep pockets. These issues are discussed in chapters 9 and 10.

There are even more basic difficulties, such as the question of how to decide the right size for a merchant project. It may be difficult to obtain information to evaluate the capability of the existing transmission system to accommodate a proposed project. The incumbent transmission owner may require excessive reinforcements that might make the project uneconomic. An analysis of the revenue-earning potential of the project is likely to depend upon an assessment of future energy prices, as well as the likely interest of market participants in acquiring transmission rights with respect to the project and the value they will place on those rights. All of this is likely to deter all but the most ardent entrepreneurs.

Different approaches to capturing market forces

We shall look at two broad approaches that could be utilized to capture some of the benefits of market forces on the basis that defaulting to a regulated solution may not provide the only or the best answer. The techniques or approaches are:

- entrepreneurial, market-based expansion
- competitive bidding for expansion projects on an institutionalized basis

Entrepreneurial Market-Based Expansion Projects

Although in its infancy, there is experience of market-based expansion

The most obvious way of achieving a market outcome in transmission is to encourage or, at least, permit entrepreneurial developers (typically market participants) to propose and implement transmission expansion projects on a pure merchant basis, where the developers take the entire usage or revenue risk.

The leading examples of market-based projects to date include the interconnectors between five states in Australia—DirectLink (in operation), MurrayLink, and Southern Link developed by TransÉnergie, Australia between New South Wales and South Australia, and BassLink, developed by NGC between Tasmania and Victoria. In the U.S. there are several projects to interconnect markets, *e.g.*, Cross Sound Cable I and II between Connecticut and Long Island, and Harbor

Cable between New Jersey and New York City, developed by TransÉnergie U.S. It is also developing the Lake Erie Link with Hydro One in Ontario. Cross Sound Cable I was a groundbreaking project and the first where the Federal Energy Regulatory Commission (FERC) gave approval for market-based rates for transmission. Its proposed standard market design accommodates merchant transmission.[4]

The four phases of the Neptune project will connect New Jersey, Manhattan, and Long Island; New Brunswick and New York City; Nova Scotia and Boston; and Maine and Connecticut. Each of these phases relies on high voltage direct current (HVDC) undersea cables. Other U.S. merchant interconnectors in the course of development include Genpower, a 1,000 mile generator lead from the Maritime Provinces in Canada to New York City; NB Power, a line from New Brunswick to Maine; and Transamerica (a Black and Veatch joint development with Siemens AG) connecting Wyoming to Chicago and Los Angeles, with possible lines to Denver and Minneapolis.[5] The interconnector between England and France is now exploited on a merchant basis even though there is a degree of regulatory oversight. Radial lines have been built in Argentina and Peru under the institutionalized competitive bidding approach described later in this chapter.

Development of mesh systems on a merchant basis perceived to be difficult

In many countries, it would be possible for a market participant to propose and implement a new transmission project, but, for a variety of reasons, there has been little built on this basis. Some of the reasons described in chapter 1 have been at play. In several cases, the key ingredients from market design and transmission pricing we are about to discuss have not been in place, so there are no signals or incentives for any developer, other than the incumbent, regulated transmission owner, to invest.

System upgrades and expansion by third-party merchant investors to mesh systems that belong to other transmission owners are seen as difficult to achieve on a merchant basis. In fact, there are relatively few systems in developed countries that are not meshed for reliability reasons. The quasi-radial parts may be the connections between generators and the rest of the system. Transmission owners and system operators are understandably nervous about entrepreneurial third parties working on their systems, because of the potential effect on transmission system performance and reliability. The responsibilities and liability of each party (especially with respect to their equipment) must be entirely separate and clear. Although there can be a remarkably low degree of performance interdependence between elements of the same system, investors and lenders are unlikely to be willing to provide the necessary finance for the investment, if they feel the performance risks associated with it cannot be properly ring-fenced or separated.

The Neptune Regional Transmission System (RTS)

The Neptune RTS (approved by FERC)[6] is motivated by an abundance of cheap energy in Canada that can be exported to the U.S., and by the problems of serving urban-load pockets such as New York, Boston, Chicago, and other cities, as well as the opportunity to arbitrage energy prices. The developers identify a number of cost and revenue drivers that justify the project:

Cost drivers

- fuel cost advantage
- capital cost (non-urban vs. urban)
- avoided local distribution charges
- avoided in-city interconnection or connection charges
- summer derating costs
- emission credits

Revenue drivers

- "urban premium" between Neptune RTS™ generation areas and Neptune load areas (Neptune RTS™ is a transmission system right or form of FTR we discuss later in this chapter)

- portfolio value—the ability to sell into multiple Neptune RTS™ markets
- meeting installed capacity obligations and in-city generation needs
- volatility value—the ability to sell energy options forward in a high volatility market

These drivers essentially represent the pool of money the market participants will use to pay for the transmission rights to the project they acquire.[7]

The Key Ingredients of Merchant Transmission

List of key ingredients

Market-based or merchant investors in transmission projects need certain ingredients to be in place if they are to rise to the challenge of meeting the needs of society for new transmission capacity. They include:

- the ability to assess who the potential users or beneficiaries will be and the ability to enter into contracts with them
- the ability to realize or deal with economies of scale
- market-based prices
- locational marginal pricing
- a mechanism established in advance for property rights (physical or financial) to be allocated to the investor, representing the additional transmission capacity that the expansion has created
- a coordinated planning process that facilitates market-based transmission and takes it into account at the earliest possible stages
- the absence of regulatory hurdles and micromanagement
- a level playing field with incumbent, regulated transmission owners or network service providers

- the ability to enter into agreements on satisfactory terms with the system operator for the relevant control area in order to cede operational control of and integrate the project with the existing system (for reliability purposes) and to enter into agreements with the relevant transmission owners with respect to the terms for interconnection and consequential reinforcement of the existing system

Identifying the beneficiaries

Given the potential size of a transmission expansion project, it might be necessary for a developer to form a coalition with other potential users or beneficiaries of the expansion to share the costs and benefits of the investment. However, given the nature of transmission systems and the laws of physics, it is not always easy to identify who the users or beneficiaries might be. The potential users could benefit in the form of lower congestion costs, access to buyers in higher-priced areas, access to low-priced sellers, and being able to hedge their energy costs and transmission charges by acquiring transmission rights.

Market-based projects can be difficult to implement if the potential beneficiaries are not easy to identify or if the beneficiaries have no incentives or clear price signals (discussed in chapter 4) to create coalitions or to purchase transmission rights that will be needed to drive investment forward. Merchant developers will always look for ways to secure the project revenues by selling rights in advance to potential beneficiaries or by forming joint ventures with them. The fact that it may be difficult to identify the beneficiaries does not mean that the attempt should not be made. The analysis used to determine whether or not to build any transmission project (merchant or regulated) involves identifying which users will benefit and by how much.

Advantages in identifying the beneficiaries

There are various reasons it is appropriate to identify the transmission users that will benefit from an expansion project. Once a cost-benefit analysis has been carried out, it would be economically efficient to assign the costs and benefits to those users that benefit. The benefits could accrue to a relatively small number of users, who may, as already explained, wish to form a coalition to fund the project. In other cases, the benefits could be completely widespread, and the costs might have to be socialized, in which case, the project might have to proceed on a regulated basis, because the regulator will have power to allocate the charges to all users. As we discussed in chapters 2 and 4, allocating costs to the beneficiaries is efficient, because it creates the necessary incentives on the part of market participants to invest in transmission.[8] By the same token, if the costs are socialized, the beneficiaries could find that they are paying more or less than they should pay, from the perspective of fairness and economic efficiency. Regulators should always try to allocate costs to beneficiaries in order to discourage free riders. This will facilitate both merchant and regulated expansion.

As explained in chapter 4, there is no perfect method for allocating these charges, and, whatever method is chosen, it often proves to be contentious. However, if it is known that the costs are going to be socialized, those users who can identify that they will receive a proportionately greater benefit than other users would have an incentive not to participate in a coalition to fund a project on a merchant basis, and would wait until their needs were met under a socialized-cost approach. Regulators and market designers should not adopt a blanket approach to transmission pricing that socializes or rolls all the costs of transmission expansion into existing transmission charges. This will deter merchant investment.

A separate, but similar argument arises with respect to the socialization of the costs of connecting or interconnecting new generation. Socialization of these costs might over-stimulate the expansion of generation capacity where a transmission solution might be more cost-effective. An example of this might be an HVDC merchant transmission project that provides an urban area with "virtual generation," *i.e.*, power from generation connected to the load center via an HVDC line or network. This raises the question as to whether this type of expansion has more the character of generation and less of transmission and exacerbates the free rider problem.

Efficiency and fairness vs. precision

There is another argument that an attempt should always be made to define the benefits and identify the beneficiaries. It will be likely to achieve a result that is more efficient and fairer than full cost socialization, even if it is not entirely precise. Unfortunately, however, the attempt may be controversial and result in a dispute, because no user will want to pay any more than it has to without putting up a fight. Perhaps the ultimate test of whether this key ingredient can be achieved should not be based on whether it is possible to identify the beneficiaries, but whether there are too many beneficiaries for the investor to be able to enter into private contracts with a substantial majority of them. It will want to be able to do so (whether in the form of transmission service agreements for physical rights or contracts granting financial rights) in order to secure the revenues, because the investment is not being undertaken on a regulated tariff basis.

Difficulties of forming coalitions

Thomas Olivier-Léautier has been relatively cynical about whether many market participants would be motivated to form coalitions to develop expansion projects. He points out that some generators may have little incentive to increase the competitiveness of the marketplace by increasing the size of the market. He argues that transmission is not within the core competence of most large users. They would prefer to sign supply contracts or install a cogeneration plant on their industrial site. Small users could be discounted, and distribution (wires) companies would have no incentives, because their profit would be based on their own performance. Independent and unregulated owners of transmission assets could not be relied upon to expand the network optimally, because their incentives were not aligned with economically optimal outcomes. Retailers would not have the financial muscle, and their incentives are not clear. They would have little competitive advantage in coordinating with other retailers to expand the network. System operators would have the necessary information, but their role is to operate the system, and long-term congestion management, while important, is secondary to them, and no incentives to expand the system are usually provided.[9] In short, he suggests that:

> *Policy-makers cannot pin too much hope on these entities for optimal transmission expansion. We are left with a for-profit grid company which has the information, the resources, and if properly regulated, the incentives to expand the network.*[10]

Although it is never easy to put together investor coalitions, various industry participants will have an economic interest in expanding the system, such as:

- generators seeking access to buyers in higher priced, constrained areas
- buyers (loads) looking for energy at cheaper prices
- transmission users seeking lower usage (congestion) charges
- transmission owners and investors interested in capturing the value of the transmission rights that are awarded following the expansion
- generators looking to protect themselves from new generating capacity being built in high price areas in which they have buyers

Methodology to identify beneficiaries

A methodology to identify the potential beneficiaries of new transmission investment has been developed in Argentina. There are no claims that the methodology is perfect, and, indeed, it will always be a challenge to identify exactly who are the beneficiaries of the expansion in economic terms. Broadly, the Argentine methodology starts by identifying the power flows and then analyzes the effect on the energy prices at each of the nodes. This provides some economic basis for deciding who will benefit. It is described in more detail later in this chapter. Chapter 8 describes an approach that identifies beneficiaries using a reference network.

The traditional view has been that it if it is not possible to easily identify who the beneficiaries are, the expansion costs should be socialized under the regulatory regime. However, as we have discussed, the approach of socializing the costs should be regarded as the last resort. As Chandley and Hogan explain:

> *Moreover, once the enhancements gained regulatory approval, regulators would need to allocate the costs to the beneficiaries, rather than simply rolling the costs into existing rates and spreading the costs across all ratepayers, beneficiaries or not. In addition to being a fair cost allocation principle, this rule would tend to discourage free riders and thus increase the likelihood of market-driven investments.*[11]

Economies of scale

Market-based transmission expansion has been challenged on the grounds that it is unable to capture economies of scale. Developers will not want to invest even a modest amount to over-size a facility to realize economies of scale, because they will be increasing their revenue and usage risk with respect to the excess capacity they cannot immediately sell. However, new technological advances mean that relatively small-scale capacity increases may be as economic as major capacity additions, allowing grid capacity more closely to match demand. Flexible AC transmission system (FACTS) devices, advances in underground and submarine cables, controllable high voltage DC lines, new super-conducting materials, low-impact cable installation processes, and other new technologies have the economies of scale that are substantially realized at those capacities at which they would be installed in a transmission system. FACTS devices reach their economies-of-scale plateau at capacities of less than 100 MW, and are comparable to modern peaking generation units. The realization of scale economies at such low levels undermines the argument that all necessary small-scale transmission investments are inefficient because they do not capture economies of scale, and therefore must be undertaken on a regulated basis.[12]

However, some of these new technologies will be significantly more expensive than traditional alternatives that have bigger economies of scale. Indeed, as a policy matter, it concerns some experts[13] that merchant transmission may result in transmission expansion consisting of a number of relatively small projects instead of a larger, more efficient project. It will make the planning coordination described in chapter 8 more difficult and may produce fewer and narrower benefits for society. There is much to recommend in this reasoning, but merchant transmission does have great promise in interconnecting regions and alleviating congestion using these new technologies, and it is clear that entrepreneurs are not daunted by the size of the project or the distances. For example, NGC and Statnett have been developing an undersea cable between Norway and England on a merchant basis.

The complex world of non-controllable power flows

As we have seen, there are powerful externalities and interactions that affect most transmission networks. Most lines are not controllable, and power flows cannot be directed to follow any designated path. If a controllable line is part of a larger system, its impact will be affected by the pattern of use of the rest of the system. As a result, if there are significant economies of scale in a transmission expansion project, the amount of power purchased and sold at source and sink might equal the expanded transmission capacity (properly defined), but the power actually flowing might not be the same. Moreover, the important linkage between expected price differences and the marginal cost of investment might be eliminated. The expected price difference between the locations might not be equal to the marginal cost of expansion. In other words, the "equations" we discussed at the beginning of chapter 4 when equilibrium prices differ at the source and sink in a competitive market would no longer be sustained. These "equations" are:

- The amount of power purchased and sold at source and sink would equal the capacity created by the transmission investment
- The expected price difference between the locations would be equal to the marginal cost of expansion
- The power flowing on the controllable transmission line would be equal to the capacity created by the investment

A simple model of charging for the flows on a new line where there are economies of scale would not support efficient investments. In extreme cases, with economies of scale and the efficient investment eliminating the price difference between locations, there might be no investment at all.

Deliberate withholding of capacity additions

As Hogan has pointed out, it would be possible for a merchant developer to install transmission facilities that were larger in capacity terms than those that were economic from the point of exploiting the congestion rents. There would be nothing to stop the developer from declaring the availability of the new facilities at a lower level than the physical, nameplate rating. Nearly every transmission system has examples of facilities that are capable of transporting or transforming higher levels of energy than the levels at which they are typically loaded. This may be for system operations reasons or because of limitations imposed by the other facilities forming part of the transmission system.

Hogan explains that:

> *The idea of allowing withholding of actual physical capacity seems at odds with the policy of open access and full utilization of the existing grid. However, on closer inspection there is no funda-*

> *mental contradiction. The market-based investment would not be like the regulated investment. In particular, there would be no rate-base protection or inclusion of the investment costs in mandatory access charges. For the capacity associated with regulated investments, there would be no withholding.*[14]

The investment might not be made if the profit incentive were removed by requiring optimal exploitation of economies of scale and immediate utilization of the full capacity. The investor would bear the risk of the market-based investment, but would require the reward of charging a price to justify the investment. Hogan goes on to explain:

> *As long as the investment did not foreclose competitive entry, the withholding should not be seen as the deleterious exercise of market power. Rather, the temporary withholding could be seen as part of the process of dynamic adjustment in the market, with the profit incentive originating in the scarcity rents. It is precisely the pursuit of these scarcity rents which motivates entry and the investment.*[15]

These arguments are powerful, but regulators may feel that it would be counter-intuitive to the principles of open access and the control of market power to allow withholding of capacity in this way, because it will be difficult from a political respective. They could be nervous of temporary changes in capacity that is withheld for profit-motivated, strategic reasons and fear some sort of gaming. A similar result could be achieved through reserve prices or minimum bids in rights auctions, but the same regulatory concerns might arise.

In a policy adopted by Australia, which encourages market-based transmission investment, there is a provision allowing the investor to control the capacity of a new transmission line. This allows the investor to limit the use of the line whenever the locational price difference is low. If this results in higher price differences, the withholding is said to be justified as the necessary cost of capturing the benefits of the expansion. This rule that is described as a "safe harbor provision" applies only to "controllable" lines, which simplifies things considerably because it mitigates or eliminates the network externalities. The use of the controllable line approximately restores the linkage between the identifiable increment in system capacity and the actual flow over the line. This allows for a payment method that is equivalent to charging a fee for using the line. This limitation is necessary, because Australia does not have a bid-based, security-constrained economic dispatch with fully locational prices. FTRs have not been implemented.

Market-based prices

A merchant expansion project is defined as one where the developer takes the revenue and usage risk of the project and levies charges that are market-based, rather than regulated. If the developer prefers the regulator to allocate the charges to users and to receive a potentially lower, regulated return with a more certain revenue stream, the project would be classed as regulated transmission. The developer could undertake the project on the basis of a long-term agreement with one or more users that secure the needed revenue stream. This would be the transmission equivalent of a traditional independent power project (IPP). The users might be an incumbent transmission owner, a generating company, or a distributor. Their prices to customers may be regulated.

In a market environment, the revenues will be derived from selling (typically, well in advance) physical or financial transmission rights. In some jurisdictions, particularly where the legal or regulatory framework is such that the regulator has jurisdiction over all transmission activities, the developers may have to obtain permission in advance from the regulator to be able to charge on a market basis. The role of the regulator is dealt with in more detail below.

The transmission equivalent of an IPP project

Ten years ago it would have taken a book to describe how to undertake the transmission equivalent of an independent power producer (IPP) project, *i.e.*, an independent transmission project. The feasibility of an independent transmission project would have depended upon a long-term (25 years or more), revenue-earning transmission control agreement being negotiated, under which operational control would have been ceded in full to the system operator of the host utility. The tariffs probably would have been calculated in much the same way as for an IPP project, based on inflation-linked construction, operations, and maintenance costs, debt service, and a return on equity. However, there are now many published examples of transmission control agreements where the operational issues have been thought through in detail. Most of the agreements have been developed in the context of a restructuring that has involved the separation of system operations from transmission ownership. For the most part, these agreements have worked reasonably well in practice if the rights and obligations of the parties have been well thought through. They have enabled the system operator to carry out its job and have not impaired reliability and security of supply. They have also enabled the transmission owner, in a more passive role, to manage its assets and, where no other mechanism applies, to earn the necessary revenues.

Those countries that are not about to deregulate electricity industries and are struggling to find the financial resources to fund needed transmission expansion should find it attractive to run a competitive procurement process for independent transmission projects. However, they will need to be prepared to live with the consequences and lack of flexibility of long-term agreements that may require government support if the credit standing of the host utility is weak. In this way, it should be possible to capture the forces of competition at the point of procurement, at least, and encourage innovative financing and technological solutions. The fact that the usage risk resides with the host utility may not matter to the utility if it is regulated and can recover the costs from the beneficiaries. Its main obligation will be to run a transparent and fair competitive procurement process that best captures the forces of competition. The role of the regulator would be to oversee that process, but not to micromanage it.

Locational marginal pricing

As explained in chapter 4, it is important for all transmission expansion, whether on a market basis or on a regulated basis, that clear locational price signals are made available that reflect congestion costs and encourage efficient use of the transmission system. Location-based marginal pricing is an important ingredient for the creation of an effective, market-based expansion environment. It creates signals as to when and where to invest. Because locational marginal prices reflect the costs of transmitted power at different nodes, beneficiaries can be identified and signals created that should allow system users to find a basis for allocating these costs among themselves efficiently and fairly. The network externalities of pricing delivered power are internalized, and, for this reason, the approach is both accurate and efficient in the context of transmission.

The voluntary offers and bids in a competitive market create the most efficient method of establishing locational marginal prices. To recap, generators submit offers to the system operator to sell electricity at their points of injection into the system, and loads submit bids to purchase power at a particular node for a maximum purchase price. The system operator then dispatches the generation in economic price order (lowest to highest) based on the bid and offer prices it has received in respect of each node. If it is necessary to depart from the economic (merit) order for reasons of system reliability, the market model is described as "security-constrained."

Transmission rights

As explained in chapter 4, a system of transmission rights, whether physical or financial, is a key ingredient in facilitating transmission expansion. The bid-based, security-constrained nodal pricing model provides what is described as "an elegant answer"[16] in the form of fixed or financial transmission rights (FTR). The FTRs "represent exactly the financial benefit that would accrue to a market participant that owned its own line."[17]

The system of transmission rights proposed for the Neptune project is specially designed and a right is called a Neptune RTS™. It is based on the concept that one transmission system right (TSR) gives the right to transmit from a specified export point to a specified import or delivery point one MW of electricity, the duration of which can range from one hour to more than 20 years. This, however, calls into question whether market participants will commit to matching 20-year power purchase agreements. TSRs can be bought and resold subject to the continuing obligation of the successful bidder to pay the contract price of the TSRs. The TSRs can be broken down into time periods as short as one hour and resold or recombined.[18]

The major advantage of FTRs is that they are tradable property rights reflecting the market value of using the transmission system at a point in time. They can be sold or acquired in advance through auctions or other allocation procedures as explained in chapter 4 and can then be traded in forward secondary markets. The prices that market participants are prepared to pay for FTRs (and at which they may trade them in advance) represent the forward valuation of transmission usage by the market. These prices represent the market participants' expectations of the value of the stream of payments that the holder will receive over the period of the FTR. They will reflect the expected value that the market places on avoiding congestion charges. In other words, FTRs provide a measure of what the market might pay to have the transmission system upgraded to relieve the same congestion.

FTR auctions in advance of implementation

It is likely that a merchant developer of a transmission project would auction rights in advance of implementing the project, and, based on the proceeds of the auction that were realized, it would make a decision on whether or not to proceed with the project. If the value that the market places on carrying out the expansion were less than the cost, the developer would not proceed. If it were more, it would clearly be economic for it to do so. This is a much more effective method of managing and mitigating the usage and revenue risks than building on a purely speculative basis and selling or auctioning rights at a later stage, unless the project were very low cost and the developer were prepared to take the risk.

Market participants may find it difficult to place a value on the future benefits of holding FTRs, particularly in the early days of a deregulated market. The initial "open season" auction carried out by Northeast Utilities Service Company with respect to the Connecticut-Long Island interconnector was cancelled because of lack of satisfactory response.[19]

One of the key ingredients for market-based expansion in the design of the wholesale market is to incorporate rules that award to those who invest in transmission expansion projects, the incremental FTRs that become possible as a result of the expansion. This means that the market rules must contain a defined method for determining the incremental FTRs made possible by a proposed expansion project and also a mechanism under which the system operator awards those incremental FTRs (after applying the simultaneous feasibility test described in chapter 4) to those who make the investment. The length of the FTRs that are awarded to the investors should ideally reflect a period that is sufficient to enable the debt to be repaid and to enable the investors to earn a return in line with profit expectations. In this way, market-driven and regulated transmission investments will be both permitted and facilitated.

Practical experience of transmission rights to facilitate expansion

As explained earlier in this chapter, the first expansion project approved by FERC on a merchant basis in the U.S. was undertaken by TransÉnergie U.S. to link the New England system with the New York system by an undersea HVDC cable from coastal Connecticut to Long Island, NY, known as the Cross Sound Cable (CSC). The project was approved by FERC on June 1, 2000. FERC described the project's negotiated rate structure as a form of opportunity cost pricing, reflecting the market's expectations of the locational price differential between New England and New York.[20] The investors will assume the entire financial risk of the project and charge location-differential prices to users of the link. FTRs that are tradable in a secondary market will be sold. The Long Island Power Authority was the successful bidder in the CSC "open season" rights auction. As explained in chapter 4, at the time of the open season, appropriately determined financial transmission rights between the New England and New York markets did not exist. However, the CSC rights can be converted into financial rights upon the implementation of an appropriate framework for New York/New England FTRs.

As we discussed in chapter 4, the question is often asked: if an optimally-sized transmission project is built, and congestion is reduced to zero, will the FTRs be worthless because there will be no congestion rents. It is unlikely that an optimally-sized transmission investment will reduce congestion to zero for all hours of the year. In any event, the developer is likely to attempt to negotiate private contracts with generators, consumers, and load-serving entities that would clearly benefit from the expansion. They should be willing to purchase rights or fund investments up to the amount that the project reduces their congestion costs. For example, if the expansion reduces congestion costs from $100 million to zero, and the cost of the project is $80 million, then the developer and the load-serving entities that have been paying for the congestion in the past should be able to negotiate a deal in which the project is funded from the reduction in congestion costs.

European experience of physical rights

There is also experience of using physical rather than financial rights that has proved to be the preferred approach with owners and users of the European interconnectors. The disadvantages of physical rights have been explained in chapter 4. However, they have been used for the Nordic interconnectors and the interconnector between England and France. Because of concerns about the potential abuse of market power and the possibility of hoarding physical rights, both the European Commission and the national regulators have scrutinized attempts at long-term capacity reservations and have looked at the mechanics of the auction or allocation processes quite closely.

The European Commission has highlighted concerns about capacity reservations on existing interconnectors, in particular where congestion occurs, such as the German/Danish, Dutch/German/Belgian, Spanish/French, and Italian/Austrian/French interconnectors and where the reservations are in favor of dominant incumbents. A senior Commission staff member, Michael Albers, issued an influential paper indicating future capacity reservations would only be permitted if they were necessary to create revenue certainty in order to assist in raising the necessary finance for a new line or interconnector.[21] However, it has taken a "favorable" approach with respect to the Viking Cable between Norway and Germany where a contract between Statkraft (Norway) and E.ON Energie (Germany) for delivery of a maximum of 600 MW/1,200 gigawatt hours (GWh) (less than 25% of the trading capacity of the link) for 25 years appears to be permitted. The rest of the capacity will be devoted to short-term exchanges.

The European Commission has intervened to reduce or eliminate capacity reservations for existing electricity interconnectors. This relates to some of the undersea Baltic interconnectors. There was a long-term reservation agreement for 60% of the total capacity of the only electricity interconnector linking Western Denmark and Norway (the "Skagerrak" cable) for the benefit of the Norwegian generator Statkraft and of Elsam, the principal generator in Western Denmark. This agreement built on previous long-term agreements and was for a 20-year term. The remaining 40% of capacity on the Skagerrak cable was also subject to a long-term reservation agreement between Statkraft and the German utility, E.ON. The second agreement also included a reservation of transit capacity through the Western Danish network and 34% of the capacity of the Danish/German interconnector towards Germany for a term of 25 years from 1998.

Following intervention by the European Commission, the 60% reservation of capacity on the Skagerrak cable in favor of Statkraft and Elsam was abandoned with effect from the start of 2001.[22] The issue was taken up in parallel in the VEBA/VIAG merger procedure where the surrender of the 40% capacity reservation on the Skagerrak cable and the 34% capacity reservation on the Danish/German interconnector were conditions of the merger clearance procedure.[23]

Physical rights have a more limited appeal to a smaller number of market participants than financial rights. The physical rights would be attractive to generators, distributors, and large consumers whose primary interest is in the physical delivery of energy. Financial rights should appeal to any market participants that are interested in physical delivery, hedging congestion costs, and generally in risk management in energy trading.

Auction of physical rights to the England-France interconnector

The auction of physical rights to use the England-France 2,000 MW HVDC interconnector was viewed as successful. Initially, there were six or seven active users of the link.

The construction costs were sunk so that it was possible for National Grid Company (NGC) to take the entire usage risk on its portion of the link. In the past, NGC's income from the interconnector was uncertain. It was based on energy transfers into or out of the England and Wales Pool by Electricité de France (EdF). In the early days of NGC's existence, EdF had not opened its market. Originally, the plan was to sell rights for a five-year period, but, after sounding out the market, a combination of three-year rights, one-year rights, and daily rights were auctioned. NGC accepted 600 MW of bids for three-year rights and rolled the rest of the capacity into the annual auction, where it sold 400 MW of rights. The remaining 1,000 MW were auctioned as daily rights. The rights were auctioned by NGC in blocks of 1 MW, but transfers were made in 50 MW blocks (although operational changes will ultimately enable the interconnector to operate in 1 MW increments). A use-it-or-lose-it rule was imposed, and a mechanism to permit trading in the rights at short notice near to real time was created.

A reserve price was put in place to mitigate EdF's market power. A variable number of MWs were auctioned with no reserve price, and the balance was auctioned with the reserve price. Each weekday, 50 MW were auctioned with no reserve price and 750 MW with the reserve price. The numbers for weekend days changed to 250 MW with the reserve price and 500 MW without it. In the winter, these numbers could be changed again. The regulator was concerned with the fact that the capacity that was subject to the reserve price was under-utilized and required information as to how it was calculated.

In Australia, the Basslink interconnector project between Tasmania and Victoria has also been financed on the basis of physical rights. Hydro Tasmania, seeking a market for its hydro energy, has decided to join the National Electricity Market and has purchased all of the export/import capacity of the link.

Is a coordinated planning process necessary for merchant transmission?

In chapter 8, we will discuss the need for and the benefits of a centrally-produced, coordinated statement (as opposed to a command-and-control style central plan) of proposed expansion projects (whether merchant or regulated), reliability projects, and transmission upgrades. The purpose of this exercise would be to enable market participants and regulated, incumbent transmission owners to assess the opportunities that will be forthcoming and to reconcile the issues that need to be resolved if all the projects are implemented. Because of the publication of useful development data and criteria, this degree of transparency in the transmission expansion development and implementation process will be extremely valuable and will help to create a level playing field between merchant and regulated transmission.

Advocates of merchant transmission are keen to avoid a process that includes coordinating plans for projects that are driven by economics and the desire to reduce congestion.[24] However, the interaction of reliability and congestion on the transmission system is such that expansion motivated by an economic desire to reduce congestion is likely to improve reliability and vice versa. The distinction becomes blurred, because congestion is often the result of limits imposed by reliability considerations.

The planning coordination process should facilitate market-based expansion

It is important that local and regional planning processes should take into account the plans of incumbent, regulated transmission owners and market-based expansion projects.

The planning coordination process will be discussed in more detail in chapter 8, but for the purposes of analyzing the key ingredients of an effective market-based expansion scheme, the important aspects of the planning coordination process to note at this stage are:

- Merchant developers will benefit from information at an early stage, in the same way as developers of generating plants and large consumers analyzing where to locate on the transmission system. For this purpose, the publication of a statement showing anticipated power flows at each node of the system over the coming 5-10 years and the anticipated new connections and interconnections will be extremely beneficial. Many transmission owners, such as NGC and its affiliates, do this either voluntarily or pursuant to regulatory rules. They do not seem to regard it as the disclosure of proprietary information they would prefer to keep secret, but rather, the publication of information that will encourage efficient use of the transmission system and help to provide the transmission service that the market needs

- In order to encourage market-based expansion and coalitions of potential beneficiaries and investors to form, it could be helpful if the system operator or other entity carrying out the coordination of the planning process were to undertake a consultation or solicitation process in order to advertise identified needs to relieve congestion, improve reliability, reduce market power, or procure ancillary services in any way that is appropriate, including market-based projects, distributed generation, and load management measures, and invite responses from the market. In this way, all of the options can be taken into account and the costs and benefits analyzed effectively. If the market fails to respond in this way, the backstop would be the traditional, regulated approach. If the generation and demand-side solutions are to be taken into account in the coordination of the transmission planning process, it will be necessary to establish a market price that reflects the value of electric power as transmitted, in order to derive a value for a transmission project that will enable comparisons leading to an efficient choice to be made among the alternatives

The relationship between merchant and regulated transmission expansion needs to be thought through at both the planning and implementation stages. Regulated or non-market-based expansion can degrade the value of other transmission investments and property rights, whether regulated or market-based. However, merchant investors need the freedom to take on and manage their own risks. Merchant projects do not need to be planned as such, but should be taken into account in a formal planning coordination process. FERC only proposes that regulated projects must go through a formal RTO/ITP planning process in its standard market design, so that the rate treatment can be reviewed.[25]

Regulatory oversight of merchant transmission

Regulatory risk is an even greater issue for market-based expansion than for regulated expansion. Prices cannot and should not be regulated, but there is likely to be some degree of regulatory oversight in terms of third-party access, anti-competitive practices, and potential market power abuse, both with respect to the transmission facilities that are built and with respect to the energy markets they serve. Therefore, it is important that the regulatory risk be minimized to the maximum possible extent, in order to create the necessary certainty upon which the merchant development of any infrastructure depends.

A pattern is beginning to emerge of the conditions that are attached to regulatory approvals. Some of them are evident from the Cross Sound Cable[26] and from the European interconnectors. The conditions include:

- unbundling (corporate or management separation and separate accounting of the merchant transmission business from other business interests of the investor)
- absence of conflicts of interest and cross-subsidies from the other businesses of the investor
- full assumption of market risk by investors
- non-discriminatory access by third parties and an absence of capacity withholdings and capacity reservation by affiliates
- availability and tradability of physical or financial rights to all potential users or beneficiaries
- approval of auction design features, for example:
    - reserve or minimum prices (if the assets are not in the regulated asset base, merchant developers should be permitted to include them)
    - transparency and fairness of any allocation process
    - preconditions for buying the rights
    - duration of the rights
    - minimum size of the rights that can be acquired
    - restrictions on assignment of rights
    - "use-it-or-lose-it" rules
    - publication of results (to facilitate secondary trading)
    - caps on total quantity of transmission rights acquired by any one entity[27]

- satisfactory governance and decision-making arrangements (to deal with rule changes, market monitoring, enforcement)
- protections against favoring affiliates (*e.g.*, cap on reservation of capacity)
- technical and operational issues:
    - degree to which the power flows are fully controllable
    - interoperability with other transmission systems
    - impact on other system users
    - the ability to provide system support or ancillary services
    - respect for pre-existing, third-party property rights
- arrangements for emergency support

Regulation in Australia

The regulators in Australia seem to be more relaxed than those in Europe and the U.S. For example, the developers of merchant links are:

- only subject to general competition law
- not subject to third-party access requirements (*i.e.*, they can withhold capacity)
- not required to hold an open auction of transmission rights
- not restricted in allowing affiliates to exploit part of the whole of the capacity of the link
- not limited to the commercial arrangements that can be entered into with generators
- not subject to "use-it-or-lose-it" rules with respect to physical transmission rights[28]

The key lessons with respect to merchant transmission for regulators are that they should:

- Resist the temptation to micromanage

- Make every attempt to avoid creating barriers or hurdles for new entrants

How far should the regulator go?

The question arises of how far the regulator should go to attract new entry. Some regulators, such as OFGEM in the U.K., have a statutory duty to promote competition. Should the regulator impose restrictions on incumbent transmission owners to facilitate new entry? This would be going too far. Even the most market-based model in Argentina does not restrict the incumbents from bidding on an equal footing with merchant developers for transmission expansion projects, even though it does restrict the degree to which the existing system can be upgraded without undertaking a competitive bidding process. The merchant developers themselves have suggested that existing transmission owners should be able to compete with them for projects, provided that the playing field is level.[29]

Level playing field

Obviously, the developers of market-based expansion projects will want the assurance that they are playing on a level playing field with the incumbent transmission owners. They are likely to feel that the incumbents will have an inherent advantage, because of the size and characteristics of the transmission systems they own and their ability to recover costs through regulated charges. By the same token, incumbent transmission owners will wish to be able to realize any latent value in their assets and to be free to compete in the marketplace.

Merchant developers will be interested in the design of the regulatory regime for incumbent transmission owners and, in particular, the nature of any regulatory incentives designed to encourage them to upgrade or expand their systems. They will also want to satisfy themselves that incumbent transmission owners are not being subsidized by consumers to compete with them for the same projects.

The issue then turns upon the degree to which incumbent transmission owners should be asked to put their hands behind their backs in order to create a level playing field or, arguably, to tilt the playing field in favor of new entrants in the form of merchant transmission developers. Alfred Kahn has argued that it would be inefficient from the point of view of society to require incumbents to refrain from market participation, because this may be the cheapest and most effective form of market entry or market participation. It is made from a platform in which the initial costs are already sunk and do not have to be paid for by consumers.[30]

Interaction of merchant and regulated transmission in Australia

Australia facilitates market-based transmission expansion by requiring all regulated expansion projects to pass a special regulatory test. As we have seen, merchant investors are free to exploit their projects in any way they wish, including trading in the energy markets to arbitrage price differences. They are not required to offer open access to third parties with respect to the lines and facilities they build, but they may contract with market participants to provide long- or short-term capacity reservations if they wish.

The regulatory test to be applied to all transmission and distribution system augmentations and interregional interconnectors is applied by NEMMCO and the Interregional Planning Committee. The test does not apply to market-based expansion. The text of the test is as follows (emphases are original text):[31]

The regulatory test

A *new interconnector* or an *augmentation option* satisfies this test if it maximizes the *net present value* of the *market benefit* having regard to a number of alternative projects, timings, and market development scenarios; and

An *augmentation* satisfies this test if:

(a) In the event the *augmentation* is proposed in order to meet an objectively measurable service standard linked to the technical requirements of schedule 5.1 of the Code—the *augmenta-*

tion minimizes the net present value of the *cost* of meeting those standards; or

(b) In all other cases—the augmentation maximizes the net present value of *market benefit*

having regard to a number of alternative projects, timings, and market development scenarios.

For the purposes of the test:

(a) *Market benefit* means the total net benefits of the *proposed augmentation* to all those who produce, distribute, and consume electricity in the National Electricity Market, *i.e.*, the increase in consumers' and producers' surplus or another measure that can be demonstrated to produce equivalent ranking of options in most (although not all) credible scenarios;

(b) *Cost* means the total cost of the *augmentation* to all those who produce, distribute, or consume electricity in the National Electricity Market. Any requirements in notes 1 to 9, [of the regulatory test] inclusive, of the methodology to be used to calculate the *market benefit* of a *proposed augmentation* should also be read as a requirement of the methodology to be used to calculate the *cost* of an *augmentation*;

(c) The net present value calculations should use a discount rate appropriate for the analysis of a private enterprise investment in the electricity sector;

(d) The calculation of the *market benefit* or *cost* should encompass sensitivity analysis with respect to the key input variables, including capital and operating costs, the discount rate, and the *commissioning* date, in order to demonstrate the robustness of the analysis;

(e) A *proposed augmentation* maximizes the *market benefit* if it achieves a greater *market benefit* in most (although not all) credible scenarios; and

(f) An *augmentation* minimizes the *cost* if it achieves a lower *cost* in most (although not all) credible scenarios.[32]

There are nearly three pages of notes on the methodology to be used in the regulatory test as applied to a proposed augmentation or interconnector. They cover the information that should be considered, including a variety of cost estimates and the construction timetable. There is a requirement that in determining the market benefit, the proposed augmentation should not preempt nor distort potential unregulated developments, including network, generation, and demand-side developments. The consultation process for determining whether a proposed augmentation satisfies the test must be an open process, so that interested parties have an opportunity to provide input. They need to understand how the benefits have been measured and how the decision has been made.

The question then arises of who should pay for a regulated interconnector project. The project to build an interconnector between South Australia and New South Wales (the SNI Project) has given rise to a debate on whether consumers in South Australia who benefit should pay some of the costs that are already sunk in New South Wales. An argument has been made that the project benefits Australia as a whole and that all consumers should pay.

The Australian regulatory test represents a transparent approach to deciding what projects should proceed on a regulated basis. It may also help to create a level playing field between market-based and regulated expansion projects by putting the regulated expansion projects through an additional test.

Application of the regulatory test in practice

The wording of any test is important, but the way in which it is applied is, arguably, more important. The regulatory test and its application have not been without controversy. In the case of the South Australia-New South Wales interconnector (SNI) project, the developers of a potentially competing interconnector (in part) between the two states took issue with the way in which the regulatory test had been applied. The developers had identified an alternative project configuration for the SNI project that appeared to provide greater interregional electric transfer capability than the proposed SNI project at a lower cost. However, this alternative configuration was not considered, and there was no calculation of the net market benefit associated with it. This failure to evaluate the alternative appeared to be in contravention with the letter and spirit of the regulatory test.[33] The point was strongly disputed and the challenge was not upheld.[34]

Test of market failure

If it is necessary to create a level playing field by allowing time for the market to respond first, the question arises of whether a formal market failure test would be needed to establish when the regulatory test should be applied to a proposed project. The test would be very challenging to establish, because it could be difficult to distinguish between a project that was simply uneconomic and which the market had rightly rejected and a project that was needed, but the market had failed to profit from the opportunity to build it. Simply giving the market a period to respond would be arbitrary and subject to political considerations as to the period during which high congestion costs should be allowed to continue without a regulated solution.

Sharing the inherent advantage of an incumbent transmission owner

A significant advantage that an incumbent transmission owner has over any new entrants is the ability to utilize or exploit the existing rights-of-way, land, and substations that it owns, given the difficulties in obtaining permits, consents, and negotiating with landowners over rights-of-way for new route corridors.

It may be possible to create a regime where a third-party merchant developer could negotiate a contract to utilize the rights-of-way belonging to an incumbent transmission owner on a basis that was entirely satisfactory from the point of view of both parties. Arrangements of this nature have been agreed to with respect to the installation of fiber optics on transmission systems. Clear ground rules have to be put in place as to which party carries out the construction and maintenance work and how liability is apportioned between them if faults or system incidents affect the performance of the assets of either party. Indeed, the ownership of all of the assets can remain with the incumbent transmission owner, and the fiber-optic lines are leased to the telecommunications company. In the case of transmission expansion, the ownership and maintenance of the assets that upgrade the system could remain with the incumbent transmission owner, and the third-party investor could be allocated the FTRs.

This approach could overcome two difficulties. First, it could address the issue that sometimes arises from the way in which rights-of-way or easements are worded, which limits their use to the owner of the transmission system. Second, if the incumbent transmission owner were, in effect, to enter into a maintenance contract with the third-party investor, there would be economies in having one company carry out the maintenance of the entire system and less scope for dispute over liability in the event of faults and system incidents.

The question then arises of whether the regulator should actively promote or facilitate market-based expansion and introduce competition in what was previously thought to be a sector characterized by natural monopolies, by requiring incumbent transmission owners to offer terms to third-party, merchant developers for the exploitation of existing land, rights-of-way, and other assets. The regulator could act as a backstop if the parties were unable to reach an agreement. This may appear to be a step too far, because it would be seen as an interference with the inherent property rights of the transmission owner.

Who owns the existing system?

Whether requiring transmission owners to make their assets available to third parties is a sustainable regulatory approach probably depends upon a broad economic analysis of who is the owner of the existing transmission assets. Any legal analysis would give the answer that the assets belong to and, therefore, should only be exploited by their owners. Any benefit that can be derived from further exploiting these assets to increase the transfer capability of the system would belong to them. However, an economic analysis undertaken from the perspective of social equity might give a different answer. Where the costs of these assets are sunk, in essence, they have been paid for over the years by transmission users directly and by consumers indirectly. Therefore, if the regulator can maintain the argument that the economic value of the assets belongs to the consumers, it could attempt to obtain the cooperation of the transmission owner in enabling third parties to exploit these assets in suitable cases, if there would be an economic benefit to the consumer to do so.

There are a number of problems with this line of thought, which is put forward on a speculative basis. The regulator may not have the legal power to force an incumbent transmission owner to allow third parties to use its assets in quite this way. If the performance risks associated with the existing and new assets cannot be ring-fenced in a manner that is acceptable to the lenders and investors, the project will not be bankable. However, although there is no precedent for such a step, there has been some thinking by regulators as to the treatment of sunk costs, where the regulator has concluded that the consumers should be given some share of the revenues derived from the increased exploitation of assets for which they have already paid. An example of this would be in relation to the income with respect to the use of the England and Wales transmission system for stringing fiber-optic lines.

Transmission owner upgrades on a merchant basis

The market-based incentives may be much more effective than any incentives that the regulator can create. Therefore, is there any reason to limit the ability of incumbent transmission owners to propose new market-based expansion projects? If they are prepared to take the usage and revenue risks, then they should be permitted to do so, because they would not be seeking to allocate the costs under regulated charges. Clearly, they should not be permitted to cross-subsidize merchant projects using their regulated business. However, regulatory ring fencing and accounting separation rules usually cope with this adequately. The question then arises of whether regulated transmission owners should be permitted to exploit their own assets on a merchant basis, with the result that part of their revenues will be regulated and part unregulated. For example, they could propose to install new technologies or to reconductor existing lines on a merchant basis, knowing that they can take the usage risk reflecting the incremental capacity, because there is no usage risk with respect to the rest of the capacity. This might be attractive from a consumer perspective, but unattractive to a new entrant.

If it proves economic for the transmission owner to declare the availability of the upgrade at a level below its physical capability (*i.e.* withhold capacity in the manner discussed earlier) in order to benefit from the congestion rent, it should not be permitted to do so at a level below the capability of the existing regulated asset base for which the consumer has already paid.

Interconnecting a merchant project

Incumbent, regulated transmission owners may be concerned about interconnecting a merchant transmission project with their systems. They may view the merchant investor as an unwelcome competitor and be reluctant to carry out system studies and agree costings for system reinforcements. It would be helpful for the regulator to act as a backstop in the event that there are delays and disagreements.

Addressing the Arguments of the Skeptics

Potential for multiple transmission charges

An immediate reaction to the prospect of market-based expansion projects has been that they would give rise to "pancaking" of transmission charges with those of the existing, regulated transmission owners (*i.e.*, the payment of multiple charges in order to cross ownership boundaries). This is not an issue in a market that has incorporated locational marginal pricing and FTRs into the design. The investor would cede control of the new facilities to the system operator in the relevant control area. The system operator would apply the simultaneous feasibility test in order to calculate the incremental FTRs that the investor would receive. The investor would receive the congestion rents from the system operator as a holder of those FTRs in the normal way. It could sell the FTRs by auction, under negotiated private contracts, or in a secondary market.

In any other case, there could be a potential problem, because it is likely that the pricing regime will involve treating fixed costs as variable costs.

As we observed in chapter 4, this is inefficient and creates perverse locational incentives to avoid the variable charges that are now higher than they would have been if some of the fixed costs had not been treated as variable. The investor could negotiate an arrangement with the system operator under which its costs were somehow incorporated into the amount collected in respect of transmission usage, in effect, utilizing the system operator as a collection agent. This would not eliminate the "pancaking." However, it does not follow that the charges of the merchant developer would be higher than if the project were built on a regulated basis or that "pancaking" would automatically increase cost.

There might also be "seams issues" at the interface between the regulated transmission system and a merchant interconnector project. For example, there is a danger that there could be different market designs and congestion management schemes at the point of interconnection of the merchant interconnector that might result in market participants having to acquire two sets of rights in order to utilize the same set of assets. This is a potential effect of the different congestion management schemes over the interconnectors between Ontario and New York. FERC's plans for a standard market design are helpful in this respect, as they will minimize the occurrence of "seams issues."[35]

Inconsistency of definition of FTRs

There has been concern that if FTRs are not consistently defined from one market to the next, the investment incentives will be inadequate. It is clearly desirable that as many of the aspects of market design as possible should be consistent between markets in order to facilitate trading between them. However, any differences in the definition of FTRs should not operate to frustrate the establishment of a scheme that permits market-based transmission expansion.

Transaction costs and negotiation of contracts

There are also those who are skeptical about market-based expansion on the grounds that the transaction or negotiating costs will be prohibitive. Potential beneficiaries and users will not be prepared to buy similar quantities of rights at similar prices. They also argue about the practicalities of entering into long-term contracts with the many beneficiaries of a large project. The amount of these costs, the time that it would take, and the ability of many to free ride by obtaining the benefits of the expansion without contributing to its costs could discourage investment in transmission on a merchant basis.

In practice, however, the developer should be able to identify a relatively small number of beneficiaries or potential users who are significant participants in the wholesale market. They will probably account for more than 80% of the revenues that the developer will need to secure in order to finance the project. An auction mechanism will allow bidders to acquire varying quantities of rights without paying more than the value they place on the rights. In any event, merchant transmission investments have been undertaken, based on the value of the transmission rights created by the project.

Does market-based expansion add cost to society?

There is an argument that the transmission expansion project might be privately profitable, but would impose additional cost on society by adversely affecting reliability. In practice, this is unlikely to be a particular problem. The system operator would not allow the connection of new transmission facilities that would adversely affect the reliability and operating characteristics of the transmission system that it controls, nor would the market participants or the regulator permit it to do so. The transmission owners would require the reinforcement of their systems with which the project would interconnect. The regulator will also be watching for signs of potential market power abuse, as we have discussed. A further idea that has been proposed is that, as an additional safeguard, and as a prerequisite for obtaining authority to charge market-based prices, a condition should be imposed that the project must not adversely affect existing property rights.[36]

As we discussed in chapter 4, if FTRs (giving the right to receive congestion rents) are awarded after applying the simultaneous feasibility test, they can eliminate perverse investment incentives to increase, rather than reduce congestion and losses that are components of efficient nodal prices. An investor could build a line to be intentionally congested so that it could profit from the congestion rents, even though the overall benefit to consumers would decline because of the congestion. If a system of FTRs were in place, the investor would be required to accept (or buy back) whatever FTRs exactly cancel the flows that are no longer feasible as a result of the lower capacity of the network caused by the increased congestion. Therefore, the concept of feasibility provides a check on this perverse incentive. Market participants can protect themselves from detrimental or irrational investments by a combination of FTRs and contracts for differences (although it is unlikely that the allocated FTRs would exactly match the market participants' actual trading in the spot market). Market participants that elect not to purchase FTRs would be exposed to spot market price volatility and would stand to lose.[37]

Free rider problem

The free rider problem is an issue that affects all types of transmission upgrades, often benefiting market participants and other users that do not have to pay for them, making it difficult to recover the costs of the improvement. As we have seen, the free rider problem can be avoided through the use of FTRs as locational hedges. Free riders would take the risk of the price volatility of access charges and would have incentives to acquire FTR cover. However, there is always the concern that the congestion and price volatility may have reduced as a result of the expansion, and the free rider could get a reasonably good deal. Over-sizing for future load growth can create a free rider problem for the transmission investor, unless it can withhold or reserve capacity.

As discussed earlier, new technologies such as HVDC links and FACTS devices can help to reduce the free rider problem because transmission service can be directed only to those who pay for it, but these technologies are not always the cheapest, nor the most appropriate for all transmission expansion needs.[38]

However, where there is a significant number of smaller market participants, each benefiting a little from a large transmission investment, the temptation to free ride on the economies of scale and scope may be considerable, even though they would all be better off by sharing the cost of the investment. Forming investor coalitions should become more attractive to smaller market participants, if they can obtain a share of the transmission rights.

A key ingredient to solving the free rider problem is the creation of a workable set of transmission rights. The point is well explained by Hogan:

> *If there is no method to allocate the benefits of transmission, the strong network interactions exacerbate the difficulty of allocating costs. Without some form of transmission rights, there could be substantial incentives to shift costs to others (i.e., free-riding), it would be difficult to define a standard of performance for transmission providers, and the normal operation of market principles would founder because of the lack of property rights. By contrast, where it is possible to define the equivalent of property rights, free-riding and these difficult allocation problems could be substantially reduced, if not eliminated.*[39]

If the free rider problem can be solved, the market could take care of many more investments.

It could be the case that the pattern of transmission use was so uncertain and the network so interconnected that no set of transmission rights would be capable of capturing enough of the economic benefits of an expansion project. This would be true for both physical and financial rights. In effect, there would be economies of scope in the transmission investment that would go well beyond the benefits of any reasonable patterns of FTR. If the beneficiaries cannot be identified or benefits cannot be assigned fairly, it is unlikely that market-based investments will take place.

Difficulty of obtaining permits

The difficulty of obtaining permits, environmental consents, and rights-of-way is often expressed to be the main deterrent to investment in new transmission facilities. This is discussed in more detail in chapters 9 and 10. However, some types of new technology cables are easier to place underground than traditional high-voltage cables, which means that roads, railway tracks, and pipelines might be more easily used for new transmission lines, as they have been for fiber-optic cables.[40] Undergrounding transmission lines obviously attracts less opposition than overhead lines strung on steel lattice towers (or even the more elegant concrete pole), but it is a much more expensive technique in many environments.

Cherry-picking

A criticism has been made of the market-based approach to transmission expansion on the basis that merchant developers will "cherry pick" the best projects, leaving the unattractive projects to the incumbent regulated utilities. Moreover, the holders of FTRs associated with merchant projects might have the incentive to resist other regulated projects, if there is a chance that the value of their rights will be adversely affected. It is likely that the developers and incumbent transmission owners will, in fact, compete for the best projects if there is a profit to be made. Other projects will rely on regulatory approvals to allow cost recovery on a traditional basis. Cherry-picking could be beneficial for consumers, as it will create incentives to act quickly to meet an obvious need.

Recipes for Market-Based Expansion

Recipes for different kinds of projects are needed

Having looked at the key ingredients of and conditions for entrepreneurial, market-based expansion to succeed as well as the arguments of those that are skeptical whether merchant projects will succeed, we must now create the recipes for this kind of expansion. Because the needs and opportunities for transmission expansion vary, we must look at the four most obvious types of projects—small and large facilities where the power flows are and are not controllable. This involves an analysis that is undertaken in Figure 5-1.

Necessary ingredients	Small scale controllable project	Large scale controllable project	Small scale network project[1]	Large scale network project[1]
Identifiable Beneficiaries (to enter into contracts)	Yes	Should be possible	Should be possible	May not be possible
Accommodate Economies of Scale	Likely to use technologies that reach economies of scale plateau	Accommodate but could withhold capacity	May be able to use technologies that reach economies of scale plateau	Need to accommodate but beneficiaries may not be prepared to pay for them
Market-Based Prices	Yes	Yes	Yes	Yes
Locational Marginal Pricing	Desirable	Yes	Yes	Yes
Property Rights	Financial or physical (if third parties will use the assets)	Financial or physical	Financial	Financial
Coordinated Planning Process	Desirable	Yes	Yes	Yes
Minimal Regulatory Hurdles	Unlikely to impose open access or auction of transmission rights	Likely to impose open access, auction of transmission rights, and limit long-term capacity reservation	Likely to impose open access and auction rights	Will impose open access and auction of rights and limit long-term capacity
Level Playing Field with Incumbents	Useful	Important	Important	Important

1 Assumes AC technologies or non-controllable flows on the new transmission factors.

Fig. 5-1 Recipes for entrepreneurial market-based expansion

Figure 5-1 demonstrates that our somewhat theoretical analysis supports the experience to date of merchant projects. A market-based approach is most likely to be adopted where the project is relatively small or where technologies that enable power flows to be controlled are used. However, as we shall see, it can be used for AC radial lines.

Competitive Bidding on an Institutionalized Basis

The primary method of expansion

There is another way to capture the benefits of market forces in transmission expansion. A market-based approach involving competitive bidding or solicitation process can be carried out on an institutionalized basis and can be adopted as the primary method of expanding the system using a market-based approach, as in Argentina, Peru, and Colombia.

There have been a number of examples where requests for proposals (RFP) for new transmission facilities have been issued by system operators or transmission administrators. For example, the transmission administrator in Alberta solicited bids for the construction of transmission substations in April 2001, having earlier issued RFPs for facilities comparable to a 144 kV transmission line. The project was awarded to a distributed generation scheme. The system planner in Victoria, VENCorp, issued an RFP in October 2001 for the construction, ownership, and operation of shunt capacitor banks to be connected to the transmission system of the incumbent utility. This process has not been tested on any scale. The U.S. Western Area Power Administration solicited expressions of interest in June 2001 for the financing and co-ownership of transmission lines in central California, and the California ISO has issued RFPs that are described in chapter 8.

Argentina—A decentralized expansion scheme

The model that was adopted in Argentina at the time of restructuring and privatization in 1992[41] can be described as a decentralized scheme, in that it is initiated by market participants and then coordinated by CAMMESA, the system operator. The scheme can also be described as institutionalized, even though it is market-based, in that all expansion projects must go through a process conducted by Ente National Regulator de la Electricidad (ENRE), the electricity regulator in Argentina, from which approvals and expropriation (eminent domain) powers to acquire land and rights-of-way also flow.

Transener, the major incumbent transmission concessionaire, is an independent transmission company that was created by combining the transmission facilities of three utilities and was privatized in June 1993. NGC is a major shareholder in Transener. There are six other private transmission concessionaires and five regional companies that own what might be described as sub-transmission facilities of 330 kV or less. These concessionaires and companies have no obligation or right to expand their systems apart from minor upgrades. Expansion projects costing less than U.S. $2 million can be undertaken by a transmission concessionaire, which may agree to the cost of amortization directly with the beneficiaries or users of the expanded facilities. A transmission concessionaire may also request ENRE to authorize the expansion and decide what portion of the costs each user must bear.

Congestion rents

The transmission expansion scheme is based on the locational marginal pricing of energy from which a surplus is derived from congestion rents and marginal losses when a line is constrained. In the usual way, the rent is equal to the difference between the prices (less incremental losses) at each end of the line, multiplied by the MWh transferred. These rents were originally rebated to the transmission concessionaires, but after the difficulties with the first proposal for the Fourth Line from Comahue to Buenos Aires described later, they are now applied to an expansion fund called the "SALEX" fund, which is used to fund new lines. The rents are no longer allocated to a single, central SALEX fund, but to the relevant SALEX fund to expand the transmission capacity in the corridor of the constrained line.

Two methods of expansion

There are two methods of major system expansion:

- public auction
- private contract

In the case of the public auction method, which is most likely to be used for costly projects, the parties wishing to expand the system submit a petition for expansion to the transmission concessionaire responsible for that part of the system to be expanded, which then refers the petition to CAMMESA. CAMMESA applies a methodology laid down by the government, known as the "area of influence" methodology to verify that the parties who have submitted or support the proposal represent at least 30% of the "benefits" of the expansion.

Area of influence methodology

The methodology measures the expected use of the new facilities by applying a load flow model together with a dispatch and simulation model. Buenos Aires, being the main load center, is the location of the swing bus (a modeling definition), that results in remote generators being identified as the "users" of the long radial lines. Although the methodology relates to usage and does not formally measure or allocate economic benefits, it was adopted as a convenient surrogate for a more economic methodology that would probably be difficult to develop and apply in practice. It works reasonably well, because the Argentine transmission lines are radial, and there is no real mesh system. The methodology has been under review by the government for some time.

The process has three steps. First, CAMMESA defines the different types of load patterns that might be expected on the transmission system during a year of the line's operation and estimates the number of hours per year the system is likely to exhibit each load pattern.

Second, for each load pattern, CAMMESA determines how much the flows on the new line are changed by a 1 MW change in the amount of power inserted or withdrawn at each node in the line's area of influence. The node's participation in the line is then calculated by weighting its marginal effect on line flow by the amount of power inserted or withdrawn at the node, and then dividing that number by the sum of the weighted marginal effects for all of the nodes. If rL_k stands for the effect of a marginal change in power generated or consumed at node k on the new line and P_k is the amount of power generated or consumed at node k, then the participation of node k in the new line is:

$$\text{participation of node } k = rL_k \ x \ P_k \ / \Sigma_{\ all \ nodes} \ (rL_k \ x \ P_k).$$

Finally, the expected participation of the node during a year was calculated as its participation in each load pattern weighted by the proportion of hours in the year that load pattern was expected to apply.[42]

The application and CAMMESA's technical study are then sent to the regulator, ENRE, for approval. ENRE is required to determine the question of convenience and public necessity from a system-wide standpoint against an economic test based on a comparison of the net present value of the line, with and without the expansion, taking into account the value of non-supplied energy[43] but without taking the SALEX funds into account. If the test is met, ENRE will then hold a public hearing within 30 days of receipt of the petition (that lasts only a for a few days). Unless stakeholders accounting for at least 30% of the identified usage make an objection, ENRE may authorize the project to go ahead and grant a certificate of convenience and public necessity.

The project is then the subject of a public auction or bidding process that is open to existing transmission concessionaires and companies. Bidders propose a 15-year "canon" or annuity that will be paid by the identified users. The canon is reallocated annually under the area of influence methodology to reflect changes in the pattern of dispatch. The amount in the SALEX fund is deducted from the cost of the line to be paid by the users, up to a limit of 70%. The successful bidder is awarded a 15-year concession at the end of which the assets revert to the relevant concessionaire and are paid for using the pricing regime that applies to all existing transmission assets.

Expansion by private contract

The second method is known as expansion by private contract, under which the parties proposing the expansion are responsible for paying for it. At least one market participant must sign a construction, operation, and maintenance contract for the project with a transmission concessionaire or an independent transmitter.

The application is made in the form of a petition to the transmission concessionaire responsible for the system to which the project will be connected, which forwards it to ENRE with its analysis of the petition. After verifying that the petition is consistent with the rules that govern the transmission system, ENRE will hold a public hearing. If ENRE does not object to the petition, it may issue a certificate of convenience and public necessity for the expansion. The amortization costs are allocated in accordance with the methodology agreed by the parties to the contract and not the area of influence methodology. ENRE will review the agreement to ensure that the expansion serves the public interest and that the transmission concessionaire is not abusing its monopoly position in the negotiations. If the project is implemented, the transmission concessionaire will grant the transmission company that builds it a license to build, operate, and maintain the project in its concession area.

Success of the Argentine expansion scheme

The Argentine model has been successful in adding four new major radial lines. There is a plan for a fifth line from Patagonia to one of the mining areas of the Andes. The costs have progressively decreased as a result of the competitive bidding process. The canon (annuitized charge) for the first and second lines was exactly the same—$24 million (U.S.)—but the first line was 300 km long and the second, 500 km. The fourth line of 1,300 km was built at a canon of a $24 million (for 15 years) and a contribution of $80 million was made from the SALEX fund.

However, the scheme has a number of difficulties. The Argentine transmission system is acknowledged to be under stress and not able to keep pace with the demands that are now placed on it by market participants and consumers. Reliability is an increasing problem. The scheme does not seem able to deliver all of the required expansion or reliability. Moreover, a number of the features of the scheme appear to act as deterrents to investment.

There was understandable logic for the approach adopted by the Argentine government in designing the expansion scheme. It wanted to get away from the concept of central planning and allow the beneficiaries of the expansion (*i.e.*, those who would pay for it) to determine when expansion was needed and whether they were prepared to pay for it. They also wanted to put a downward pressure on the cost of construction of new facilities through a process of competitive bidding.

The story of the fourth line

As we have seen, three lines were built under the Argentine scheme successfully decreasing cost, but in 1994 a proposal to build a fourth line from Comahue to Buenos Aires ran into difficulty. A coalition of generators in the Comahue area, burning cheap gas and claiming to represent 30% of all the users or beneficiaries, had a strong incentive to propose the line to give them access to much higher prices in Buenos Aires. In the hours when power could not be exported, the nodal prices in the Comahue region would fall to $12/MW while prices in Buenos Aires rose to $23/MW. At the public hearing in 1995, more than 70% of the identified beneficiaries (identified using the area of influence methodology) objected and the petition failed.[44]

There were a number of reasons the original petition failed:

- There was potential for free riders on the line, and the generators clearly did not want to pay for others to use it free of charge
- There were no incentives on distributors to support the expansion, because they were permitted to pass all their transmission costs on to consumers
- The area of influence methodology did not identify any wholesale buyer in the Buenos Aires area as a beneficiary, because the market reference node was also in Buenos Aires, which meant that any simulated change in power consumption or production at a node in the same area would be unlikely to affect long-distance transmission
- Some of the generators along the route of the proposed line that were identified as beneficiaries did not think the expansion was in their interests and voted against it. They had higher marginal costs than the Comahue generators, which meant their revenues would decrease because the wholesale price of energy would decline as a result
- There was no system of transmission property rights for investors

In response to these events and in recognition of the fact that the line was clearly needed, the government established the initial, centralized SALEX fund to subsidize the cost. The logic for this was that the surpluses from the congestion rents and marginal losses reflected the cost of the lack of transmission capacity and should be used for investment. The fund grew rapidly (it was expected to amount to $90 million by the time construction began in 1998) and was sufficient to overcome objections at a second hearing in 1996. Indeed, it was sufficient to reduce the cost to the beneficiaries by one third. An innovative package of minor upgrades, which increased the capacity of the existing three lines from 2,700 MW to 3,300 MW, was also developed. The government decided that the package was small enough that Transener should implement the program without competitive bidding and be reimbursed from the SALEX fund. ENRE put the project out to bid in early 1997, and Transener was selected as the lowest bidder in October of that year. In accordance with the time limits imposed by the terms of the concession and despite a four-month delay caused by litigation initiated by one landowner that resulted in the rerouting of 40 kms, the line was placed in service in October 1999.

Lack of redundancy in the Argentine transmission system

There is little or no redundancy in the main transmission system, and its loading has been increasing at a rate of between 5-6% per annum. The fourth line has attracted even more generation, and the free rider fears of the original generators have been realized. In 2001, there was an excess of generating capacity over transmission capacity of 1,100 MW. As a result, CAMMESA has been forced to load the lines up to and, in some cases, beyond their limits in order to respond to market needs. Transener has power to approve or prohibit loadings above nameplate ratings or loadings that might compromise the safety of personnel or plant. Otherwise, Transener must follow CAMMESA's instructions. However, because of the strain that the needs of the market place on CAMMESA, it is under pressure to not allow Transener to take circuits out for routine maintenance. Reliability is a serious issue. There is a thin line between optimizing economically and safe loading. If there is an outage, Transener may foot the bill in the form of penalties or compensation.

Federal transmission plan

It is acknowledged that the time has come for the radial system in Argentina to be improved, both from the standpoint of reliability and adding needed capacity and flexibility, by the addition of links between the transmission lines that radiate from the center of the Buenos Aires region. This would enable CAMMESA to transfer power from one side of the region to the other if one of the lines were to suffer a reduction in capacity. It would also reduce the need to pay for standby generating capacity to be available in the region.

The government has published, but so far not acted on, a federal transmission plan to build this transmission ring. There would be difficulties similar to those encountered with the fourth line, in using the current market-based expansion scheme. Arguably, there would be a large number of broadly-based beneficiaries, although the area of influence methodology would not identify everyone as a user. In any event, there appears already to be an easier way to pay for this expansion by using the SALEX fund and a government trust fund derived from the retail tariff.

The penalty regime

Transener's revenues are affected by a system of fines or penalties for outages on any part of its system. The transmission concession contains a schedule of penalties related to the type of line or transformer that is out of service and the duration of the outage. The penalties are paid to the customers affected by the outage by way of compensation.

The design of the penalty scheme did not take into account the lack of redundancy in and the linear nature of the existing system that Transener inherited. At the time of privatization, bidders were required to factor the risk that outages would be caught by the penalty regime into the price they bid. It was possible to carry out only a limited amount of due diligence as to the state of the system, given the length of the radial lines and the lack of accurate asset registers. It certainly was not possible to anticipate the developments in the wholesale market and the strain under which the system would be placed throughout the term of the concession. Accordingly, factoring in a discount to the offer price in a competitive bidding process to cover the risk was extremely difficult.

Transener has responded to the incentives that the regime created by utilizing most of the innovative techniques that NGC had developed under its transmission services scheme (TSS) in England and Wales (see chapter 2). Indeed, it uses more live-line working than NGC in the UK and has installed as much new technology as it is permitted to improve the existing capacity. However, many of the causes of outages for which it is penalized are outside its control, and although it accepted that it must be incentivized to restore service as quickly as possible, it finds the penalties capricious and unfair in some cases. Although the system of penalties has been renegotiated, it is clearly acting as a disincentive to invest.[45]

Outages caused by tornados used to incur the heaviest penalties, but it is now agreed that they are classed as sabotage (curiously) where there is a more lenient regime. Another cause of outages outside Transener's control is the burning of sugar cane under its lines, which can cause flashovers. No account was originally taken of the fact that the same storm that brought the lines down might also impede or prevent access to restore the lines because of flooding. Disputes over the causes of the outages have resulted in payments to Transener being delayed by 8-10 months after the event. In the meantime, interest on unpaid penalties continues to accrue and is payable by Transener to the affected customers. ENRE does not allow Transener to recover the costs of repairs, replacement, and additional maintenance.

Changes to the penalty regime have been negotiated recently. For example, a short period is now allowed to find and then to repair the faults created by storms, before the penalties apply. The penalties paid have dropped from $15 million to $3 million per annum. However, Transener needs to be extremely efficient and carry a large inventory of spares.

Because of the closed nature of the concession and the restriction on its ability to expand its own system, Transener cannot easily undertake investments that would increase the efficiency of its own system and its ability to restore outages, and protect itself from the imposition penalties.

Transmission tariff in Argentina

The transmission tariffs are regulated by ENRE and have three components:

- variable charges for energy transported
- connection charges for connection of generators and consumers to the system
- line capacity charges covering maintenance costs

The transmission business of Transener also comprises rights to conduct certain unregulated businesses including new generation connections.

Transener's rate of return is based on the price the government thought should have been paid for the shares in the company, and not what was actually bid. This price is fixed for 15 years. The argument for this approach for valuing the asset base is that there would otherwise be an incentive to increase bids in order to increase revenues throughout the concession period (which seems to ignore the competitive pressure of the bidding process to acquire Transener in the first place). The rate of return in the 1999-2003 review period was set at 10.54% (post tax). Devaluation of the peso at the end of 2001 reduced this rate of return to –15.6% (without taking into account Transener's international debt).

There is no periodic review of the asset base, and the fixed costs are similarly set for the same period. A complicating factor is that part of the price is attributable to unregulated rights or business. Before the difficulties with the fourth line and the creation of the SALEX fund, the congestion rents and marginal loss surpluses under the nodal pricing system were paid to Transener. Because of the volatility of this element, Transener's annual variable energy revenues were guaranteed at a certain level, and any shortfall or surplus was made up by a surcharge or discount on the capacity charges in the 1994-1998 review period and on the connection charges in the 1999-2003 review period. The guarantee was utilized to compensate Transener for the application of the surpluses it had previously received to the SALEX fund.

However, because of the reduction in nodal energy prices in response to market forces in the wholesale market, the variable charges have reduced by approximately 45% since 1993, and CAMMESA's forecasts of energy prices show further reductions. However, Articles 40 and 41 of Law No. 24065 of 1992 guarantee transmission owners a reasonable rate of return, although they do not indicate on what the return is to be based. After taking the unregulated rights into account, the price of which was assumed to amount to 30% of the total price, the rate of return was based on 70% of the price the government considered should have been paid for the shares in Transener. The general provisions of Articles 40 and 41 (which prevail over the concession) were applied, following negotiation to restore Transener's revenues to a level that reflected a reasonable return.

Rights-of-way

Although the time scales in Argentina appear to be swift by comparison with some common law jurisdictions, obtaining rights-of-way has proved problematic because the general law appears to give landowners effective remedies that can cause delay or even block expansion. The general law also appears to be able to override an administrative order, *i.e.*, the certificate of convenience and public necessity that ENRE effectively issues when it approves a project and its route. The certificate of convenience and public necessity, according to law, granted Transener the right to expropriate the necessary rights-of-way along the route for the fourth line. Tranesener was forced to dismantle towers that had been erected along part of the route of the fourth line when a landowner obtained an injunction prohibiting entry on to his land on grounds of "visual impairment." ENRE had not held the required public hearing or published the project in the local newspapers before issuing the certificate and did not intervene to uphold its order. Transener suffered a four-month delay and had to bear all the costs involved. ENRE did hold a hearing after the event. It is important to note, in this context, that the route was established by ENRE as a result of the process, and not the successful bidder, Transener.

Criticisms of the Argentine scheme

There is widespread acknowledgement of the shortcomings of the scheme:[46]

- The signals to market participants as to when to invest are not particularly strong
- The area of influence methodology is relatively crude and does not identify the true economic benefits, particularly to wholesale buyers, or the disadvantages to other users
- Generators may delay proposals for a new line that is already economic, either because they prefer to wait until the SALEX

fund can make a significant contribution or because the congestion rents and marginal loss surpluses are flowing to the SALEX fund, rather than to the investors in the new line

- The SALEX fund may be used to subsidize lines that would otherwise be uneconomic
- The absence of transmission rights will discourage investment
- There are inherent free rider problems—*e.g.,* the generator might look to find a location where there is an already congested line (because the funds in the SALEX fund build up quite rapidly) and locate close to that line in order to benefit from a subsidy that its competitors have paid for
- The benefits that accrue to generators could exceed the benefits to consumers that may over-incentivize generators to propose expansion, because they will receive a transfer of wealth from the SALEX fund
- There may be inefficient location decisions as a result of the scheme

Congestion rights

Many observers of the Argentine scheme recommended that a system of tradable financial transmission rights (FTR) would assist in completing the design and make it more effective from the point of view of identifying and financing needed investment.

In October 1998, the government proposed the introduction of transmission congestion contracts (TCC) or FTRs (that were more akin to physical rights than conventional FTRs, because they were defined by reference to the physical capabilities of a line, rather than by the power flows). The proposal was discussed at a public hearing in November of that year. There was considerable support for the concept of congestion rights and the general approach of allocating the rights to new lines to the investors that built them.

However, there was strong opposition to the proposal with respect to the auction of transmission rights associated with new lines that might be built in the future. Under this scheme, the bidder would specify the portion of the construction cost or canon of the line that it proposed to construct and the average percentage of the total capacity of the line that the bidder believed would be utilized during the 15-year amortization period. If the bids did not cover all of the construction cost, the government could cover up to 30% of the cost out of the SALEX fund. The successful bidders would pay the relevant portion of the cost and have the right to charge other users a toll or fee for using the line. The annual toll would be the canon times the ratio of the actual to the expected utilization of the line, which the bidder had specified in the bid. In this way, the bidder would assume part of the risk of whether the line was actually needed. The objections were based on difficulties in raising finance, given the uncertainties of the revenue stream and the fact that the bidders had no say in whether the expansion should be undertaken.

As far as the existing system was concerned, the government proposed to retain all the rights and to continue to apply the congestion rents and marginal loss surpluses to the SALEX fund. There was objection to this as well, and many of the concerns included the perverse incentives created by the fund already discussed.

The proposals were withdrawn at that time, but were included in a draft decree (Decree 604) in 2001. There was strong opposition to many of the provisions, which included broad reforms of the wholesale market, but no objection to the concept of congestion rights. The decree was withdrawn. One of the reforms would have allowed transmission expansion "by free initiative." The transmission companies objected, because there would have been no requirement for the investor to operate under the technical supervision and in accordance with the standards applied by the transmission owner to which it would connect its assets. Under this scheme, a developer could have built a new line within an existing system and required its interconnection, without meeting any of the interconnection standards or conditions that were necessary to protect the existing system or other connected plant or equipment.

Recipes to solve Argentine problems

Clearly, the government is anxious to avoid central planning and to harness market forces to determine need and timing. With this in mind, we can recommend some immediate, simple steps towards solving the difficulties that have been encountered with this imaginative and, in several respects, successful scheme:

- The SALEX fund should be abolished. The surpluses derived from the congestion and loss schemes should be rebated back to the owners of the lines
- A well-designed system of FTRs should be implemented where the rights with respect to new facilities are allocated to the investors
- The area of influence methodology should be reviewed, if only to remove the difficulties we have identified
- The distribution companies should be given incentives to participate actively in transmission expansion decisions
- Some of the operational arrangements between the transmission companies and CAMMESA require attention, at least to enable routine maintenance to be carried out
- The system of penalties for outages that is clearly acting as a deterrent to investment needs to be revised
- The regulatory and revenue risk should be reduced by creating more certainty as to how the tariff will be calculated from one review period to the next, without having to rely on the general language of the 1992 law
- Transmission companies should be allowed and incentivized to upgrade and maximize the capacity, capability, and flexibility of their existing assets, overcoming the problems created by lack of redundancy in the system. Greater use of the existing transmission owners could be made to build the smaller schemes quickly
- The restrictive nature of the scope of the concession and the objects of the transmission companies should be reconsidered, because they appear to act as a disincentive to investors.

They are unable to make investments in other infrastructure industries from which they could derive beneficial economies of scope and scale

- ENRE or CAMMESA should issue coordinated, indicative statements of future transmission needs and connection opportunities based on the statements of opportunities that the transmission companies produce, in order to give stronger signals to market participants on where to locate and when to propose expansion. This would fall a long way short of the central planning that Argentina is keen to avoid
- A backstop is needed in case the market fails to propose some expansion projects that are clearly needed, so that they can be implemented on a regulated basis, while still preserving the benefits of the competitive bidding process that has done so well to bring construction costs down

Peru and Colombia

Both Peru and Colombia have instituted a transmission expansion regime that captures market forces on an institutionalized basis.

The schemes might be described as hybrid in nature, in that they involve a government planning unit that develops an indicative central plan. While the plan is fully discussed with all stakeholders, the planning unit decides which projects will be implemented. However, in Peru, anyone can request or propose a new transmission line. There are two different types of transmission facilities:

- *Main transmission lines*—the costs of which, by law, are to be socialized. The concept is that they are primarily for reliability purposes and for the benefit of all users. The costs are recovered under postage stamp prices. A proposal for a main transmission line has to be presented to the regulator, and, if it is approved, the regulator will arrange for a competitive solicitation process in which public or private sector entities can compete. A concession period is specified in the RFP that has allowed 3 years for construction and 30 years to recover the investment, for example. The shorter the period, the greater the impact on the prices that the consumer is charged, and therefore there is flexibility in considering longer periods. At the end of the period, the assets revert to the government under the concession. A rate of return of 12% has been allowed for the first 10 years

- *Secondary transmission lines*—are paid for by a particular beneficiary or group of beneficiaries rather than by all users. A proposal is presented to the regulator that has to establish how the line is to be used and will determine the cost allocation between the beneficiaries, if the promoters and the beneficiaries cannot agree among themselves. Once the beneficiaries have been identified, the allocation methodology essentially requires beneficiaries to pay in accordance with their use of the new facilities. If the line is over-sized because of economies of scale, the beneficiaries will still have to pay their shares of the over-sizing. As in Argentina, there is a review of the cost allocation every year, because the beneficiaries may change. While this creates flexibility, it adds some uncertainty as to the costs that the beneficiaries will have to bear. The major purpose of secondary lines is to connect new generation and large consumers

Needless to say, in many cases, there is pressure to classify a line as a "main transmission line." The regulator uses two or three different congestion models to establish who the beneficiaries are, with or without the line.

Peru has successfully used this approach to build new lines, and the results of the competitive procurement process have been satisfactory. TransÉnergie was the successful bidder for a major, radial AC line.

The central planning process has been accepted by stakeholders, who feel that a "unified vision" is helpful, otherwise each industry participant would present a partial view of its own problems. The planning is carried out on an indicative basis by the Ministry of Energy and

Mines in Peru and the Ministry of Development in Colombia. The planning process in Colombia is required to take place by law, and if a proposal is not included in the program of the planning unit, it will not be built. However, the planning unit will look for private sector initiatives, and if none are forthcoming, they will put a proposed line out for bid.

The successful bidder will be awarded a concession, but then must fulfill all the environmental requirements, including public consultations and hearings. The environmental hearings can take about six months after the submission of the environmental impact assessment and remedial or mitigation proposals. Once the necessary approvals have been given, the powers to expropriate the right-of-way and land will automatically follow. Full commercial value must be paid for land and rights-of-way expropriated.

The planning unit in Peru is able to place a maximum limit or cap on the price that can be bid, so that the regulator is not obliged to accept any price. The cap has been set fairly high, and turned out to be approximately 50% above the price bid by TransÉnergie with respect to the AC line that it bid for successfully. A minimum of three bidders is required for the process to be regarded as competitive.

Recipes from Latin America

Although there are some flaws in the models that have been implemented in Argentina, Peru, and Colombia, there are many strengths. They have successfully captured market forces through competitive bidding and achieved some significant expansion, using innovative financings. Argentina clearly has provided us with some excellent experience from which we can draw some conclusions for our recipes. For example, we know that:

- Market-based expansion can and does work
- Central planning (command-and-control style) is not necessary, but there would be benefits from more formalized and regular planning coordination of the type described in chapter 8. The Federal Transmission Plan in Argentina demonstrates this need

- It may be difficult to identify beneficiaries with perfect precision, but it can be done. It is a much better approach for market-based expansion than socializing all the costs
- There must be some upside for transmission investors if they are to be attracted. An aggressive system of penalties for actions over which they have no control and the inability to upgrade existing assets will deter investment

Peru and Colombia have embraced central planning on an indicative basis in a way that Argentina has not. This is logical in jurisdictions where the transmission systems are relatively weak, and the economy (and, therefore, demand) is growing. Particularly where local investment funds are lacking and the need for expansion is bound to materialize, the models that Peru and Colombia have adopted of actively combining market-based and regulated expansion with indicative planning are sensible. We shall watch with interest their experience of implementing them further.

CHAPTER 5 APPENDIX

The European Internal Market in Electricity

The origins of the single market

For more than 10 years, the European Commission (EC) has been attempting to complete the "single market" in electricity. The single market has its origins in Article 95 of the treaty that established the European communities. The treaty incorporates the concept of subsidiarity, which essentially means that the implementation of EC policy directives is left to individual member states. EC directives, therefore, do not guarantee an identical outcome in all member states. The EC is not a federal regulator in the same way as the Federal Energy Regulatory Commission (FERC), but it has power to issue regulations that do have direct effect in member states.

The IME Directive

The so-called internal market in electricity (IME) Directive creating common rules for the internal market in electricity[47] has, inevitably, led to 15 rather different outcomes. It was originally designed to achieve:

- competition in generation
- retail competition, with a progressive market opening to around 35% of consumers
- the unbundling of transmission system operators (and distribution system operators) not necessarily into separate companies, but so that they become "independent at least in management terms"[48]

- negotiated third-party access based on a published "indicative range of prices" or regulated third-party access (which was adopted in most member states apart from Germany)

Amendment of the IME Directive

The EC was disappointed with the implementation of the Directive.[49] It tabled proposals to amend the Directive and to issue a regulation having direct effect in all member states (the first in the energy sector).[50] The proposals failed to reach political agreement in Stockholm in the spring of 2001 but broad agreement was reached in Barcelona in March 2002.

The proposals for amendment involve:

- full corporate separation of transmission system operators (TSO)
- the requirement that TSOs should exercise full control over all assets necessary to operate, maintain, and develop the network
- a requirement that TSOs should meet certain minimum levels of investment
- further retail market opening to cover all commercial consumers by 2004
- regulated third-party access to transmission systems for all potential users

The Directive implies only regulated transmission

The amended draft implies that all system operators are regulated transmission asset owners—in other words they are transcos. The assumption is not unreasonable, because many of the European utilities remain vertically integrated to a considerable extent. The IME Directive only required them to functionally unbundle their transmission system operations. For the most part, the European transmission systems are strong, and what is lacking is sufficient interconnector capacity to make the internal market in electricity a reality. Given the lack of experience with merchant transmission, the European policy makers did not focus on the market as a mechanism to strengthen the interconnections of Europe.

They may also have concentrated on the very considerable problems of loop flows between the systems of the various member states. It was not obvious to them that new controllable technologies could be integrated into the complex AC networks of Western Europe to enhance the interconnection capacity.

The regulator in Italy has inadvertently provoked another structural change in the Italian industry by only allowing the ENEL a low rate of return on its transmission business. Having argued to retain the transmission assets, ENEL appeared to lose interest in transmission and agreed with the Italian government that its transmission subsidiary, Terna, should be merged with the transmission system operator to create a transco. It is not clear what incentives the new transco will have to invest.

Key ingredients are not in place

The key ingredients for cross-border trade in electricity and for transmission and interconnection expansion both on a merchant and regulated basis are not in place. It is very difficult for a consumer to choose a supplier in a different member state. There are a wide variety of different pricing structures, and pricing is not cost-reflective. There have been complaints over the difficulty of negotiating agreements for the use of transmission systems in Germany in the absence of a regulator. The national regulatory regimes are all different. The European regulators do meet together twice a year at the so-called Florence Forum under the auspices of the EC, as the Council of European Energy Regulators (CEER), so there is some degree of coordination, but the regulatory regimes are still evolving.

Getting the prices right

In the same way that we began by looking at the list of key ingredients from the perspective of getting the prices right, the EC has taken a courageous step in proposing the first energy regulation on transmission pricing that would have direct effect in all member states. The broad objectives of the regulation are to achieve:

- cost-reflective transmission charges avoiding "excessive transaction costs"
- transparent, non-discriminatory rules for allocation of scarce interconnector capacity to avoid long-term reservation contracts
- efficient, locational, economic signals to market participants and TSOs

The regulation on cross-border networks contains a set of general principles that may be difficult to implement in practice.[51] In essence, the regulation:

- allows the EC to set the level of compensation payments between TSOs for the use of their systems for cross-border trade
- empowers the EC to issue guidelines detailing the relevant principles and methodologies, *e.g.,* for cost calculations or power flow measurements
- compensates TSOs for hosting transit flows (including loop flows), based on long-run average incremental costs
- requires access charges to be cost-reflective and not distance-related
- deals with the recovery of access charges, mainly from consumers
- requires tariffs to provide locational signals as to losses and congestion
- allows the EC to specify general principles of congestion management
- requires congestion rents to be applied to reduce expansion costs
- prohibits "pancaking," *i.e.,* payment of different transmission charges to multiple transmission owners

The EC focus on seams issues

The approach of the EC is different from that of FERC. Whereas FERC has focused on a standard market design for all RTOs in the U.S. in order to avoid issues at the "seams" between markets, the EC is not empowered to propose such a radical solution, which would be likely to be inconsistent with the concept of subsidiarity embodied in the EU treaty.[52] It must therefore focus on the "seams" issues between member states and is right to pay particular attention to getting the prices right. However, the broadly-worded regulation will give plenty of scope for interpretation and dispute over its implementation.

Problems for merchant transmission

Article 22 of the amended IME Directive presents a problem for merchant transmission and interconnectors in that member states are required to establish national regulatory authorities that have sole responsibility to fix or approve tariffs for transmission and interconnection facilities and to define the rules on the management and allocation of interconnection capacity, in conjunction with the regulatory authorities of the neighboring member states. The assumption that all transmission and interconnection development will be carried out on a regulated basis was apparently not intended by the EC to preclude merchant transmission. However, Europe has a long way to go before the necessary ingredients to stimulate market-based transmission and interconnection development that were discussed in chapter 5 are in place. To recap, the key ingredients are:

- identifiable beneficiaries
- accommodating economies of scale
- market-based prices
- locational marginal pricing
- property rights
- a coordinated planning process
- minimum regulatory hurdles
- a level playing field with incumbents

Regulation will be a useful tool

Regulation will be a useful tool to create a number of the key ingredients. Because it requires tariffs to provide locational signals, locational marginal pricing could be adopted that will assist in identifying beneficiaries and act as a platform for property rights. A coordinated planning process could probably be put in place through ETSO (the association of European Transmission System Operators), which is an active organization with a problem-solving culture, despite the differing political views with respect to the pace of progress towards the internal market in electricity among member states.

Well-designed regulatory regimes will be vital

However, the attitude of the regulators will affect the achievement of the key ingredients for merchant expansion. For example, they may find it difficult to allow capacity to be withheld on a large scale, controllable, merchant project in order to maintain congestion rents at a certain level. They may be unwilling to allow market-based prices, preferring all assets to be regulated as part of the interconnected system. There is no particular reason that they should have anything to fear by allowing market-based prices on facilities where the investor is using its own money and taking the whole of the usage and revenue risk. However, there may be a fear of losing jurisdiction, to some extent. Regulators have no role in micromanaging merchant expansion. Their roles must be limited to the broad issues of open access and potential market power abuse through long-term capacity reservations. The regulators will also have a strong role to play in creating the level playing field between merchant and regulated transmission investors in Europe, given the strength of the incumbent transmission utilities.

Status quo is not sustainable

The temporary cross border trade (CBT) system that came into force on March 1, 2002, involves a one-off €1 per MWh standard declared export fee that grants access to all the main West European grids, removing the "pancaking" of national charges. However, since the charges only began by actively being levied in Belgium, Switzerland, and Germany, the system is scarcely uniform. There is a CBT compensation fund of €200 million, around half of which will be derived from a socialized charge spread across all users, and the other half of which will come from the €1/ MWh fee on declared exports. The €1/MWh fee would make the majority of trades more expensive, since they involve crossing only one border, for the most part the German border, where the previous export charge was €0.64/MWh. All of this indicates that the initial arrangements were unsatisfactory.

Six Swiss grid operators have taken a complaint to the Supreme Court in Rome over the allocation of import capacity volumes between Italy, France, and Switzerland. Because the transmission systems are so closely meshed in the region to the northwest of Italy, there is no straightforward way of establishing how much technically belongs to each country. The Italian regulator had an unenviable task of determining how much France and Switzerland should be allowed to import.

Agreement that status quo is not an option

All this may sound rather negative and the objectives seem difficult to achieve, but there is consensus that status quo is not an option. There may be differences of opinion within ETSO and CEER on how to proceed. There is also the potential for the transmission owners to delay agreeing to anything that removes the transit fee component in the temporary system of charging for CBT. There is agreement to put in place a permanent CBT mechanism that has no transit charges by 2003, but it is subject to a proviso that appropriate and efficient locational signals should be in place. It is not clear that this will happen in the time available, particularly in France, where the regulator intends to spread the CBT charges uniformly across generation. There is no regulator in Germany at present, which means that the government and the industry associations will have to agree what the charge should be and what locational elements should apply to it as a replacement for the much criticized transit or export charge. The transmission charges within national boundaries will also need to be reexamined.

It is apparent that there are many different ideas as to the way to create locational signals that may fall short of the locational marginal pricing that we have discussed in chapter 4. Policy makers may fail to go the necessary distance to put charging systems and other ingredients in place that will enable market-based transmission and interconnector projects to flourish. This would be a pity because market-based investment could offer the best chance for interconnector expansion in Europe, even if the loop flow problems are challenging. Merchant investors may be much more innovative in using new controllable technologies to help solve these problems. As we have seen, locational marginal pricing and FTRs also help solve the same problems.

There is an active dialogue to solve problems

The good thing is that all the member states are engaged in an active dialogue in an attempt to solve the problems of cost-border trade in electricity between member states. The meetings of ETSO and CEER at the so-called Florence Forum twice a year and the work done between the meetings indicate that there is a will to move forward, and where there is a will there is a way.

NOTES

1. William W. Hogan, "Market-Based Transmission Investments and Competitive Electricity Markets," Center for Business and Government, John F. Kennedy School of Government, Harvard University (August 1999) available from Professor Hogan's web site, www.whogan.com

2. Paul L. Joskow and, Richard Schmalensee, *Markets For Power* (Cambridge, Massachussetts: The MIT Press, 1983), p. 31

3. Paul L. Joskow and Richard Schmalensee, *Markets For Power* (Cambridge, Massachussetts: The MIT Press, 1983), p. 32

4. FERC Notice of Proposed Rulemaking, 18 CFR Part 35, Docket No. RM01-12-000 (July 31,2002)

5. FERC has approved the following merchant transmission projects and their docket numbers: (1) Cross Sound Cable Interconnector Project (TransÉnergie U.S., Ltd), Docket No. ER00-1-000, (2) Harbor Cable Interconnector Project (TransÉnergie U.S., Ltd), Docket Nos. ER02-552-000 and ER00-1-000, (3) Lake Erie Link Transmission Interconnector Project (TransÉnergie U.S., Ltd.), Docket No. ER02-406-000, (4) The Neptune Project (Neptune Regional Transmission System, LLC), Docket No. ER01-2099-000, (5) Connecticut-Long Island Cable Project (Northeast Utilities Service Company), Docket No. ER01-2584-000

6. The Neptune Project (Neptune Regional Transmission System, LLC), Docket No. ER01-2099-000

7. Edward N. Krapels, "Merchant Transmission: The Neptune RTS Case Study," presentation to Harvard Electricity Policy Group (November 30, 2001)

8. José Rotger and Frank Felder, " Promoting Efficient Transmission Investment: The Role Of The Market In Expanding Transmission Infrastructure" (November, 2001), pp. 32-34

9. Thomas-Olivier Nasser, "The Hidden Value of Transmission Assets," *The Electricity Journal* (June 1999), pp. 69-78

10. Ibid, pp. 75-76

11. John D. Chandley and William W. Hogan, "Independent Transmission Companies in a Regional Transmission Organization" (December 20, 2001), p.10, available on Professor Hogan's website, www.whogan.com

12. José A. Rotger and Frank Felder, "Promoting Efficient Transmission Investment: The Role Of The Market In Expanding Transmission Infrastructure" (November, 2001)

13. The author has received verbal comments from Larry Ruff and Daniel Kirschen to this effect

14. William W. Hogan, "Market-Based Transmission Investments and Competitive Electricity Markets," Center for Business and Government, John F. Kennedy School of Government, Harvard University (August 1999), p. 32, available from Professor Hogan's website, www.whogan.com

15. William W. Hogan, "Market-Based Transmission Investments and Competitive Electricity Markets," Center for Business and Government, John F. Kennedy School of Government, Harvard University (August 1999), available from Professor Hogan's website, www.whogan.com

16. José A. Rotger and Frank A. Felder, "Promoting Efficient Transmission Investment: The Role of The Market In Expanding Transmission Infrastructure" (November 2001)

17. Karen Lyons, Hamish Fraser, and Hethie Parmesano, "An Introduction To Financial Transmission Rights," *The Electricity Journal* (December 2000), pp. 31-37

18. Edward N. Krapels, "Merchant Transmission: The Neptune RTS Case Study," presentation to Harvard Electricity Policy Group (November 30, 2001). It is essentially a project to build a DC backbone for the northeast RTO

19. Northeast Utilities Service Company, 97 FERC ¶61,026 (2001). FERC issued an Order on March 18, 2002, conditionally accepting the proposal. Docket No. ER01-2584-000

20. FERC Order Approving Proposal Subject To Conditions. Docket No. ER00-01-000 (issued June 1, 2000)

21. Michael Albers, "Compatibility of Long Term Capacity Reservations with EC Competition Law," discussion paper prepared for the 6th Regulatory Forum for Electricity in Florence (November 2000)

22. European Commission, IP/01/30

23. European Commission, Comp/M.1673 VEBA/VIAG (June 13, 2000)

24. Roy J. Shanker, "New Transmission Investment Policies-Implementation Issues," presentation to Harvard Electricity Policy Group (May 22, 2000)

25. FERC Notice of Proposed Rulemaking, 18 CFR Part 35, Docket No. RM01-12-000 (July 31,2002) p. 347

26. Petition of TransÉnergie U.S. Ltd for Order Accepting Tariff for Transmission Interconnector and Granting Related Authorizations and Waivers, FERC Docket No. ER 00-1-001

27. Reinier Lock, "Transmission Pricing: Are We Finally On the Brink of a Breakthrough," *The Electricity Journal* (November 2000), pp. 47-56

28. Philip Gall, "Regulated and Merchant Transmission Investment-Lessons From Australia," presentation to Harvard Electricity Policy Group, San Diego (January 23, 2002)

29. José A. Rotger, "Reconciling Market-Based Transmission and Planning," presentation to Harvard Electricity Policy Group, San Diego, California (January 24, 2002). See also José A. Rotger and Frank A. Felder, "Reconciling Market-Based Transmission and Transmission Planning" (November 2001), pp. 31-43

30. Alfred E. Kahn, "Electric Deregulation Defining and Ensuring Fair Competition," *The Electricity Journal* (April 1998), pp. 39-49

31. The initial language in the Regulatory Test had to be changed, because it required projects to be evaluated in terms of customer benefits, where customer was defined as a retailer that purchases wholesale energy and on-sells to end-consumers

32. Australian Competition and Consumer Commission, "The Regulatory Test for New Interconnectors and Network Augmentations" (December 15, 1999)

33. National Electricity Tribunal, "In the Matter of an Application for Review of a NEMMCO Determination on the SNI Interconnector," Dec. 6, 2000, Application No. 1 of 2001.

34. Stephen C. Littlechild, letter report to Dr. Tony Cook, TransÉnergie Australia (November 23, 2001), available on the TransÉnergie Australia website, www.transenergie.com.au

35. FERC Docket No. RM01-12-000, "Working Paper on Standardized Transmission Service and Wholesale Electric Market Design," available on the FERC website, www.ferc.gov

36. José A. Rotger and Frank Felder, "Promoting Efficient Transmission Investment: The Role of The Market In Expanding Transmission Infrastructure" (November 2001)

37. James Bushnell and Steven Stoft, "Electric Grid Investment Under A Contract Network Regime," *Journal of Regulatory Economics* 10(1) (July 1996), pp. 61-79

38. Eric Hirst and Brendan Kirby, "Key Transmission Planning Issues," *The Electricity Journal* (October 2001), pp. 66-67

39. William W. Hogan, "Market-Based Transmission Investments and Competitive Electricity Markets," Center for Business and Government, John F. Kennedy School of Government, Harvard University (August 1999), available on Professor Hogan's website, www.whogan.com. This article provides a comprehensive and readable account of the issues affecting market-based transmission investments

40. José A. Rotger and Frank Felder, "Promoting Efficient Transmission Investment: The Role of The Market In Expanding Transmission Infrastructure" (November 2001)

41. The legal and regulatory framework was created by Law No. 24065 of 1992. The main transmission system was privatized in 1993

42. For a more complete description of this process, see José A. Gómez-Ibáñez, "Regulating Coordination: The Promise and Problems of Vertically Unbundling for State and Local Government, John F. Kennedy School of Government, Harvard University (December 1999). Omar O, Chisari, Pedro Dal Bó, and Carlos Romero, "High Tension Electricity Network Expansions in Argentina: Decision Mechanisms and Willingness to Pay Revelation," Centro de Estudios Ecónomicos de la Regulación, Universidad Argentina de la Empresa, Buenos Aires (1998)

43. Procedimientos, Article 19 of Title III of Annex 16

44. New Zealand has proposed a system where 75% of the voting participants must support a project involving a service change (affecting nodal prices). Commerce Commission Application, Part F: Transport (December 6, 2001), pp. 16-24

45. Other disincentives to invest include local taxes and stamp duty in the usual way, but there is also a provincial tax on economic activity (a gross revenue tax). The application of this tax has been challenged on the grounds that it relates to inter-provincial commerce that can only be taxed at federal level. Proof of payment must be submitted to ENRE before the tax can be passed on to users

46. Omar O, Chisari, Pedro Dal Bó, and Carlos Romero, "High Tension Electricity Network Expansions in Argentina: Decision Mechanisms and Willingness to Pay Revelation," Centro de Estudios Ecónomicos de la Regulación, Universidad Argentina de la Empresa, Buenos Aires (1998); or see José Gómez-Ibáñez and Rodríguez Pardina, Transener (C), p. 7, fn. 13. See also Kent Anderson, Sally Hunt, Hethie Parmesano, Graham Shuttleworth, and Stephen Powell, "Analysis of the Reform of the Argentine Power Sector: Final Report," prepared by National Economic Research Associates (NERA) for the Secretary of Energy of the Argentine Ministry of Economy and Public Works (January 1998), pp. 58-68

47. Commission of the European Communities, Directive 96/92/EC Concerning Common Rules for the Internal Market in Electricity, *Official Journal* L 027, 30/01/1997 pp. 0020-0029

48. Chapter IV, Article 7(6), Directive 96/92/EC of the European Parliament and of the Council of December 1996 Concerning Common Rules for the Internal Market in Electricity, *Official Journal* L 027, 30/01/97, pp. 0020-0029

49. Proposal for a Directive amending Directives 96/92/EC Concerning Common Rules for the Internal Market in Electricity (2001/0077 (COD), EU Council Presidency text (March 14, 2002), published on the EU Council website, www.europa.int

50. Proposal for a Regulation on Conditions for Access to the Network for Cross-Border Exchanges in Electricity (2001/0078 (COD) (March 14, 2002), published on the EU Council website

51. Yves Smeers, "Towards a Common European Electricity Market-Interconnections," presentation to the Regulation Initiative Executive Forum, London Business School (January 3, 2001)

52. FERC Notice of Proposed Rulemaking, 18 CFR Part 35, Docket No. RM01-12-000 (July 31,2002)

CHAPTER 6

The Setting Agent I: Approaches to Regulation

Ingredients of this Chapter

A setting agent

Regulation is a good setting agent, because it is capable of achieving greater certainty of outcome, which consumers, market participants, asset owners, new entrants, and investors need. It provides the gel to reconcile social welfare objectives with the profit motivation of investors. As with all setting agents, regulation must be appropriately chosen to mix with the other ingredients of our transmission expansion recipes. If regulation is not well designed and handled, it will give rise to a very fluid situation, far from well set, that may deter the kind of behavior and investment that the regulator is seeking to achieve.

Also a raising agent

Good regulation can also act as a raising agent as well as a setting agent. In other words, it can raise the game of the regulated entities to achieve the desired regulatory objectives. Through the use of incentives, it can lighten the degree of regulatory burden and minimize any tendency on the part of the regulator to micromanage. It can create more freedom for the regulated entity to make innovative and efficient management decisions. It can also increase the responsibility and accountability of utility executives for their decisions. They can derive more job satisfaction if they meet and beat performance targets and gain some upside from a regulatory incentive. Good regulation can also help to remove barriers and hurdles to new entry, new investment, and market-based activities. In short, it can simulate market forces to some extent.

Organization of the chapter

This chapter is based both on literature and practical experience. We shall begin by looking at why economists regard markets and private contracts to be superior to regulation, and then we will consider the unique, obvious, and not-so-obvious features that make the regulation of transmission necessary. After an explanation of the principles of good regulatory design, the objectives, trade-offs, and the special concerns of transmission investors with respect to regulatory risk, we shall investigate the asset base, cost of capital, and rate of return issues that face transmission investors. The chapter will then describe rate-of-return regulation and explain the general dissatisfaction with its incentive properties and other features.

Market Forces Are Superior to Regulation

There is no doubt that market forces and competition are superior to regulation, not the least because the regulator is being asked, essentially, to simulate market forces with less than perfect knowledge.

Stephen Littlechild, the first U.K. electricity regulator, explained in his original proposal of the RPI-X formula:

> *Competition is indisputably the most effective – perhaps ultimately the only effective means – of protecting consumers*

> *against monopoly power. Regulation is essentially .... a means of holding the fort until competition arrives.*[1]

Similarly, Daniel Yergin and Joseph Stanislaw are often quoted:

> *The objective is to move away from governmental control as a substitute for the market and toward reliance on competition and the marketplace as a more efficient way to serve the public.*[2]

Markets may not be perfect

As we discussed in chapter 5, the market for transmission expansion may not be perfect. The key ingredients that we have looked at may not have been in place to bring forward the necessary transmission investment. There may have been a market failure with respect to a particular need. Governments and regulators cannot allow consumers to suffer high congestion costs for long periods in the expectation that market participants will eventually bring forward the necessary investment. Regulated transmission expansion will be the only hope. Therefore, we must look at how a regulatory regime can be designed to encourage transmission investment, because, for the foreseeable future, merchant and regulated transmission must exist side by side.

Regulation of natural monopolies

In New Zealand, there has been a belief that industry-specific regulation of the electricity industry is not necessary. The New Zealand policy makers have preferred to create the threat of legislation if things go wrong. This runs contrary to the belief that is expressed in many textbooks on microeconomics, regulation, and competition law that regulation is necessary in the case of a natural monopoly to protect consumers, both as to prices and quality of transmission service, because small consumers will be unable to rely on meeting their needs through private contracts with the transmission owner.[3]

Private contracts are better than regulation

José Gómez-Ibáñez maintains that:

> *All things being equal (which they seldom are), private contracts are better than concession contracts and concession contracts are better than discretionary regulation.*[4]

The stronger the exposure to market forces, the greater the incentives to improve services and reduce costs. Private contracts will directly involve consumers in the design of the quality of service and the price for themselves. In that way, the regulator will not have to guess what they need and what they are prepared to pay. However, not all consumers will have enough technical knowledge to know what service to accept and what price to pay. They will not be sure what compensation to seek if service is not provided. However, the regulatory system may find it difficult to simulate the effect of private contracts and market forces, because the regulator will never have the information or innovative and commercial skills of a facility owner or market participant. In theory, the transmission owner and the consumer could enter into a long-term contract that would protect each of them from opportunism by the other, but there are two major difficulties with this approach—transaction costs[5] and unequal bargaining strength.

Contract negotiation with regulatory backstop

Many large customers are able to bargain satisfactorily on their own without any assistance from the regulator. However, in England and Wales, which have a regulatory system that depends upon NGC offering terms to enter into contracts for connection and use of system, the regulator is empowered to determine disputes that have arisen in the course of the negotiations between NGC and some of the large industrial users of its transmission system. Most long-term contracts need to include reopeners, because the parties cannot provide for every eventuality that might arise over a term of, say, 30 years. If it is difficult for the parties to decide what the contract will say, it will be even more difficult for a regulator to determine what it should provide. The regulator (or a hired consultant) does not have the commercial skills and information that are available to the parties themselves.

Long-term contracts can also be expensive for consumers. They tend to move from being fixed price to "cost-plus" towards the end of the contract term, in order to deal with uncertainties. This may mean that the cost to the consumer will be higher than it would have been under a short-term contract. The longer the term, the greater the risk and the premium that a facility owner may demand.

As the experience in New Zealand has shown, it is not always the customer that needs protection. The transmission company is at a disadvantage in reaching agreement as to prices and standards for connection and service, if there are no ground rules or minimum requirements that have been agreed to by the regulator that form the basis upon which it is to provide service. Indeed, a user has no incentive to reach agreement with the transmission owner if it can take service anyway. The user will have every incentive to dispute the charges on the basis that it is being allocated too high a proportion of the revenues that the transmission owner is seeking to recover.

Concessions

Some civil law countries, such as Argentina and Chile, use concession contracts as a means of regulating transmission, as well as allowing private sector participants to exploit and operate transmission lines or systems for relatively long periods (*e.g.*, 30 years on a renewable basis). The assets remain in the ownership of the state for policy reasons, because of the strategic importance of the transmission system to the public.[6] However, the concessionaire enjoys "virtual ownership" through its right to keep an agreed level of revenues from its exploitation of the assets and its maintenance and repair obligations. The concession has its origins in French administrative law and has been used extensively in the water industry.

Under most legal systems, the concession is typically not sufficient on its own to deal with all aspects of the regulatory regime for transmission, but it is a useful tool to increase the level of commitment by the government and to set out clearly what is expected of the concessionaire. In Argentina, the transmission concession was supplemented by a plethora of regulatory rules embedded in primary and secondary legislation. This included a provision in the law that assured concessionaires of a reasonable return on their investments. In theory, a long and detailed concession contract could be drafted in an attempt to cover all eventualities, but there are many difficulties for both parties in agreeing to a long-term contract that cannot be changed. Regulators cannot accept a very long-term commitment. They need some flexibility to make mid-course adjustments to carry out their duties, particularly in order to protect consumers and to oversee the market. The chances are that a concession would have to be supplemented by other rules or incentives and be renegotiated at some point.

Why Is Regulation of Transmission Necessary?

Practical problems

In addition to the fact that transmission is a natural monopoly, there are a number of practical reasons that regulatory oversight of transmission will be necessary:

- For small and medium-sized transmission users, the lack of knowledge and transaction costs involved in negotiating private contracts will be too extreme to make private contracts a workable option (unless, perhaps, there were a sufficient degree of retail competition to allow distributors or retailers to negotiate them on behalf of their customers). Even then, they may need a regulatory backstop to determine what the contract should say if they fail to reach agreement. There would be obvious concerns over unequal bargaining strength and abuse of monopoly power

- The transmission system must serve all users on a non-discriminatory basis and be maintained and expanded to meet all of the needs of the market or the trading arrangements. Therefore, for both market and technical reasons, it is inappropriate and, in practice, extremely difficult to give effect to individual arrangements with particular users
- Notwithstanding that many aspects will be dealt with under the rules and agreements of the wholesale market, consumers will still need regulatory protection as to prices and quality of service
- The transmission owner may need protection, because there will be little incentive for a user to enter into a contract or agree to a price if it knows that it can take service anyway. The transmission owner is not in a position (technically or politically) to interrupt service or de-energize or disconnect the user's facilities as a result of its other commitments
- Transmission planning and performance standards are an important element of maintaining reliability, reducing congestion, and facilitating the working of the market. There are trade-offs between the level of the standards and the cost to the consumer of implementing them that require the regulator to take a stand
- Relaxation of standards for individual users (to minimize the costs they have to pay) may increase congestion and compromise reliability, affecting the ability of the transmission owner to meet its other obligations. Costs could be shifted to other parties. The transmission owner will need regulatory authority to impose these standards on users
- As we discussed in chapter 4, one of the most difficult issues that a transmission owner faces is how to allocate its fixed charges among the different users. It will require the regulator to approve the charging methodology and decide the allocation

of costs among users, because it will always be a contentious issue among them who bears how much, and they will never reach agreement

- The investment programs will require regulatory oversight, because they will result in costs that will be reflected in charges that will have to be allocated to users
- The regulator must make decisions as to inherent contradictions and tensions between the regulatory objectives—incentives are needed that will encourage investment, but that will also achieve the most efficient use of the transmission system to minimize the need for expansion
- The central role that the transmission system plays in the successful working of the wholesale and retail markets means that even with the maximum use of private contracts, market forces, and incentive regulation, continuing regulatory oversight is justified. Competition is, after all, not a state of nature. It needs regulatory oversight
- Maintaining the security of the system is paramount, because of the catastrophic impact of major failures. There is much more to security than preventing local interruptions. In short, transmission is fundamental to the welfare of society. Electricity has become a social and economic necessity. Issues of access to electricity and equity may arise. Society will require and expect transmission to be regulated

The Principles of Regulatory Design

What constitutes good regulation of transmission

There is no definition of "good" regulation. Often, good regulation is defined by its objectives, but they are never easy to achieve, and there are, inevitably, trade-offs between them. There is seldom much attempt to assess the success of a regulatory regime by objective and measurable criteria. Regulatory failures are usually observed, however, only with the benefit of hindsight.[7] Only in the long term can one judge the quality of the regulation. One of the roles of government is to take the long-term view. In that respect, regulation is an extension of the role of government.

In the context of transmission expansion, the test of good regulation must be that it protects customers both as to prices and quality of service by creating a climate (through incentives and, if effective, rules of conduct) in which efficient performance is achieved, the benefits of market forces are simulated, to the extent possible, the necessary investment is encouraged, and implementation of expansion is facilitated.

Good regulation is not easy to achieve

The difficulty is that the alternative forms of regulation that are discussed later—performance-based regulation and rate-of-return regulation—are not as distinct as they appear in theory, when applied in practice. Performance incentives do not remain in place indefinitely, and incentives for cost reduction are blunted to the extent that they are reset in line with costs.

If there is uncertainty about the nature and scope of the regulatory regime, the utility will tend to under-invest in the provision of service quality. If there is a danger that the regulator would set a price that covers operating costs, but does not give the firm a return on sunk investment costs, the firm may foresee this, take the view that it will not gain an adequate return on an investment project, and may decline to invest. Any regulation that forces the utility to maintain prices strictly in line with costs may diminish incentives for cost reduction. It is always difficult for a regulator to obtain good information about minimum cost levels. This is a good reason to adopt incentives designed to improve efficiency and to achieve cost cutting. In order to provide incentives for short-term efficiency, it is important that tariffs must be fixed, so that the regulated firm profits from its improved efficiency and cost reduction.

Achieving the right balance between the interests of the consumer and the interests of investors and between attracting investment and achieving efficiency gain is never easy, but it needs to be done.

Regulatory objectives and trade-offs

There are a number of basic regulatory objectives, of which the goals are to:

- improve efficiency, taking both cost and quality of service into account—the supply-side efficiency goal
- attract investment to expand capacity to include new facilities and to invest in maintaining existing facilities—the capital attraction goal
- enable the transmission owner to finance the carrying on of its activities—the firm viability constraint
- share the benefits of cost savings and efficiencies with consumers in the form of lower prices—the rent extraction goal
- create incentives for market participants to utilize the transmission network efficiently—the utilization efficiency goal[8]

Designing regulatory incentives and rules of conduct that achieve good performance and that penalize or make poor performance unprofitable involves trade-offs between potentially conflicting goals. Inevitably, it is not possible to achieve all of these goals to their fullest extent, simultaneously.

The most obvious trade-offs that are inherent in regulatory goals include:

- maintaining incentives for cost reduction while ensuring that prices are not too far out of line with costs
- ensuring that regulated firms have a reasonable assurance of cost recovery and a reasonable return on long-term investments, while limiting the scope for significant over-investment created by guaranteed returns on investment
- limiting the scope for regulatory discretion or opportunism, but giving the regulator discretion to respond correctly to new information (the exercise of which creates uncertainty and

blunts the incentives that are designed to promote efficiency and investment)
- creating incentives to invest, but at the same time requiring efficiency in operations and utilization of the transmission system so as to minimize the need for investment
- encouraging regulated firms to take a short-term view with respect to efficiency gain and a long-term view with respect to investment

There needs to be some means of assigning appropriate weights or priorities to these regulatory goals at different times, without creating uncertainty or deterring investment or efficiency gain. Incentives that are reset after a fixed period are more useful as a tool to achieve the necessary weightings than broadly-drafted regulatory rules. Even so, there is seldom much transparency in resetting regulatory goals, and investors are deterred by the uncertainty. They will want to invest for short-term gain, because they know that the regulator will not make a long-term commitment upon which they can rely.

Weighting the goals without diluting the incentives

In those countries with weak transmission systems and interconnections, regulators may need to give more weight to attracting investment, achieving supply-side efficiency, and monitoring utilization efficiency, in that order of priority. Sharing the benefits with consumers (the rent extraction goal) will be less important at this moment than the capital attraction goal.

However, investors will be wary of the ever-present rent extraction goal, which has been a dominant feature of the regulatory landscape for transmission in the past. If a regulator introduces incentives to invest, the regulator will need to give assurances that the investor will be fairly treated when performance under the incentive formula is reviewed. There will be a fear that the benefits of the efficiency gain or cost reduction and return on investment will be affected or, at worst, clawed back for the consumer when the performance is reviewed and the formula is reset. In other words, they will be afraid that incentive formula and prices for the following period will be reset— in a manner that does not allow the investors to retain these benefits, but passes them back to the consumer. At best, this will act as a disincentive to make efforts to achieve any efficiency gain as the end of the review period approaches, and at worst, to invest.

Transparency

The regulator's objectives may become overtaken or influenced by political objectives or government policy. It is also possible for the regulator to set policy on an ad hoc basis in areas that are really a matter for government, partly because it has specialist knowledge in that area or because it spots a policy vacuum. This adds to the uncertainty that investors have to deal with.

The most important step in the recipe for a good regulatory regime for transmission is for the regulator at the outset to decide, and make transparent, what it wants to achieve, when, and why. This will enable regulated transmission companies to know what is expected of them. It might also put some pressure on the policy makers not to act capriciously.

For example, if the primary objective is to encourage increased investment in transmission and improved efficiency rather than price reductions to consumers, the regulator could put incentives in place to achieve this for a defined period (*e.g.*, 10 years), making it clear at the outset that other objectives will be taken into account at the end of the period. Arguably, this is what may happen at the end of each review period, but the process lacks transparency, structure, and any certainty of outcome.

Commitment and flexibility

Regulators are often criticized for lack of commitment to the regulatory rules and incentives that they impose. The criticism can be unfair, because regulators also need flexibility to cope with changing circumstances in order to protect consumers. The key ingredient in the recipe for the right balance of regulatory commitment and flexibility is a mechanism that allows the regulator to give notice of its changing priorities, without creating the uncertainties and fears that might deter investment and slow productivity or efficiency gain. There is need for more focus on this ingredient than it normally receives, because it will act as a setting agent that will reduce the regulatory risk we are about to discuss.

Ratchets

In order to reduce regulatory risk and attract investment, a pre-agreed formula or "ratchet" could be applied to an existing scheme at the end of the review period, although it may serve to dilute the incentive properties in some cases. Flexibility could be achieved by utilizing a test to reopen a scheme that either the regulator or the regulated utility felt was not operating fairly or as intended. The test would need to be both strong (*e.g.*, with a materiality limit) and very specific, so that it was not open to interpretation and dispute. Joskow has suggested a ratchet that is discussed in chapter 7.

Special regulatory concerns of investors in transmission

Regulatory risk is a key concern of investors in any sector, but is particularly important in transmission investment because of the size of the investment, the length of the asset life, and the fact that significant transmission expansion will be the focus of attention of both market participants and regulators alike. The key concerns that transmission investors have in relation to regulatory risk are not simply the usual concerns regarding certainty, regulatory commitment, regulatory discretion, and changing the rules of the game. There are many others:

- the length of review periods. Perhaps, some investors would like no review period and no reexamination of their regulated asset bases, once the regulator had approved the costs of the new facilities. Some might prefer one that reflects the period of their financing (*e.g.*, 15 years). If they are subject to an aggressive, performance-based incentive scheme, they may welcome a relatively short review period (*e.g.*, less than 5 years), in case it operates unfairly against them

- downward adjustment of the value of the regulated asset base, particularly if the cost or value can be written down or disallowed for reasons that do not relate to the efficiency or prudence of the investment decision at the time it was made
- the calculation of the cost of capital and the lack of certainty that the investor will be given reasonable rates of return and depreciation for a period that enables it to recover the full cost of its investment, within a timescale that is considerably less than the life of the asset
- an allowance that is insufficient to cover the cost of investment to accommodate increased power flows and to maintain reliability created by load growth. (In Norway, Statnett receives only 50% of the costs that it incurs for this purpose)
- the inability to obtain funding or recover costs on an extraordinary basis, should the need arise (*e.g.*, to implement new connection or interconnection requests), because of the perceived level of regulatory risk
- where benchmarking or yardstick competition is used, lack of confidence as to how benchmarks are set and interpreted (and their relevance to the transmission system in question)
- incentives that do not create an adequate upside for the transmission owner or which expose it to excessive downside or risk
- the inability to retain a significant portion of its cost savings or efficiency gain when performance is reviewed (clawback)
- regulatory penalties for inadequate performance, especially if they are punitive or are affected by actions or circumstances outside the control of the transmission owner
- costs of, delays in, and risks posed to a project by regulatory, siting, and permitting agencies in the process of obtaining permits, approvals, consents, siting certificates, and rights-of-way (particularly, if the jurisdiction is fragmented)
- the costs of and time taken in meeting the requirements of the regulatory regime

The Asset Base and Cost of Capital

Market-based incentives to make efficient investment decisions

There is an ever-present risk of disallowance by the regulator of the whole or part of the value of an asset in the asset (rate) base of a regulated transmission owner on which prices are calculated. This can deter investment. If the transmission owner were a firm participating in a competitive market, the profit that it would be able to earn would depend on its ability to make efficient investment decisions. These decisions would be based on information available to it at the time and on the costs, demands, and prices that are directly realized as market conditions change over time. An efficient competitive firm would expect, on average, to earn what might be described as a normal return on its investments when they are made, and, over time, the average firm will earn a competitive rate of return. In the absence of any long-term contracts, competition provides incentives for firms to make efficient investment decisions. Even if the investment decisions were made efficiently at the time, the firm would expect to run the risk that it may be earning more or less than a competitive return as prices fluctuate with changing market conditions.

The regulated monopoly will not make investments unless it expects to earn a competitive return, and the role of regulation is to allow the utility to do just that with respect to investment decisions that are efficient, provided that the regulated monopoly operates efficiently as well.

Many regulators approve investment plans in detail. They require transmission owners to justify the need for and value of the assets that are included in the asset or rate base from which prices are derived. They may also review whether the facilities are "used and useful" and whether the costs have been "prudently incurred." Regulators might disallow capital and operating costs if they found them to represent expenditures that were "imprudent or unnecessary." This is a common feature of U.S. state legislation under which electric utilities have been regulated. State regulators do not always allow the full cost of installing the assets to be included. There have been examples in California where the cost of meeting conditions imposed by the state siting agencies has not been allowed.

Interestingly, Steven Stoft has shown that a line can be useful for mitigating market power despite the fact that no power flows on it. The size of a line might need to be much larger than the flow on the line. Therefore, he rejects the traditional requirement of "used and useful" in some circumstances.[9]

In some jurisdictions, such as Australia and the U.S., the regulatory approach to reviewing the asset base is designed to achieve a result where the utility covers its costs, but not the costs of inefficient investments and excessive operating expenses. If, for example, the utility incorrectly forecasts demand and adds capacity to meet the demand, such that part of the capacity is not necessary, it would not be penalized for events outside its control, as a competitive firm would be, so long as it can defend its forecasting procedure. In this way, regulation punishes only bad decisions, and not bad luck.

Asset revaluation

Under some systems of regulation, the quality of the investment decision is assessed only once—at the moment when the asset is first included in the asset base. Although, in theory, it could be reviewed again, it is unlikely to be revalued downwards or disallowed if the investment was thought to have been efficient at the time and the circumstances that might cause the regulator to question whether it is still "used and useful" were outside its control.

The Australian approach to asset "optimization"

In Australia, the entire asset base is reviewed at the end of each five-year regulatory review period and will be "optimized." The value of the asset base can only be optimized downward. For example, the revenue requirements for Transgrid in New South Wales are based on the value of its assets, using a technique called "optimizing the deprival value" that gives a "depreciated optimized replacement value (DORV)." Although the methodology is not yet entirely settled or clear, it is taken to represent a methodology that a prudent planner would undertake today. It is not about building new assets from scratch and looking, with the benefit of hindsight, at whether a line was sized correctly, with the result that it might be revalued downwards if it were found to be oversized. The exercise would take into account the potential for load growth and economies of scale. Inevitably, there are some elements of subjectivity about the approach, and it is still an immature regime. In one case, the asset base has been devalued by 15%. The regulator brings in consultants to look for obvious areas of excess engineering that should be removed.

As far as replacement of assets is concerned, the modern equivalent asset (MEA) valuation basis that is utilized exposes Transgrid to new technology risk, because it results in redundant assets or write-downs. The consultants and the regulator are fairly pragmatic and look to see whether Transgrid has been prudent and done a good job.

Depreciation was originally on a straight-line basis over the life of the assets. The regulator is exploring the concept of "economic depreciation", which has not yet been defined. It could be calculated by reference to how quickly the asset becomes redundant in the marketplace. Transgrid could decide to accelerate the depreciation of assets that are at risk of optimization, but it is not in a good position to see what the market is going to do and, therefore, when to do so.

In the view of Transgrid, its optimization risk is an essential part of an effective incentive-based regulation framework for transmission, but there needs to be greater clarity as to the rules, how they will be applied, and the linkage to the rate of return. If there is no certainty as to the basis upon which assets will be considered underutilized and therefore optimized, not only will the planning process become very difficult, it is likely to deter investment entirely.

Transgrid is also subject to a system called "optimized depreciation replacement cost" that places the technical risk on Transgrid as well as the asset optimization risk. It is subject to the same five-year review period.

In short, Transgrid accepts the optimization risk as a discipline placed upon it to make efficient investment decisions as if it were in a competitive market. It cannot justify being protected from its own bad decisions by passing all of its costs on to consumers. Optimization of the assets is one means by which the regulator can counter a tendency on the part of a regulated utility to over-build or make inefficient investment decisions, but the rules must be clear and applied intelligently.

Low rates of return are not an effective incentive

Another less effective technique for creating incentives to invest is to keep regulated returns close to the cost of capital. The cost of capital will be higher when the optimization risk is present. However, optimization is a better incentive to invest prudently than low rates of return. Low rates of return simply deter investment in transmission, whereas the optimization risk forces regulated transmission businesses to seek out the most efficient investments. This would be analogous to the effect that would be provided in a competitive market. The approach creates incentives for innovation, which reduces costs and improves the quality of service to the consumer in the same way as a competitive market would. Transgrid is only allowed a rate of return equal to its weighted average cost of capital (WACC). This creates little incentive for it to invest. However, like many state-owned transmission utilities, it makes an effort to do the right thing.

In order to encourage greenfield site development of gas pipelines in Australia, higher rates of return and regulatory holidays have been allowed. For example, one gas pipeline was allowed a return on equity of 15% on the basis of "greenfield risk." The Australian Productivity Commission has adopted a policy that it is better to run the risk of a moderate amount of over-investment than to run the risk of under-investment. The Australian Commerce and Competition Commission (ACCC), which regulates the electricity sector on a national level, is understood to be considering the adoption of this policy with respect to transmission.

The New Zealand approach to asset optimization

By contrast, New Zealand uses a different optimization deprival value methodology, which is imposed as part of the market rules and under the contracts between Transpower and system users, because there is no formal regulatory system. It also seeks to create some degree of competition and is part of the "light-handed approach" to regulation for which New Zealand is so famous.

This approach allows an asset value to be challenged by virtually anyone as if competitive market conditions applied on the basis that a cheaper alternative might be available. If, for example, the replacement cost of a line reduces over time, Transpower's asset value reduces, and, therefore, its transmission charges are also reduced. If it can be proved that some other solution can transport the power more cheaply, Transpower must reduce the asset value. A user can maintain an argument to this effect and would have every incentive to do so continually. It can be challenged by users who dispute their terms of service, for example, on the basis of the uncertainties of demand forecasts. If the user has a lower cost alternative, such as distributed generation, the economic value of the assets that serve the customer will be reduced to the lower cost, but the user does not have to build the alternative.

The New Zealand approach begs the question of how far a user must go to prove that the alternative really is viable in the long run. This game for Transpower becomes one of calling a bluff. There are likely to be few occasions, for example, when it is clear that a user will actually build a captive generating station if Transpower does not reduce its prices.

This approach also affects the computation of the rate of return in a manner that is not entirely clear. An accounting net profit (after tax) is determined that is applied over shareholders' funds. This gives a figure for earnings before interest and taxes (EBIT). It is not relevant whether the funding involves equity or debt. The calculation derives a figure that becomes the numerator, and the asset value becomes the denominator, leading to a return on assets.

A target cost of capital is determined using a model with interest added in, which is derived ex-ante and that should result in a proper return. The review after the fact determines what the return actually is. If Transpower has earned more than it should have, it has to hand the surplus back. If it earns less than it should have, it recovers the shortfall by an adjustment in the following year.

The optimized deprival value has a large impact on future revenues, and market participants have a natural incentive to mount challenges to the asset value. This can be done virtually at any time. It creates considerable uncertainty for Transpower and is a deterrent to investment.

Cost of capital and rates of return

Transmission owners all agree that allowing increased rates of return and accelerating depreciation over shorter periods than 40-year asset lives will be the quickest and most effective way to incentivize them to invest on a regulated basis. They will argue that because the cost of transmission service represents only about 4-11% of the average customer's bill, a significant increase in the rate of return on transmission investment, designed to achieve needed expansion, would result in an increase in the average retail price of less than 1%, even if the increase in the rate of return on assets were, say, 5%. They argue that regulators should not be tempted to reduce the rate of return on transmission, because it would deter investment and increase the associated indirect costs that could harm consumers more. They also argue that regulatory risk and uncertainty increase the cost of capital that the consumer must bear.

Figure 6-1 shows a range of quoted rates of return (differently stated and based on different asset valuation methodologies). It demonstrates that none of the transmission owners is particularly well rewarded for what it does.

Transmission owner	Regulated rate of return
Transgrid	8.3% (pre-tax nominal WACC)
Powerlink	8.83% (pre-tax nominal WACC)
NGC	6.25% (pre-tax return on assets) and 6% (pre tax) on post-vesting connection assets
Transpower	6.8% (target rate for year to June 2002), 7.2% (indicative rate for year to June 2003)
Southern California Edison	9.68% (in a FERC ruling in 1999)[1]
ESB	6.1% return on capital expenditure
Transalta	9.5% return on assets (2000), 9.75% (2001)
National Grid, US	10.25% return on equity
Pacific Gas & Electric	11% return on equity (2000)

[1] Southern California Edison Company, 86 FERC P 63,014 (1999) Docket no. ER97-2355-000.

Fig. 6-1 A selection of rates of return

The cost of capital, including the thorny issue of the rate of return, is always a contentious issue between the utility and the regulator, because it is at the heart of achieving the balance between the interests of the consumer and the interests of the investors. It was never easy to deal with this issue under any form of regulation. It is important for the regulator to decide what it is trying to achieve in deciding what is a reasonable or fair rate of return. This is not something that can be done without information, analysis, and modeling.[10]

Transmission is not risk-free business

Regulators have tended to take the view that transmission is a risk-free business, not subject to competition, revenue risk, usage risk, or payment default risk and that it justifies only a low rate of return. This thinking has been an important deterrent to investment in transmission. However, there are several reasons that transmission should not be regarded as risk-free.

- Distributed generation and load curtailments present real competition with and alternatives to transmission
- Technological changes and innovation mean that assets may become obsolete long before the end of their physical or accounting lives, and innovation and retirement of generation can decrease the value of transmission assets associated with particular generation investments
- The obtaining of permits, consents, and rights-of-way has become more complex (with many more objectors), so that the optimal design, siting, and building of transmission facilities become more risky, and project development costs and delays are increasing
- A transmission system was never designed to be used as the marketplace for wholesale competition. Market participants who put the system under stress and place new demands on it, in turn, increase system performance and maintenance risks faced by transmission owners
- Transmission owners are running increased risk of stranded assets that are non-revenue earning, because generators and consumers may be able to build plants rapidly or decide to retire facilities on little or no notice
- There is increasing uncertainty regarding the future usage patterns.
- Counterparty and payment default risks will arise from the diversity and relatively small size of a larger number of users
- In addition to the demand risk, a transmission company will be affected by fluctuations in market prices and fuel prices—competition makes supply and demand conditions less certain and will increase the risks and costs faced by the transmission company, both in terms of its investments and its operations
- The regulatory risk has increased, particularly in the early stages of a new market, as the regulator grapples with overseeing a market that is settling down, and where both market rule

changes and regulatory rule changes that may affect transmission charges are a common feature
- Unbundled transmission companies are smaller in financial terms than the vertically integrated utilities from which they were created, and their financing or leveraging ability will be less
- Transmission companies have less revenue diversity, being primarily dependent on revenues from their transmission assets[11]

Cost of capital for a stand-alone transmission company

Michael Cragg, William Lehr, and Ron Rudkin argue that traditional techniques for estimating rates of return on equity, such as the discounted cash flow (DCF) model, the two-stage discounted cash flow model used by FERC, and regression-based techniques, such as the capital asset pricing model (CAPM) and arbitrage pricing theory (APT) models, are likely to produce estimates for a return on equity that are too low for an unbundled stand-alone transmission company operating in a deregulated market.[12] The reason for this is that all of the methodologies are based on historical data from comparable firms, and, in the early stages, there will be no comparable stand-alone, publicly-traded transmission companies in a single jurisdiction among which comparisons can be made.

All estimation techniques underestimate the return on equity

All of the estimation techniques will understate the return on equity to the extent that the debt percentage, or leverage of the stand-alone transmission company, is expected to be higher than that of the industry comparators used in the models. The more debt, the greater the required return on equity. The authors' analysis assumed that the debt: equity ratio will be 60:40. This is higher than that of vertically-integrated utilities for which comparable figures were available. In order to adjust for the higher leverage of the transmission company, data were used to restate the return on equity as if it had no debt; each company's return on equity was then "re-levered" at the expected debt level of the transmission company before taking the average for the industry. After adjusting for the new leverage, the estimated rate of return increased to 11.8%. However, even with this adjustment, the estimate was inappropriately low using traditional discounted cash flow methodologies.

Gearing

Increasing the gearing of a firm up to an optimal level (the level of borrowing or debt: equity ratio) can (in circumstances when the assumptions of the Modighani-Miller theory we looked at in chapter 2 do not apply) lower the cost of capital. Increasing the gearing beyond the optimal level will increase it again. One of the reasons it achieves this beneficial effect is that it increases the pressures for managerial efficiency, and the share price will be more sensitive to changes in performance.[13] However, increasing the level of debt to 100% would give the utility no balance sheet to enable it to withstand unforeseen, adverse events.

NGC's cost of capital

There is always considerable debate over the cost of capital in the U.K. when the price cap is reset at the end of each five-year review period. The regulator applies a capital asset pricing model (CAPM) that has four elements in it:

- risk
- risk premium
- optimal gearing
- beta (the measure of systematic risk of the company)

The weighted average cost of capital (WACC) consists of the cost of debt and equity that is weighted to reflect the gearing for the regulated asset base. Assessing the debt might be relatively straightforward, because it is a more factual exercise of assessing rates in the market. However, the cost of equity is more difficult. These concepts were described in more detail in chapter 2.

The cost of equity formula involves adding the risk-free rate of return to the beta and the equity risk premium. Because the value of the regulated asset base is indexed to reflect inflation, this is carried out in real terms, rather than nominal terms (which would be the case if historic cost were used, as in the U.S.). The equity risk premium is a rather soft number and varies between 2-9%. It is compared with the yield on a portfolio of long-term index-linked gilts. The beta rises (as the portfolio of gilts rise) and is a measure of the risk to the particular company. Unfortunately, there are not many comparators for transmission companies, and it is difficult to look to a particular market as a whole for a yardstick. If the beta is set very low, this gives an implausible view of the cost of capital. In NGC's case, the beta for the England and Wales natural gas pipeline owner is regarded by the regulator as the best proxy. However, the two companies have now merged.

Rates of return and depreciation are key ingredients

It is clear that the cost of capital in general and the rates of return and depreciation, in particular, are the key ingredients in creating incentives to build new transmission facilities. The allowed rate of return determines the average return on incremental investment that investors expect to earn on invested capital, including the market-determined opportunity cost associated with that investment.

An unbundled transmission company will lack the revenue diversification and the ability to cross-subsidize its business activities enjoyed by an integrated firm. The transmission business is, as we have seen, becoming increasingly risky. The old, regulated returns, based on an assumption that transmission systems have been overbuilt, will not be sufficient to attract investment. The social costs of setting the rate of return too low will be greater than setting it too high, because it will deter needed investment.

Leonard Hyman asked the question: "What would it take to get investors to enthusiastically embrace and expand the transmission system?" He has calculated the impact on price of raising allowed returns and accelerated depreciation rates (assuming that an incentivized owner will reduce costs using realistic estimates). He concludes that:

- Raising the return from 9% to 11% and accelerating depreciation from 40 to 16 years would increase the delivered price of electricity by 2-3%
- Raising the return from 9% to 12%, accelerating depreciation from 40 to 20 years, and passing operational cost savings of 10% to the consumer would increase the delivered price by 1.5-2%
- If capital expenditure needs to be doubled to keep up with demand, the return on new capital is 20% (after tax), and depreciation is accelerated to 5 years, the rise would be only 1% p.a. during the program. It would not, however, encourage the utility to use existing assets more efficiently or to look for the lowest cost solution[14]

The Classic Approaches to Regulation

Although there are many variants and hybrids, the two classic approaches to economic regulation are:

- rate-of-return regulation
- performance-based or incentive regulation

Components of rate of return regulation

Rate-of-return regulation (also called cost-of-service regulation), at its simplest, is designed to ensure that the prices paid by consumers will be sufficient to create a specified (but regulated) return to a firm on investments approved by the regulator. Under this approach (as developed in the U.S., for example), the revenue requirements of the regulated utility are based on three components—operating expenses, depreciation, and the permitted rate of return, multiplied by the depreciated value of the asset base. In order to make the calculation, the regulator needs to make a number of decisions, for which it will require accurate information. It needs to understand what the operating expenses of an efficiently run utility would be, the appropriate depreciation rate, and whether it was prudent to make the investment the assets. The regulator then needs to decide (within the relevant legal framework) what the minimum rate of return should be that the capital markets require in order to finance investments of comparable risk. The treatment of sunk costs is a controversial issue. If a return on sunk costs is not permitted, or is too low, it may deter future investment. These issues are not straightforward and were often the subject of long regulatory proceedings and debate.

Historic cost and straight-line depreciation

Under rate-of-return regulation, prices are set to cover both operating costs and capital costs. Operating costs can be derived directly from the utility's accounting system, and capital costs are taken to be depreciation plus a "fair return" on the utility's actual or estimated investment. Many regulators calculated the cost of capital by multiplying an estimate of the utility's nominal cost of capital by the depreciated original or historic cost of its assets. Straight-line depreciation was employed with asset lives that were long and, to some extent, arbitrary. The more modern accounting approach uses a current cost or modern equivalent asset cost approach that takes inflation into account.

In an inflationary situation, the historic cost accounting rules (*e.g.*, as used in the U.S.) overstate the costs in the early years and understate them in the later years. As a separate but related issue, when there is technical progress (that results in constantly declining real prices for the same service), the rules generally understate the costs in the early years. Current cost accounting rules (used in the U.K. and elsewhere) correct for the first error by adjusting for inflation.

This traditional approach would give the utility a revenue stream for each asset that had its present value (using the cost of capital as the discount rate) equal to the original cost of the asset. The market value of the utility would then exactly equal its book value. However, other approaches to computing capital costs would probably achieve similar results.

It is unlikely that the capital cost charged at a particular moment would represent the true, economic cost of using the utility's capital at that moment, because the depreciation rules are, to some extent, arbitrary. Joskow and Schmalensee point out that, in general, straight-line depreciation is equal to economic depreciation only under very special conditions unlikely to be encountered in practice, so that accounting and economic capital costs generally differ.[15]

Regulatory lag

Under rate-of-return regulation, prices were not continuously adjusted over time as the costs changed. They were only changed at the request of the company or the regulator after a lengthy regulatory process or hearing. Indeed, in the U.S. during the 1950s and 1960s, prices and the provisions of filed tariffs (terms of service) remained the same for years in many cases. This phenomenon is referred to as "regulatory lag." As a result, the actual rates of return would be above or below what the regulator determined to be fair at the time.

Rate-of-return regulation for transmission

Rate-of-return regulation for transmission would, typically, involve estimating the amount of electricity that the transmission system would transport during a year, estimating how much it would cost the transmission company to transmit the quantity, and then dividing the second number by the first to derive a transmission price. This approach presented some immediate difficulties. It required accurate information about forecast demand and costs. Demand was easy to forecast only if it was relatively insensitive to price. If the demand forecast was affected by unexpected changes, there was a downside risk that there would be insufficient demand to cover transmission costs and an upside risk that capacity would be insufficient to meet demand. The transmission owner had an incentive to exaggerate its costs, but had no incentive to minimize its operating costs or use its assets efficiently. It had no incentive to adopt technological innovations that would reduce costs or benefit consumers. Essentially, rate-of-return regulation required the regulator to micromanage the operations of the transmission owning company. At worst, the cost data reported by the transmission owner would have to be audited, creating the potential for disallowances and protracted dispute.

Criticism of rate of return regulation

Rate-of-return regulation has been severely criticized since the early 1960s, when a number of economists drew attention to lack of improvements in performance by regulated monopolies in comparison to unregulated monopoly outcomes.[16] However, rate-of-return regulation apparently served utilities and consumers well for 50 years. Its greatest contribution was that it provided a reasonably acceptable means of balancing the interests of consumers and the utility. However, this was only the case so long as inflation was low and changes in technology were not dramatic. It also depended upon volumes growing steadily. It did not seem to matter greatly if the utility failed to achieve its expected rate of return in a particular period, because it would be allowed to recover the return in future periods. Prices could be kept low by the straight-line depreciation rate based on the long asset lives.

The effect of inflation, technological change, and competition

The 1970s and 1980s saw inflation, rapid technological change, and competition. All of this made it difficult to maintain price or rate stability. Technological change, particularly in communications technologies, became quite dramatically rapid, and with entry by competitors, it was difficult for the traditional regulatory contract to be maintained. The regulator could not sustain exclusive franchises, free from competition and the various cross-subsidies that were common in rate-of-return regulation. Depreciation over long asset lives based on little technological progress began to result in capacity deficiencies. Charging lower than economic depreciation rates represented a form of cross-subsidy. In essence, future customers paid for under-recovery of capital by current customers. This would not be sustainable in a competitive market environment. Moreover, the shareholders would ultimately pay for the shortfall that would affect the utility's future cost of capital. Apart from rapid technological change and competition, the regulatory processes had become cumbersome, particularly, the formal rate cases.

Insufficient incentive for internal efficiency

There were other attacks from economists, such as the criticism that rate-of-return regulation blunted incentives for internal and dynamic efficiency. In other words, rate-of-return regulation created insufficient incentives for internal efficiency as well as product and process innovations over time. After all, if the utility could recover all of its costs eventually, there was no incentive to cut costs.[17] The notion that the regulator could simply order a utility to cut costs was clearly unrealistic. Rules of conduct and regulatory micromanagement could not work as well as profit-motivated incentives.

Inherent shortcomings of regulation in general

Some of the deficiencies of rate-of-return regulation were caused by the deficiencies of any regulatory system, rather than the lack of profit-motivated incentives. The literature was critical of rate-of-return regulation and did not always make a distinction between the inherent problems of regulation, in general, and the shortcomings of rate-of-return regulation, in particular. For example, there were complaints that regulators were not in a position to distinguish efficient from inefficient performance and that they had imperfect information with which to analyze costings. The difficulty that utilities always have better information than regulators is a problem existing in all forms of regulation, not simply rate-of-return regulation. The fact that regulators face in monitoring the performance of regulated utilities imperfectly is also a problem for all forms of regulation.

Average cost pricing

The legitimate criticisms included the lack of incentives that a competitive market provides for the kind of decision making that benefits both the consumer and the regulated firm, even though regulatory lag did provide informal incentives that resulted in cost cutting. The average cost pricing that was a feature of rate-of-return regulation resulted in prices that did not properly track changes in short-run supply and demand conditions. Any average approach will lead to prices that are sometimes too low and sometimes too high, however efficiently decisions are being made. If prices are based on average cost principles rather than true marginal cost, consumption decisions will tend to be socially inefficient.

Lack of marginal cost data

The traditional accounting data that regulators used did not directly reveal the marginal costs that were essential for efficient price setting. It was difficult for the regulators to disentangle decisions that were efficiently or inefficiently made and circumstances beyond the utility's control. Where the regulators set non-marginal linear prices based on total costs and consumption, not only would consumption be inefficiently encouraged or discouraged, but the utility would have little incentive to minimize costs and would be punished if it set prices below cost in order to encourage consumption. According to Joskow and Schmalensee, it would be better to provide weaker incentives for efficiency in the case of linear pricing than to levy fines or make payments directly to reflect the trade-off between cost minimization and optimal pricing. If ordinary or linear prices are used to provide incentives, greater efficiency can only be achieved at the cost of less efficient consumption, while this additional cost is absent if direct payments are possible.[18]

Dissatisfaction with rate of return regulation

In short, the deficiencies of rate-of-return regulation were seen to be many. It failed to provide adequate capital recovery and to create incentives for internal efficiency and innovation. Its rigid, cross-subsidized price structure and the high regulatory transaction costs created serious problems that made the search for alternative regulatory structures an imperative.

From a transmission expansion perspective, rate-of-return regulation could lead to under-investment for a number of reasons, including lower returns and depreciation over long asset lives as well as over-investment if the rate of return was attractive. It is usually based on historic cost accounting, which inhibits innovation if it is coupled with low depreciation rates. The approach also tends to allocate fixed operating costs to multiple outputs through fully-distributed-cost (FDC) pricing incorrectly, leading to cross-subsidies that are problematic on unbundling and deregulation. However, the imperfections of the administration of rate-of-return regulation are chiefly to blame for its unpopularity. Regulators tended to micromanage utilities and intervene frequently in a variety of ways, including "prudence" reviews of investments. This gives the utility little or no freedom to use its core competencies and to develop its entrepreneurial skills using its superior information for the benefit of the consumer.

To quote Daniel Yergin and Joseph Stanislaw:

> *For participants in the electric utility industry, deregulation requires changing practices and long-held cultural beliefs. Traditional utility regulation centred on a laborious, legalistic, highly ritualized process called rate*

> *hearings, in which – Kabuki-like – lawyers, lobbyists, corporate officials, experts, intervenors, environmentalists, consumer activists, and regulators all performed their stylized roles.*[19]

We shall now look at other regulatory approaches and strategies in our search for a better recipe than rate-of-return regulation. In chapter 7, we shall discuss performance-based or incentive regulation.

Other Approaches to Regulation

The output standards concept

The "output standards" approach is based on the concept that utilities should be paid by results, *i.e.*, the output of their efforts that society needs. The approach requires the regulator to define what consumers need, because they are not able to do so for themselves.

Large customers may be able to define the level of security (interruptibility) or service they want for themselves and may be in a position to negotiate a reasonable contract. Small and medium-sized consumers will not be in that position. In both cases, the fact remains that the transmission utility will be a natural monopoly. Accordingly, this approach involves the concept of the regulator standing in for consumers to negotiate with the transmission owners on their behalf as "an informed buyer of transmission services," as if it were entering into a contract or series of contracts as agent for consumers.[20]

It is difficult to imagine that it would make sense to allow different groups of consumers to negotiate separately for different levels of security. Problems in the transmission network affect large numbers of consumers, and it is presently impossible to discriminate between those that have paid for a higher level of security and those that have not. In terms of investment, this could lead to significant free riding.[21]

The challenge

The contract is challenging to strike because it is more difficult to translate the concept of what consumers need (over and above what the wholesale market requires) into key performance indicators (KPI) with respect to transmission (as opposed to distribution) and to define the permitted or allowable cost, which may only be in the order of 4-11% of the total cost of service to consumers. The issue of performance standards is discussed in more detail in chapter 8.

Emphasis on output rather than input cost

If a mechanism could be created to decide what delivery of the performance the customer wants, then this approach could be a means of finding out what it will cost and also of putting a downward pressure on those prices. Arguably, this is what NGC's transmission services schemes (TSS) created and also the Information Incentives Project 2002 (IIP) that OFGEM has created for the electricity distributors in England and Wales. The IIP approach is designed to overcome the fact that the RPI-X price cap places more emphasis on cost (input) than on performance (output) of transmission services. A major objective of the IIP is to establish a framework under which quality of services will be valued and incentivized.[22]

The concept is based on the philosophy that the regulator should not deny higher standards at lower cost to the consumer simply because it does not fit the regulatory model or have the tools available for regulatory monitoring. The approach is not substantially different from setting performance standards or benchmarks. Indeed, the KPIs that reflect what the customer wants can be used as benchmarks for comparisons between utilities.

The difficulties of this approach are no different from those traditionally involved in regulating transmission and include:

- setting the KPIs and performance targets in a manner that is relevant both to the function that the transmission system plays in delivering service to market participants and consumers and its

relationship with system and market operations, generation, and distribution system performance
- measuring and evaluating the trade-offs between cost and quality of service
- establishing future performance and cost targets, because past measures are an unreliable indicator of future needs and performance
- the avoidance of perverse incentives that will result in transmission owners spending unwisely for short-term gain
- designing key performance indicators and targets that reflect the long lead-in times to expansion and the long asset lives

Costings and the capture of market forces

Having set the performance targets and KPIs on behalf of the consumer, the role of the regulator is then to create a mechanism that both determines the allowed cost of delivery of the required performance and puts a downward pressure on the prices that the consumer is charged. This could be done through an economic purchasing obligation requiring the transmission owner to harness the forces of competition in deriving the price, *e.g.*, through requesting bids or quotations from contractors and qualified asset managers (if there is a market for these services). In the alternative, the regulator could use consultants to establish whether the costing proposals represent value for money, although value-for-money audits are difficult to carry out effectively.

The approach of looking at output standards rather than input costs is reflected in some aspects of performance-based or incentive regulation that we shall discuss in chapter 7.

Asset management concept

The concept of asset management,[23] involves placing the risks associated with management and maintenance of the assets with a separate asset manager. The asset manager could be a business unit within the transmission owner or a separate professional asset manager. If there were a separate asset manager, it would enable the assets to be owned by an investor that had no experience in the industry at all, *e.g.*, an investment fund or an institutional investor. The idea is that the asset manager would be hired under a competitive procurement process, and only asset ownership would remain a local monopoly, thereby reducing the regulatory burden.

The required performance of the asset manager would be dictated by the output standards or performance required by the regulator and also by the transmission owner, particularly in terms of quality of service and service levels agreed with transmission users and the regulator. The asset management contract would specify the price at which the output standards would be delivered. The asset manager could be incentivized to manage the assets on a basis that it would share some of the cost savings with the transmission owner. This approach would be entirely consistent with the "output standards" approach (that we have just discussed), which a regulator could adopt in defining or agreeing to the quality of service and prices.

A specially-tailored asset management contract

The fact that a single asset manager may manage the assets of more than one transmission owner does not mean that the same performance standards and KPIs need to be set for each transmission owner or system. Incentives could be set differently, if need be, from one system to another. The terms of the asset management contract can be tailored to the special characteristics of the transmission system and the regulatory goals.

The asset manager could also be tasked under the contract with producing investment plans on a rolling basis and required to keep them regularly updated. The plan would be required to meet basic reliability, safety, and other industry and regulatory standards as well as the technical and market rules. It should also be possible to factor in the need to accelerate or vary expansion projects in anticipation of developments on the system. Incentives could be created for the asset manager to minimize the need for expansion and for saving money. Penalties for delays and cost overruns could be incorporated.

Asset manager incentives

The asset manager would be required to establish a series of options and identify the trade-offs. It could be incentivized to capture the main issues or problems, which should, in theory, deal with at least 80% of the expansion planning needs. Another incentive could be put in place to give the asset manager a share of any cost savings that it can achieve in implementing new investment more economically than budgeted, or in avoiding the need for new investment at all, as a result of improved operations and maintenance. Overall, it would be incentivized to improve performance, meet KPI targets, and reduce costs.

This scheme could result in innovative systems to achieve the necessary performance and a highly professional and focused approach to asset management, which should result in maximizing asset availability and flexibility to achieve better use of the existing assets, minimizing the need for expansion. It could also capture the forces of competition through the way in which the asset manager is hired, which should result in improved efficiency and innovation, provided that there are enough qualified contractors with reputations at stake.

David Currie observes that: "it is important to note that it is far from automatic that healthy competition will emerge in this way, particularly in view of the legacy of the old monopoly structure." He also makes the point that long-term contracts may have better incentive properties than short-term contracts that introduce agency incentive conflicts between the interests of the asset owner and the interests of the asset manager. However, long-term contracts may inhibit the development of a competitive market in asset management services.[24]

This leaves the asset owner with a choice of whether to retain a small staff to manage the contract or a larger staff to act as an informed buyer of asset management services and expansion projects. Outsourcing these key activities does require a number of skills, and the asset owner cannot abdicate its regulatory and market responsibilities and pass on to the asset manager the risks that arise from its own errors of judgment. There are important decisions to be made with respect to profitable trade-offs between capital and operating expenditure, and which of several potentially-competing transmission expansion options to choose.

Advantages and disadvantages of the asset management model

Stephen Littlechild has said:

> *I'm confident that there are significant efficiencies which can be achieved by this greater specialization in running the network, and by the discipline of commercial contracts. This is the way to go to achieve better service and lower costs.*[25]

He felt that if there were benefits in longer operating contracts, then the regulator might want to look at setting longer price-control periods.

The asset manager may be one step removed from interaction with the market participants and the regulator (depending on how its tasks are defined). The chances are that it will not take into account stakeholder concerns in the way that the transmission owner would, because it will not care about the long-term future of the transmission business. The asset manager may focus on making profit in the short term. Unless the asset management contract is carefully designed, the asset manager may not care about the long term, because it may be able to find another client and if things go wrong, may not be discontented if the contract is terminated.

It is possible to take the concept of asset management to an extreme. Railtrack plc in the U.K. was criticized for outcontracting large parts of its responsibilities to other companies. As a result, it was difficult to establish "who was in charge", and the institutional memory of the assets was lost. The danger is that once all the experienced staff of an asset-owning company have left or retired, the people coming in have no way of knowing what is actually there and the potential of asset deterioration.

Although the approach could relieve the regulatory burden with respect to current and future operating expenditure, the regulator will still want to be sure that the management of the asset-owning company had made the right decisions and had not abdicated responsibility. Clearl,y the use of the output standards approach that we have just discussed would help in this regard. The regulator would still have to set performance targets and standards and monitor performance.

The asset management approach is being debated in the U.K. It reflects a new paradigm in competency-focused management structures and business models for utilities. It is not clear whether there would be a really competitive market for asset management, and whether it has advantages over the performance-based, incentive schemes we shall look at in chapter 7.

Management-based regulatory strategies

Many regulatory approaches are designed to command or incentivize regulated firms to adopt specified standards or achieve specified levels of socially desirable outputs or performance. Performance-based and incentive techniques give the regulated firm flexibility in deciding what technologies and management processes to use to achieve that performance. An approach described as management-based regulation commands firms to engage in the planning and decision-making needed to identify both the technologies and management processes as well as the performance targets needed to achieve socially desirable goals. This management-based approach to regulation is currently being pursued for application in the areas of food safety, occupational health, and environmental protection. It is an approach that can be described to some extent as the "privatization of regulation."

Under management-based regulation, utilities will be required to prepare plans to comply with general criteria outlining how to achieve the policy goals in question.[26] These plans may be subject to approval by regulators, and sometimes have been developed with their assistance. For example, fish inspection was transformed in the 1990s by introducing a hazard analysis approach requiring firms that handle fish to do an analysis of where likely points of contamination of fish would arise and to develop plans to avoid contamination. This approach was rapidly adopted around the world. These plans generally require firms to produce documentation of subsequent compliance and third-party auditors to certify compliance. The concept is similar to that of ISO 9000.

Maximizing the use of available information and utility buy-in

The advantage of this approach is that it places the locus of the regulatory decision-making at the level with the most information about how to achieve the desired result. The behavior that utilities adopt under the management-based approach has the potential to be less costly and more effective than that under a regulator-imposed regime of standards and penalties for poor performance. By placing the locus of the standard-setting authority at firm level, there should be greater "buy-in" from the management of the firm, which should lead to greater compliance with the standards. (Enforcement resources of regulators are always inadequate to ensure thorough oversight of regulated utilities. It can be argued that compliance is, to a large extent, voluntary in any event because the probability of a violation being detected is low. This approach involves the assistance of private, third-party certifiers in the monitoring and enforcement process.)

A tool for regulating planning

The management-based approach operates at the planning stage, which is appropriate for the regulation of transmission expansion. Some jurisdictions, such as Australia and the U.K., already regulate transmission-planning processes by the use of rules requiring the production of plans. This looks like a management-based approach, but where the standards are imposed by a regulator or through legislation. In other jurisdictions, such as the U.S., there is no requirement for the utility to produce an investment plan or capital expenditure budget to the regulator, unless it needs regulatory approval to recover the costs from market participants or consumers. The plans may be reviewed as part of the process for obtaining certificates, permits, and environmental consents, where the need for the project and alternative solutions may be examined in detail.

The management-based approach, however, could be utilized to elaborate on the regulatory requirements relating to the production of upgrade and expansion plans as well as the planning processes, in order to gather information for the regulator as to the appropriateness, or otherwise, of the current industry standards and plans to achieve the socially-desirable outcomes.

However, it is unlikely that a pure management-based approach would be either necessary or appropriate, because, as we shall discuss later, it would be unacceptable to leave the standard setting and determination of the necessary performance levels entirely in the hands of the regulated utilities. Nor would it be necessary to adopt a full management-based approach, which is more suitable where an industry is characterized by large numbers of regulated entities operating in a heterogeneous sector and where ease of measuring output is low.

The New Zealand Model—No Regulation

Living under the threat of regulation

As we saw at the beginning of this chapter, the culture in New Zealand does not allow for a formal regulatory regime. Instead, market participants and asset owners operate under private contracts (even though some of the major participants are state-owned). They live under the threat that the government might legislate to bring in a regulatory regime if they do not behave or perform in a manner that is acceptable. The government has issued good Government Policy Statements[27] but they are open to interpretation, and there is no formal monitoring and enforcement mechanism. It is clear that this approach has produced a variety of difficulties, and a new governance structure for the industry has been necessitated that appears to be intended to create a surrogate for a regulatory regime in some form. It creates a mechanism for dealing with changes to the market rules, taking the public interest into account.

Difficulties for the transmission company

The transmission company, Transpower, has encountered a variety of difficulties in reaching agreement with transmission system users. Some users have disputed the transmission system standards that Transpower applied, because they did not want to be charged with the costs of installing the necessary equipment at the point of connection in order to maintain those standards. However, as we discussed in chapter 4, it has been in relation to the allocation of transmission charges that the greatest difficulty has arisen.

Disputes over charging methodology

Inevitably, the disputes over charging methodology were difficult to resolve, and, in 2001, Transpower took a state-owned generating company generator to court. It had not paid any transmission charges since 1999, when the industry was restructured and had never signed a contract with Transpower because of the disagreement over the charging methodology. Eventually, the dispute wound up in court, because there was no other mechanism to deal with it. However, the court was unable to provide relief to either party, because it came to the conclusion that there was no agreement or meeting of minds that could be said to amount to a contract over which the court could claim jurisdiction and provide relief.[28]

The sad consequence of this tale is that even where there appears to be equal bargaining strength, adopting the private contract route may not work. In this case, it is the monopolist that appears to need the protection most because it cannot deny service (for technical, system, and political reasons). In the case of transmission, the parties to private contracts need a regulatory backstop and also regulated service standards.

Transmission expansion in New Zealand

It is not clear that the present industry arrangements are effective to encourage or facilitate transmission investment. Although New Zealand has a nodal pricing system, it has not yet introduced financial transmission rights (FTR). Transmission expansion is perceived to create "winners" and "losers", with some market participants benefiting from the investment, while others are disadvantaged by it. Although the transmission system was relatively robust in the past, expansion is becoming an issue. Under its legal framework as a state-owned enterprise, Transpower must seek a commercial return on any investments in the transmission system that it makes. However, under this private contract regime, it will be taking a significant usage and revenue risk, unlike its regulated counterparts. In essence, it is being asked to own and expand the system on a purely merchant basis, with no means of securing its revenues or a reasonable rate of return. The key ingredients for a market-based approach that we looked at in chapter 5 are not in place, particularly FTRs, and there is a very evident free rider problem.

Before the Commerce Amendment Act 2001 (No. 2), Transpower had no means to enforce the recovery of its costs. The legislation gave Transpower interim powers of enforcement for its current sunk-cost pricing methodology. It did not give powers to enforce a methodology for the efficient recovery of new investment costs. Under Transpower's present economic reporting framework, coupled with the application of the optimizing deprival value rules discussed earlier, any investment in transmission has an expected net present value (NPV) equal to or less than zero. In other words, the investment would not be expected to return its cost of capital.

As we have seen, there is no process that is supported by the industry to identify efficient investments and agree on the cost allocation for them. Transpower accepts that these processes should facilitate commercially-based investment decisions by market participants, but points out that if Transpower is to be used as the transmission provider of last resort, it will need some regulatory backing to facilitate and enforce decisions to carry out investments that yield a "net public benefit" as well as to enable it to allocate and recover the investment costs. It also observes that the introduction of financial transmission rights is necessary in order to facilitate investment.[29]

In the next chapter, we will discuss the design of a better recipe for regulated transmission expansion than the "no regulation" approach that we have just observed and a recipe with superior incentive properties to rate-of-return regulation. We shall bear in mind the other regulatory approaches that we have discussed.

NOTES

1. Stephen C. Littlechild, "Regulation of British Telecommunications' Profitability," Report to the Department of Industry (1983), para. 1.11

2. Daniel Yergin and Joseph Stanislaw, *The Commanding Heights: The Battle Between Government and the Marketplace That Is Remaking The Modern World* (London: Simon & Schuster, 1998)

3. See for example Paul L, Joskow and Richard M. Schmalensee, *Markets For Power* (Cambridge, Massachusetts: The MIT Press, 1983), chapter 3 and W. Kip Viscusi, John M. Vernon, and Joseph E. Harrington, Jr., *Economics of Regulation and Anti-Trust* (Cambridge, Massachusetts: The MIT Press, 1998), chapters 11 and 12

4. José Gómez-Ibáñez of Harvard University, *Flexibility and Commitment* (forthcoming book). He argues that even small, domestic consumers have made both tangible investments (homes and businesses) and non-tangible investments (schools and jobs)

5. According to Ronald H. Coase, all the potential gains from trade could be realized through negotiation in a world in which there were no costs to negotiating, monitoring, and enforcing transactions. Ronald H. Coase, "The Problem of Social Cost," *Journal of Law and Economics,* vol. 3 (October 1960), pp. 1-44. He also argued that the solution would be efficient no matter how the property rights that were involved were defined

6. The concession also protects the consumer from insolvency of the concessionaire. A liquidator or administrator would not be permitted to dispose of the assets.

7. Paul L. Joskow, "California's Electricity Crisis," Massachusetts Institute of Technology (September 28, 2001). Jerry Taylor and Peter VanDoren "California's Electricity Crisis: What's Going On, Who's to Blame and What to Do," *Policy Analysis No. 406* (July 2001)

8. Regional Transmission Organizations, Notice of Proposed Rulemaking: Comments of Professor Paul L. Joskow, Docket No. RM99-2-000 (August 13, 1999)

9. Steven Stoft, "The Effect of the Transmission Grid on Market Power," University of California, Berkeley (May 1997)

10. Alfred E. Kahn, *The Economics of Regulation: Principles and Institutions* (New York: John Wiley & Sons, 1970-71), vol. I, pp. 45-54. Colin Meyer and John Vickers, "Profit Sharing Regulation-An Economic Appraisal" (September 1995) mimeo. They describe profit measures as "subjective, open to manipulation, and prone to circularity," p. 1

11. Joseph S. Graves and John D. Clapp, "The Future of Electric Transmission," *The Electricity Journal* (December 2001). See also Shimon Awerbuch, Leonard S. Hyman, and Andrew Vesey, *Unlocking The Benefits of Restructuring: A Blueprint for Transmission* (Vienna, Virginia: Public Utilities Reports, Inc., November 1999), chapters 1 and 2

12. Michael Cragg, William Lehr, and Ron Rudkin, "Assessing the Cost of Capital for a Standalone Transmission Company," *The Electricity Journal* (January/February 2001)

13. David Currie, "Mutualisation and Debt-Only Vehicles: Which Way for RPI-X Regulation?" The Beesley Lectures on Regulation, Series XI, The London Business School (November 13, 2001)

14. Shimon Awerbuch, Leonard S. Hyman, and Andrew Vesey, *Unlocking The Benefits of Restructuring: A Blueprint for Transmission* (Vienna, Virginia: Public Utilities Reports, Inc., November 1999), chapter 2 supplement

15. Paul L. Joskow and Richard M. Schmalensee, "Incentive Regulation for Electric Utilities," *Yale Journal on Regulation*, vol. 4:1 (1986), pp. 1-49

16. Harvey Averch and Leland L. Johnson, "Behavior of the Firm Under Regulatory Constraint," *American Economic Review* 52 (December 1962), pp. 1,052-69 George Stigler and Claire Friedland, "What Can Regulators Regulate? The Case of Electricity," *Journal of Law and Economics* 5 (1962), pp. 1-16

17. As Joskow and Schmalensee point out, the literature somewhat overlooks the incentive properties of regulatory lag, which is the term that described the tendency of regulators to respond slowly to changes in costs. As a result of regulatory lag, the rates of return earned by utilities may be above or below the permitted rate. When prices are fixed, the utility can increase its profits by cutting costs. Paul L. Joskow and Richard M. Schmalensee "Incentive Regulation For Electric Utilities," *Yale Journal on Regulation*, vol. 4:1 (1986), pp. 1-49

18. Paul L Joskow and Richard Schmalensee, "Incentive Regulation for Electric Utilities," *Yale Journal on Regulation*, vol. 4:I (1986), pp. 1-49

19. Daniel Yergin and Joseph Stanislaw, *The Commanding Heights: the Battle Between Government and the Marketplace that is Remaking the Modern World* (London: Simon & Schuster, 1998), p. 350

20. I am grateful to Richard Morse of Goldman Sachs for discussing this approach with me

21. I have Professor Daniel Kirschen of the University of Manchester Institute of Science and Technology to thank for this observation

22. Goran Strbac and Ron N. Allan, "Performance Regulation of Distribution Systems Using Reference Networks," *Power Engineering Journal* (December 2001)

23. It is based upon the experience of 24/Seven Utilities Limited, which is the asset manager of the electricity distribution assets in London and East Anglia. David Jefferies and David Owens kindly explained 24/Seven Utilities Limited to me. See also Robin Cohen "Regulating energy: a new era?" *Economics Business,* n.d. (Arthur Andersen publication)

24. David Currie, "Mutualisation and Debt-Only Vehicles: Which Way for RPI-X Regulation?" The Beesley Lectures on Regulation, Series XI, The London Business School (November 13, 2001), pp. 6-7

25. Supplement to *Utility Week* on 24 Seven Utilities Limited, n.d.

26. Cary Coglianese and David Lazer, "Management-Based Regulatory Strategies," Regulatory Policy Program, Center for Business and Government, John F. Kennedy School of Government, Harvard University (July 2001)

27. The Government Policy Statements with respect to electricity can be found on www.med.govt.nz/ers/electric.html (February 2002)

28. Transpower New Zealand Limited vs. Meridian Energy Limited, Reserved Judgment of Fisher J. (June 6, 2001), in the High Court of New Zealand, Auckland Registry, CL45/99. "The simple fact is that Meridian never agreed to Transpower's terms"

29. Transpower New Zealand Limited, "Submission to the Post-Winter Review of the New Zealand Electricity System" (October 2001)

CHAPTER 7

The Setting Agent II: Performance-Based, Incentive Regulation

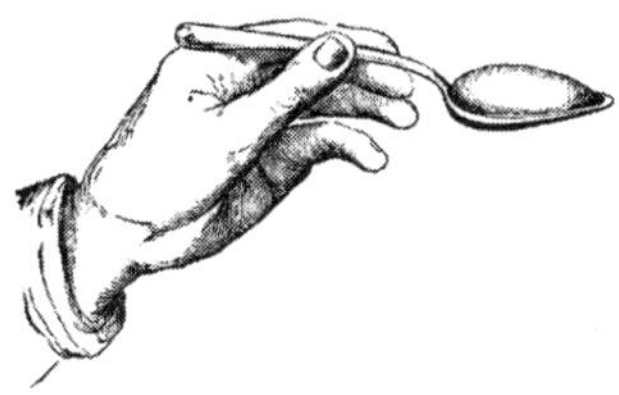

Ingredients of this Chapter

This chapter will discuss the concept, objectives, and guidelines to be followed in using performance-based or incentive regulation. We shall then look at the experience with several different performance-based approaches. After discussing the review periods and transitional issues, we will look at the experience of different approaches in the U.S. and England and Wales. The chapter will end with a description of the best available recipe for a regulatory approach to incentivize transmission expansion. The supplement to this chapter describes the different approaches to regulating for transmission expansion in the U.S. and the U.K.

The concept of incentive regulation

The shortcomings of rate-of-return regulation in general, and its inability to stimulate efficiency gain, in particular, led to the search for alternative methods of regulation in order to simulate some of the effects of market forces. Despite the incentive properties of rate-of-return regulation that were felt to have encouraged investment (over-investment in some cases) and the incentives to cut costs as a result of regulatory lag, more effective incentives were sought. The result of the search was what we now know as incentive or performance-based regulation.

This is taken to mean that the regulator delegates certain performance-related decisions to the firm and the profits of the regulated firm depend upon performance measures made by the regulator. The regulator does not attempt to control or micromanage behavior, but, instead, rewards outcomes.[1]

Incentive regulation was designed to overcome, to some extent, the regulator's inherent lack of information and to make use of the utility's information advantage and profit motivation. The regulator inevitably has less information than the regulated utility, despite its wide information-gathering powers. It needs a range of analytic and modeling skills that its budget will not often allow. It is particularly difficult to obtain information as to the potential for cost savings and efficiency gain. The economists hoped that incentive regulation would reduce the regulator's need for information. However, as we shall see, the reality has proved to be somewhat different.

Incentive or performance-based regulation, by contrast to rate-of-return regulation, is primarily designed to create performance indicators and incentives to reduce cost, achieve efficiency gain, and improve performance. However, it can be used to attract investment.

Part of the trend toward deregulation

Although incentive regulation can be broadly applied in different scenarios, it has been part of the trend toward privatization, liberalization, and deregulation. It has been taken to represent a move toward competition, because it frees regulated firms from some major constraints. In the U.K., incentive regulation has been associated with privatization and the regulation of many of the public utilities, whereas, in the U.S., it has facilitated liberalization and partial deregulation, particularly in telecommunications. Because of the existing rate-of-return regulation, the shift to incentive regulation was more difficult in the U.S. and led to more compromises than in countries that were starting with a blank sheet of regulatory paper. It is also more diverse in the U.S. than in other countries.[2]

The movement to incentive regulation was also the result of technology and demand developments, driven by the adverse results of rate-of-return regulation and the observation and belief that markets could perform better. There was also a conviction that incentives based on the profit motive would work best of all.[3]

The objective of incentive regulation

The objective is to achieve an improvement in performance over the status quo, converging to an optimum over time. This process may, however, be affected by externalities to which the mechanisms can only react with a lag. The most common form of incentive regulation is price cap regulation, which was based on Stephen Littlechild's report on British Telecommunications in 1983. It was consistent with earlier literature.[4] There are other techniques, such as the revenue cap, the sliding scale (earnings sharing), menus (in which the utility is given a choice), the discounted cost pass-through, yardstick regulation, performance target schemes, rate freezes, and the rate case moratorium. As none of these are mutually exclusive, it is always possible to invent a hybrid model that utilizes more than one of these techniques and that can incorporate rate-of-return regulation to some degree.

No single incentive scheme is likely to be optimal

In 1986, Paul Joskow and Richard Schmalansee observed that the theoretical literature had not yet produced a neat set of cookbook rules that could be readily applied with the benefit of available empirical information to develop optimal or even good incentive mechanisms for electric utilities.[5] There was a clear indication that the relationship between the regulator and the regulated firm was dynamic and that they were playing a game over many periods. The firm had an inherent incentive to try to fool the regulator, perhaps even raising costs and sacrificing profits in order to make the future reward or penalty structure more favorable.[6]

They concluded that no single incentive scheme was likely to be optimal in all circumstances and that the appropriate incentive scheme for any particular utility might change dramatically over time, as circumstances changed and the regulator's information improved.

Guidelines for the design of incentive schemes

Joskow and Schmalensee offered a number of general guidelines for the design of incentive schemes:[7]

- It is useful, at least partially, to decouple the revenue that the utility is allowed to earn from actual accounting costs because pure cost-plus regimes are never satisfactory (although regulatory lag automatically accomplishes decoupling to some extent). The optimal decoupling will depend directly upon whether the regulator has the ability to determine efficient cost and, indirectly, the economic and technological uncertainties the utility faces
- The design requires careful definition of the regulator's objectives and priorities, what information it has and will need, and the nature of the uncertainties about cost and demand
- The revenues that the utility is allowed to keep under an incentive scheme should, ideally, be based on comprehensive measures of performance. If cost minimization is the objective, it is not helpful to tie an incentive to the performance of a particular facility or to the costs of a particular service element rather than total costs. Otherwise, the utility will improve only the performance measure on which it is monitored or graded at the expense of other aspects of its performance[8]
- Rewards and penalties should be tied closely to outcomes or performance that are subject to managerial control, because incentive schemes work best by inducing management to make efficient decisions. There is no point in rewarding or penalizing management for events they can do nothing about
- The utility should be allowed to earn profits above its cost of capital (permitted rate of return) in some circumstances and less

than its cost of capital in others. In this way the firm can expect, at least, to recover its costs on average over time[9]

- For political reasons, incentive schemes that allow utilities to earn very high or very low profits will not be acceptable or credible. However, the utility may make no effort to improve performance if it believes that it will not be allowed to earn high profits for superior performance. It is important, then, to convince the utility that the incentive scheme will actually be followed (*i.e.*, there will be strong regulatory commitment, even if the rewards and penalties are limited to politically-acceptable levels)
- Where there is a regulatory regime already in place, incentive schemes must mesh well with current regulatory accounting principles, because, at least in the transitional stages, comparisons with traditional procedures will be made. However, the move to incentive regulation often involves improving on the existing accounting system, particularly in the case of transmission, for which separate accounts may not have been kept prior to the unbundling or restructuring of the industry
- Even in theory, optimal incentive schemes cannot produce perfect performance. Regulation is inherently inferior to competition, and poorly designed incentive schemes may result in outcomes that are worse than traditional regimes. Incentive schemes should be evaluated in the context of, and integrated with, unavoidable regulatory lag and disallowances of imprudent expenditures

Low- and high-powered incentives

The power of the incentive can be low or high. A high-powered incentive is one in which the firm bears a high fraction of its cost at the margin. A fixed-price procurement contract is an example of a high-powered incentive scheme. A cost-plus procurement contract would be an example of a low-powered incentive scheme, because the firm is not made accountable for cost overruns or savings. Accordingly, the higher the power of the incentive, the greater the value of the outcome to the consumer or to society, but on the other hand, the higher the payment to the firm to achieve it. High-powered incentives may be attractive to utilities, or they may find them very challenging. Much will depend upon the availability of information to the regulator and whether it is in a position to set a high-powered incentive. If it has limited information, it may only be able to set a low-powered incentive, because of the danger of being criticized for allowing the utility to earn excess profits from a poorly-set incentive. Sometimes incentives are described in terms of being "soft" or "hard."

There are three main disadvantages of high-powered schemes:

- It is costly for the firm to supply quality. As a result, service levels may deteriorate
- The regulatory commitment is generally much shorter than under other forms of regulation (*e.g.*, five years under price cap regulation, two years under a performance target scheme, and it may be even shorter if the firm makes substantial profits)
- They create a large benefit for the firm from capturing its regulator[10]

Penalties vs. incentives

There is often an immediate reaction that penalties and incentives necessarily go together—like salt and pepper. If there are incentives to perform, there must also be penalties for non-performance. However, we know that carrots work better than sticks. Appealing to the profit motive in the form of a financial incentive that could involve accepting a financial downside[11] as well as an upside is more likely to result in the desired performance than a system of penalties.

Penalties are notoriously difficult to set at the right level and tend to result in dispute. As we have observed with respect to Argentina, if they are set too high, they will be seen to be unfair and will deter investment. If they are set too low, it may be cheaper for the utility to pay the penalty than to perform appropriately. Then there is the question of what is to be done with the money that is collected in the form of penalties. A penalty is a form of punishment—not compensation to those who have suffered loss. It is a complex matter to return it to the consumer, not simply because of the administrative aspects, but it is not always easy to determine which consumers have suffered what loss as a result of the failure to perform. As a result, the mechanism may be quite arbitrary. Some jurisdictions have established rules for the payment of compensation on a single price or fixed basis for the duration of interruptions in transmission service (Argentina and Norway).

Price Cap Regulation

The concept of price cap regulation

In simple terms, the price cap or RPI-X formula operates by setting an opening price for a specified period (typically 5 years), after which it will be adjusted. The formula allows the utility to increase its prices each year by a rate equal to the increase in the rate of inflation over the previous year, less a percentage designed to create an incentive to increase efficiency or cut costs. Performance standards are usually set so that quality of service is maintained. "RPI" refers to the English-published consumer retail price index that is used to reflect the economy-wide level of input prices in preference to an index reflecting increased costs to the utility. It is an expression of the increased wealth of consumers and, therefore, the prices they can afford. A "Y" factor can be added to allow for the pass-through of specific cost items that may be outside the control of the utility or that reflect required investment.

Elements of the price cap formula

In order to put in place RPI-X regulation, it is necessary to establish the following:

- the opening price (P_0) (which can be rebased, at the time of each review)
- the efficiency deduction (X) factor
- the cost pass-through (Y) factor (if any)

Contrary to the original hopes of governments and regulators that applied this model, in order to carry out this exercise, a significant amount of data and analysis is needed, both initially and when the periodic review is carried out. It involves looking at cost data (*e.g.*, of operating and capital expenditure) and asset registers, which utilities may not initially have where transmission is being unbundled. Even where the data are readily available, some cost items may prove controversial (*e.g.*, cost of capital), and there are various approaches to the setting of the X factor over which the regulator and the transmission owner may well disagree. Given the long lives of the assets, it may be difficult to assess the difference in the value of an asset that is 25 years old and one that is 40 years old.

Effect of a price cap formula

The approach should result in the setting of an X that reflects a level of efficiency gain that the utility has a reasonable chance of beating (so that it is not demotivated to make the attempt by the fear that it will never come close to the target and, therefore, have nothing to keep for itself). Ideally, it should also be set at a level that passes a reasonable percentage of the efficiency gain (*i.e.*, equal to the percentage represented by the X factor) to the consumer, in the form of the limit on price increases.

As Michael Beesley and Stephen Littlechild explain, the RPI-X formula does not assume that costs and demand are given or known—the objective is to provide adequate incentives for the regulated company to discover them.

> *The aim is to stimulate alertness to lower cost techniques and hitherto unmet demands. The regulator's priority is to ensure that the laggards improve to match the (observed) performance of the leaders, while providing sufficient incentive for the leaders to stay ahead and blaze the way for the next round of target setting.*

However, they go on to point out:

> *Conversely, where there is only one company in an industry, the regulator is more dependent upon the company for information, and his effectiveness in bargaining for productivity improvements is thereby reduced.*[12]

This will be relevant in the context of transmission.

Attempting to beat the X factor involves changing management and operating behavior. The utility must be prepared to relax unnecessary or bureaucratic constraints on management and to give the staff the freedom to act more commercially and with a greater degree of responsibility. Under a rate-of-return regulation, the activities of the regulator are sometimes characterized as micromanagement of utility operations.

The review period is an important element of price cap regulation, because the utility will be unlikely to be incentivized to cut costs, increase efficiency, or invest if it fears that the benefit it achieves may be clawed back or given to the consumer by virtue of resetting the X factor within a short period of time. On the other hand, the regulator needs to be able to measure performance and check periodically to see that the utility is not achieving an easy benefit for itself that the consumer paid for in the past and in all fairness ought to be returned to the consumer.

Success of price cap regulation

Price cap regulation has been highly successful in achieving cost reductions, particularly in telecommunications, where it has been extensively used. The cost-reducing incentives of price caps appear to be both stable and viable. In order to be effective, the incentives have to hold over a period of time (at least five years). The strength of the incentive is usually measured by the percentage of cost reductions the utility is allowed to keep. However, in the long run, stability tends to interfere with viability, because of unforeseen events such as economic downturns and technological advances. If the regulated firm becomes insolvent or makes unconstrained monopoly profits, the political fallout will be considerable. However, the utility should be protected, provided the regulator abides by its commitment not to interfere during the relevant period.

Price flexibility

In principle, RPI–X allows any price charged by the utility to be changed at any time, subject only to the price cap on the price in question or the average price within a specified range of prices for different services, usually called a "basket". This will create an incentive to use efficient pricing methodologies, such as Ramsey (also known as Ramsey-Boiteux) pricing, which allows certain costs to be recovered from the most inelastic demand.[13]

There is theoretical support for the claim that price caps result in efficient price structures.[14]

Price caps are also attractive because they create the flexibility to change relative prices in the regulated basket of services and can be combined with a weighting scheme that promotes price rebalancing towards more efficient price structures. The use of baskets has been helpful in enabling the utility to rebalance its prices in the direction of allocatively superior prices, and has enabled regulated utilities to compete with new entrants.

Setting the X factor

As Beesley and Littlechild put it: "There is nothing unique, optimal or mechanical about the initial choice of X."[15] Although it appears to have been set and, indeed, reset quite arbitrarily, in practice, there were extensive audits of the potential for efficiency gain in the U.K. electricity industry at the time of restructuring, and a great deal of information is analyzed and modeled in the review processes. Even with the introduction of benchmarking in setting the X factor,[16] from the inside, the exercise feels to be a negotiating art rather than a regulatory science.

However, the price cap formula will also involve using an inflation adjustment that will require the X factor to include both the productivity adjustment and the adjustment for the expected difference between the relevant input prices and inflation.[17] Littlechild and Beesley explain that, essentially, the RPI–X embodies "forecasts of what productivity improvements can be achieved and what future demands will be and is set on the basis of predicted future cash flows."[18]

Typically, the X factor will be seen as a measure of efficiency gain in the form of cost cutting. It is less often taken to mean a measure of productivity improvement. Setting the X factor involves a number of issues beyond productivity. It is based upon judgment.[19] If the X factor is simply coupled with total firm productivity, the potential benefits from flexibility in price setting and revenue growth would be lost.

A political and commercial bargaining process

The level of the X factor is the product of negotiations between the regulated company and the regulator, which can be wide-ranging, and the negotiations are often used by the regulator to obtain detailed information. Littlechild and Beesley go on to explain:

> *In short, X may be thought of as one of several variables in a political and commercial bargaining process. Our claim here is simply that the UK approach offers greater and more direct scope for bargaining, with a correspondingly more active role for the regulator.*[20]

The dilemma the regulator faces is that if the X factor is set too high, it will result in confiscation of the utility's property that will be both unfair and inefficient, particularly if the maintenance and expansion of the infrastructure are important. Setting the X factor too low will dilute the incentive effect, and easy efficiency gains may not be achieved.

It may be necessary to take into account externalities and events over which the utility has no control in setting the X factor. For example, the pass-through of costs of repairing hurricane damage, undergrounding of transmission lines imposed by a siting or permitting agency, or the effects of changes in accounting or tax law could be permitted.

The utility will focus on the rate of return

Although the regulatory effort will focus on the setting of the X factor, the utility will focus on what it means as far as the rate of return is concerned. The negotiations with the regulator will cover many of the issues we discussed in chapter 6. It could be said that price cap regulation involves an implicit rate of return and an explicit review period, and that rate-of-return regulation involves an explicit rate of return and an implicit review period. Transmission owners will be incentivized to invest by a negative X or a Y cost pass-through. However, they will also argue for higher permitted rates of return, if those rates pass the capital markets test of being sufficient to attract new capital for transmission investment, taking into account the risks involved and the current business environment. Shorter depreciation lives would reduce risk for investors, because they cannot predict what future events over the decades of the transmission asset lives might ultimately affect them. They will argue that shorter depreciation lives on a present value basis would leave transmission customers indifferent. Although the regulator should have in mind the capital attraction goal and the firm viability constraint we identified in chapter 6, it is primarily focused on the efficiency goal and the rent extraction goal.

Review Periods

Institutionalized regulatory lag

A key ingredient to the success of the price cap approach was seen to be the relatively long period for which the formula would remain fixed. The idea was not only to minimize the regulatory burden, but also to give the utility enough time to utilize whatever commercial freedom it needed to achieve efficiency gains and maximize profits. In essence, the approach institutionalized regulatory lag.[21]

Longer review periods

The typical review period that regulators have applied in the case of transmission price caps is five years. Shorter periods tend to reduce the incentive properties Many transmission owners would argue that there is a case for longer review periods, given the long lead in times and the need for regulatory certainty in order to minimize the cost of debt and equity. Their reasoning would probably not be motivated by a desire to respond to the properties of an incentive, but more by the desire to avoid a review and the risk that the regulatory deal might not be so good after the review. However, if the review period were significantly longer, they would be uncertain of the commitment of the regulator not to reopen the formula, particularly if there was a prevailing view (political or otherwise) that they were doing "too well" and making excess profits.

However, most regulators will feel they are under duty to make periodic reviews in order to monitor performance and assess whether the regulatory regime is delivering the outcome the consumer needs. The regulator may be justified in making adjustments to reflect the rent extraction goal, if it feels the balance is too much in favor of the transmission owner, and to apply the firm viability constraint and redress the balance in favor of the transmission owner if the regulator feels, for example, that the required level of investment is not forthcoming.

Regulatory commitment and flexibility

The commitment of the regulator to a fixed review period is of fundamental importance to investors and a key element of the regulatory risk that affects the cost of capital. The electricity regulator in the U.K. reset price caps and X factors in August 1994 for distribution utilities for the usual five-year review period. In response to public pressure, because the companies sustained higher earnings and because the regulator had insufficient information at the time, they were reset again in March 1995 just before they would have come into effect.[22] Although the action caused a great deal of controversy and debate, it created relatively little impact on the market value of the utilities as evidenced by the share prices and the fact that the distribution utilities proved to be attractive acquisition targets for the U.S. and other investors.

The answer to the problem of regulatory commitment and regulatory flexibility could lie in allowing the pass-through of costs outside of the control of the utility and in creating an effective mechanism for intermediate or emergency reviews initiated by the regulator or the utility. Some strong objective test would need to be applied so that the regulator could not utilize the possibility to renege on its regulatory commitment and the utility did not make capricious applications for relief. The kinds of tests that come easily to mind are that the regulator could be required to prove a material adverse economic effect on the consumer, and the utility would have to establish that its cost of capital had materially increased. However, these are very broad in their scope, they lack precision, and they would be open to interpretation and dispute, but they might be better than nothing.

Coupling price caps with profit sharing

Ingo Vogelsang with Crew and Kleindorfer suggest that the uncertainties surrounding price caps and review periods could be mitigated by combining the price cap with a profit-sharing scheme. If profits rose to a level regarded by the regulator or consumers to be excessive or if the transmission owner considered a loss or cost to be unfair, a formula could be devised to enable the transmission owner to share the upside and downside equitably with the consumer. We shall discuss the general principles of profit sharing under the heading "sliding scale regulation" later in this chapter. They point out that raising the consumers' profit share reduces the transmission owners' risk and also its efficiency incentives.[23]

Slowing of efficiency gain

As the end of a review period approaches, the behavior of the regulated firm can be expected to change. There is discernible behavior in some countries indicating that utilities will slow their efforts to achieve cost savings and efficiency gain, fearing that it will be clawed back by the regulator (either in resetting the X or Y factors or, indeed, in rebasing the P_0) and be passed over to the consumer. The U.K. energy and water regulators have started using what is known as a "glide path," enabling utilities to retain the benefits of cost savings and efficiency gains under programs initiated before the price cap review. The retention is permitted for a period of five years from the time it was achieved. Operating costs are being taken out of the price review, and a moving average is used instead.

Information requirements

The review process requires a significant amount of information and is also a renegotiation between the regulator and the utility. It is as much a negotiation process as a review of the price cap formula and the prices charged. The whole idea of the review period was intended to lighten the burden of regulation. Experience has shown that it involves considerable effort to analyze performance potential, to assess the appropriate performance indicators for the following period, and to negotiate them with the utility. The review can begin between 18-24 months before the end of the review period. As a result, there is not much time between review processes. The regulator sends out consultation papers on the key issues for comment by the utilities and stakeholders. The utilities submit business plans and investment programs and explain their special circumstances. Everyone engages consultants. This state of affairs appears to argue for two things—longer review periods (which are unlikely to be put in place for the reasons already explained) and some pre-agreed ratchet or negotiating parameters within which the new performance indicators (and the opening prices) can be reviewed and set, which we discussed in chapter 6.

Ratchets and negotiating parameters

It is always difficult to agree ratchets and formal negotiating parameters some years in advance, because it is never clear what the circumstances and needs will be at the time. It is unlikely to be relevant to utilize past operational experience as a guide. Past performance is seldom a good indication of the potential for future performance and efficiency gain, particularly when the easy costs savings and efficiencies have been achieved in the early stages. However, as negotiators of long-term contracts have experienced, it may be possible to establish some negotiating objectives and statements of intent that could even be given binding force in order to increase certainty of outcome.

Application of price cap regulation to vertically integrated utilities

Price cap regulation was only intended to apply to monopoly services, but it does have useful properties in dealing with a competitive environment where the regulated utility is vertically integrated to some extent. It allows the competitive outputs of the regulated firm to be separated from monopolistic outputs. Price caps can be restricted to a subset of the firm's outputs, such as transmission.

As already discussed, price cap regulation could drive down the costs of incumbent transmission owners to such an extent that other investors may not be able to compete. In this case, price cap regulation could have the effect of deterring new entry if the circumstances are such that incumbent and new entrants are in direct competition to expand the system. Vogelsang concludes that:

> *the reduction in entry and market share of rivals caused by the aggressive behavior of an incumbent utility under price cap regulation could well result in a better market outcome than under unregulated competition.*[24]

Experience from the England and Wales water industry

In the context of transmission expansion, we need to focus on two main goals. One is to create incentives for long-term investment, and the other is to achieve shorter term efficiency gain. Therefore, the challenge is how to set and reset the X factor to do this without being reliant upon measures of past performance and, thereby, undermining the price cap's efficiency incentives. These were exactly the regulatory goals when the water industry in England and Wales was privatized, and the experience is instructive for our purposes. Because there were a number of comparable water companies, it was possible to develop independent "yardsticks" that were outside the direct control of each company, but still reflected actual or best practice among roughly-comparable water companies. The X factor could be based on the average for the industry as a whole, so that each company would have a natural incentive to beat the average. In the U.S., where there are many transmission owners, this approach would be possible. Where there is only one, international comparators would have to be used, which, as we shall discuss later, is more difficult.

Relevance of the U.K. water industry to transmission expansion

The experience with the England and Wales water industry is interesting from the point of transmission expansion because, at the time that it was privatized, new European, mandatory water quality standards had been brought in. They required a significant investment program on the part of each of the water companies, if they were to be able to meet those standards within the required periods of time. Some of the coastal and bathing water standards were quite onerous, particularly for a company that had a very small percentage of the country's population, but a substantial proportion of its coastline. For example, South West Water had 3% of the population and nearly one third of the coastline.

The water companies embraced the investment programs enthusiastically, because they thought that they could build new water and sewage treatment plants for less than the government's estimates. The new plants would be included in their regulatory asset base in the future that would be used as a basis for calculating prices. The water companies were given an RPI+K formula that incorporated the usual X element creating an efficiency incentive and a Q element (Y factor) allowing the pass-through of the costs of mandatory investments to improve water quality and environmental protection. The regulator took the view that the larger the scale of the investment obligations under the Q element, the more severe the incentive to achieve efficiency gain should be under the X element, if it was to put a downward pressure on prices.[25]

Use of benchmarks in setting the X factors for U.K. water industry

The water regulator could not simply base the X factor on the average rate of cost reduction for the industry as a whole. It was necessary to take into account that the potential for efficiency gain might vary, because of circumstances beyond the control of a particular water company, such as topography and population density. As a result, it established the benchmark of an "efficient firm" by adjusting costs to reflect the relevant local circumstances and using the "efficient firm's" performance as a benchmark for others. It was not particularly useful for regulating the most efficient firm (another technique described later was used for that), because it was not clear how much potential there was for greater efficiency in this particular case, but the regulator was able to assess how much the other companies could improve just by catching up. This exercise involved collecting more data, in more detail, and ensuring that the data were submitted in a form where comparisons could be made. It also involved building sophisticated models and statistical techniques.

The resulting analysis was then circulated to the industry for comment. This process created more credibility for the X-setting process and the degree of objectivity than had been the case in the past. This was regarded as helpful, even though the benchmark analyses were extremely controversial at the time they were introduced or revised. However, the regulator was able to justify the detail of all the variables that were used, and the degree of sophistication and understanding of the issues affecting costs in the water industry was impressive. Changes were made to the variables, equations, modeling, and analysis in response to industry comments. This approach certainly did not reduce the burden of regulation.

The benchmarking was originally applied only to operating expenses in the 1994 review, but in the 1999 review, capital maintenance and construction activities were benchmarked as well. This was important, because those activities accounted for approximately half of the water industry costs. The variables used in the analysis were kept under close review in order to reflect changes in the industry and to improve the accuracy of the cost models.[26]

Benchmarking against other comparable industries

The water regulator also attempted to estimate the potential for efficiency gains in the best performing companies by making comparisons with the best companies in other, comparable industries. If the potential for efficiency gain was greater in water than in the average of the other industries, then the "efficiency frontier" component of the X would be negative. If not, it would be positive. The estimate was made from trends in real cost per unit of output in other, comparable industries. The scope for disagreement on how unit costs should be measured, over what periods of time, and for which comparator industries was considerable.

Regulation of investment plans

The water companies were subject to the detailed rules regarding the preparation and approval of investment plans, and the regulator was sometimes accused of micromanagement, although the general feeling was that the regulatory regime for water was being well handled,[27] even though there were controversies over the shareholder dividends being too high. Because of the introduction of the Q element that was designed to incentivize investment to meet the new European water standards, prices obviously rose in the early days as the investment plans were implemented. However, the price cap did create an overall incentive for water companies to procure the new plants competitively and economically, because, under the overall price cap formula, they were allowed to keep the cost savings they were able to make.

Conclusions from the U.K. water industry that are relevant to transmission expansion

The conclusion from the England and Wales water company experience is that the price cap formula can be used to incentivize investment and to minimize the cost of that investment. However, it is fair to say that it requires the regulator to adopt an active approach in monitoring and approving investment plans. In both industries, lumpiness and economies of scale must be dealt with. The experience of the use of benchmarking or yardstick competition in setting the X factor is also instructive for those jurisdictions where there are multiple transmission owners. It clearly helps to achieve greater credibility and accuracy in setting the X factor.

The water industry gives us two insights that are also apparent from the electricity industry. It is not the case that the use of price cap regulation makes asymmetry of information less of a problem. It is important to set the incentives as accurately as possible. It is clear that excess profits or losses are tolerable, and may even be necessary or desirable in the short run, because they have encouraged innovation and price reductions in the long run.

Advantages and disadvantages of price cap regulation

There is both theoretical support for and practical experience of the advantages of price cap regulation over rate-of-return regulation. It is more likely to help promote competition, protect consumers from monopoly power, reduce the burden of regulation, promote innovation, simulate some of the effects of market forces, improve the expected profitability of the regulated firm, and improve efficiency.[28]

The disadvantages are that the primary focus of price cap regulation is on acheiving efficiency and productivity gain. It needs to be supplemented by performance standards in order to avoid deterioration in quality of service and regulatory approval of investment plans. The review period encourages the utility to take a short-term view, because of the negotiating strategy and uncertainties over the outcome of the next review. Transmission investment requires a long-term view.

Revenue Cap Regulation

The nature of revenue cap regulation

Revenue cap regulation allows the total revenue of the utility simply to increase in line with an index of prices. Typically, this will be an exogenously-set retail or consumer price index (RPI or CPI). It may also incorporate increases in total revenue to reflect load growth. The technique was used in U.S., but much less so in the U.K. It is used for transmission owners in Australia. The incentive properties are such that it should encourage internal efficiency, because the utility can keep the difference between its total revenue and its total costs. Revenue cap regulation can be used to encourage investment by allowing prices to rise above the limit set by the price index. This can result in some intrusive regulation to deal with adjustments that need to be made if a project does not materialize or if an event occurs necessitating unforeseen investment.

Revenue caps in Australia

Transgrid in New South Wales is revenue capped. During each year of the five-year review period, the total revenue is adjusted by the CPI adjustment. The X factor is a negative figure of -1.3% per annum, so that its revenues can increase by more than CPI, which creates an incentive to invest. There is a significant capital expenditure program over the five years in question. Powerlink in Queensland has a negative X factor of -8%, because of an even larger capital expenditure program.

Revenue cap in Norway

Statnett is also subject to revenue cap regulation, which, from 2001, introduced a scheme where it also suffers a deduction in permitted revenues, because it has to pay compensation for interruptions. This is designed to encourage investment to improve reliability. The arrangement covers all interruptions (both planned and forced outages) in all grid components of more than 1 kV that result in any interruption lasting more than three minutes. The arrangement does not involve the direct payment of compensation to all affected customers, but Statnett can enter into agreements with a large customer (that consumes more than 400 MWh per annum) to pay compensation directly, if the customer's interruption costs can be estimated in advance. This cost will then replace the average interruption cost fixed by the regulator that is deducted from the permitted revenues. The idea is that the payment of compensation directly to a customer or through a general deduction in Statnett's permitted revenues should have the same incentive effect. The intention is that a permanent increase in the quality of service will result in increased permitted revenues, and a permanent decrease in quality will result in reduced permitted revenues.[29] It is not easy to utilize the scheme to produce clear investment signals. In order to encourage investment, Statnett needs an assurance as to the minimum rate of return it will be allowed to earn on its investment, but it does not know what that minimum would be.

Disincentive properties of the revenue cap

However, Crew and Kleindorfer argue that revenue caps can be damaging to efficiency and might be more appropriately termed "disincentive" regulation.[30] They demonstrate that where the incumbent can freely set output, it will do so in a manner that may drive price above the monopoly level.

If service quality standards are put in place and monitored, the revenue cap should provide incentives to minimize costs, including the indirect costs of transmission. However, it can be argued that it would be preferable to define the transmission utility's output more precisely in terms of the services it provides, and use it as a basis for applying price cap regulation, rather than defining an aggregate revenue cap that masks, rather than verifies, the detailed value of the utility's service offerings. The service-based incentive scheme, in theory, promotes selling more and better services to customers, rather than focusing on maintaining the asset base of the utility. This focus arises because the revenue cap is set by calculating the required return on investment for the asset base at the beginning of the period and translating it, with operating and maintenance expenses, into a revenue requirement.

Sliding Scale Regulation

The nature of sliding scale regulation

Sliding scale regulation has its origins with the London gas utilities in 1875 and was utilized both in the U.K. and the U.S. It was also applied to electricity companies. It allowed dividends to rise if prices were reduced, and required dividend reductions, if prices were increased. The fall or increase in unit costs was shared between the consumers and shareholders. It is also known as profit- or earnings-sharing regulation. It lets consumers directly participate in excess profits or profit shortfalls achieved by the utility. This participation could take the form of ex-post refunds or price reductions for future purchases, which would happen automatically when the actual rate of return exceeded the rate of return that had been predetermined to be acceptable.

Shortcomings of the sliding scale

On the face of it, this approach has some obvious efficiency and fairness properties. It has incentives to reduce costs and appears to share the savings reasonably fairly between the utility and the consumer. However, it proved to be administratively complex, because the profits had to be agreed at the end of each period. It was also difficult to operate where only some of utility's services were regulated and where there were multiple cross-subsidies.

The sliding scale was discarded

However, the sliding scale created substantial difficulties in both the U.K. and the U.S. and was largely discarded in the 1950s, when a number of utilities became insolvent or incurred losses. It operated effectively when there was no inflation and before electricity competed with gas in the U.K. It was unable to deal with the price decline of the late nineteenth century and the inflation of World War I. After World War II, nationalization put an end to sliding scale regulation in the U.K., and it was taken out of use in the U.S. shortly afterwards.[31]

The sliding scale had a number of shortcomings. It resulted in prices that were persistently too low or too high when there was inflation or recession. If input prices rose because of inflation, the rate of return would obviously decline, even if the utility was efficient in its operations. If technological change reduced the costs, the opposite would occur. Incentive schemes must take account of observable changes in input prices as well as changes in technology, and increases in demand that are beyond the control of the utility. The sliding scale was unable to do this. It was also unable to minimize the real, economic cost, because it rewarded the utility for minimizing total accounting cost. It failed to recognize the multi-product character of some utilities. It determined only the average level of prices. Increases or decreases in prices for individual services can either be linked with the average change or left entirely to the utility. Neither approach is satisfactory on its own.

Limited efficiency incentive properties

Under price cap regulation, the shareholders benefit from any productivity increases in excess of the X factor and suffer the costs of any failure to achieve X, until the formula is reset at the next periodic review. The consumers benefit from forecast productivity increases of X, whether or not they are realized. Under the sliding scale approach, the benefits of whatever productivity increases realized were shared between the consumers and the shareholders on a formula predetermined by government. The incentive to the firm to achieve the productivity increases was somewhat less than under a price cap formula. The task of specifying the price and dividend limits that the government took upon itself was more difficult than the task of predicting achievable productivity growth or efficiency gain that is required in order to set the X factor.

Recent earnings-sharing schemes

The concept behind the sliding scale does not appear to be dead. An earnings sharing arrangement coupled with rate freezes, rate case moratoria, price caps, and revenue caps has been observed over the last decade in the state regulation of U.S. electric utilities in 29 cases.[32] The advantage is that this arrangement can improve regulatory commitment, because consumers (and therefore, regulators) have a positive interest in profits. There should also be less slowing of efficiency gain before the commencement of a review period. However, if the profit sharing is reviewed in every period, the review process will be very similar to a rate of return rate case that is costly and time consuming. If the transmission owner bears the risk of long-term losses, there will need to be an adjustment mechanism to deal with inflation and input price changes. Sharing losses with consumers to a significant extent will be politically difficult. Profit-sharing schemes need a number of safeguards to protect the utility and the consumer. The right design can involve as much effort as rate-of-return regulation with incentive properties that offer no real improvement on the price cap. In the U.K., regulators have thought profit-sharing schemes to be ill conceived.[33]

Menu Regulation

The concept of menu regulation

The idea behind menu regulation is that by offering firms a choice among the two or more incentive schemes, the regulator can increase consumer welfare by implementing more demanding incentives, without raising concerns about adverse effects on firm viability.[34] The menu of options is structured so that it is in the self-interest of the regulated firm to choose the option that reveals the firm's own performance expectation. This has given rise to the name "revealed expectations" (REx) incentives.[35] The desired outcome from the point of view of the regulator is that the firm will select the highest realistically achievable performance target that should reflect the consumer interest, but also coincides with the firm's self-interest. It is difficult for traditional regulatory approaches to achieve this coincidence of private and public interests.

Under menu regulation, the regulator could, for example, set a menu of rates of return that are permitted according to the cost target or X factor the regulated firm chooses. In simple terms, the greater the cost reduction or increase in efficiency the firm agrees to, the higher the rate of return it is allowed to earn. In effect, it gives the firm an incentive to choose a higher productivity or cost-saving target. The approach is designed to enable the regulator to overcome the difficulty it faces in not having information that is as good with respect to the potential for efficiency gain as that of the regulated firm. It also gives the regulated firm an incentive to choose the highest-powered efficiency factor or cost target it is capable of achieving.

Economic literature on menu regulation

A theoretical model was analyzed by Jean-Jacques Laffont and Jean Tirole in 1986 that was designed to incentivize a utility to avoid over-stating costs and to supply accurate cost estimates and information to the regulator that would be used to set prices.[36] The utility would chose a deduction factor reflecting its level of effort. The regulator would choose a bonus factor for operating cost estimates that reflected the regulator's expectations and the fraction of the cost over-runs the utility was allowed to recoup. This would provide a means of sharing the risk of cost over-runs or under-runs with the consumers.

As a result, the utility would only recover a fraction of the difference between its actual costs and the initial estimate of that cost.

Laffont and Tirole observed that it is never efficient to put in place a pure cost-plus arrangement, and that the higher the utility's cost estimate, relative to the regulator's expectations, the lower the reward to the utility should be, but the higher the fraction of cost over-runs the utility would be allowed to recoup. Accordingly, firms that estimated in advance that costs would be relatively high would get a relatively small share of the risk of cost over-runs and under-runs. In short, the utility's revenue would depend almost entirely on its ability to estimate costs accurately and not on the actual costs. If there were significant uncertainty, the utility and its consumers would generally share the risk that costs will depart from expectations. This work supports the general principle that good incentive mechanisms involve some sharing of risks of cost over-runs between the utility and consumers. However, it also illustrates the complexity of designing good incentive schemes. The formula they analyzed required the regulator to have good information about demand conditions and the probabilities of the alternative values of the cost parameters, as well as the cost of various levels of effort to the utility.

In 1992, Crew and Kleindorfer made another proposal to couple the level of the X factor with the rate of capital recovery.[37] The menu was designed so that the consumer benefit was held constant across all the menu alternatives. The utility then chose its preferred option by assessing its own ability to carry out efficiency improvements. It was implemented by allowing the company some flexibility in choice from the menu. They proposed a relationship between the X factor chosen by the company and a sharing factor for excess profits. On the face of it, there was a combination of price cap and sliding scale regulation. However, the authors did not feel that the scheme removed all the problems of asymmetry in information between the company and the regulator.

The Crew and Kleindorfer scheme involved:

- setting initial prices prior to the application of price cap regulation

- setting the allowed rate of return including upper and lower limits upon which the rates were based and eliminating cross-subsidy
- linking the X factor with a capital recovery factor—the Z factor—which allows the company increased capital recovery if it agrees to a higher X factor
- the company being required to propose the Z factor and in return being required to accept a higher X factor that, in effect, gave the utility increased pricing flexibility in the early years, in return for guaranteed lower rates in later years
- the company assuring itself that it could reduce costs, improve service, or otherwise increase demand by achieving higher capital recovery and putting it to good use in the form of profitable investment. However, it had to back up its judgment by providing lower prices and at least the same service quality. If it failed to achieve its rate of return it had limited recourse to the regulator for rate (price) relief, which provided strong incentives for the company to examine how it could best achieve the efficiency gain

Once the agreement had been reached between the regulator and the utility, the regulator would still be involved in monitoring performance, quality, and prices, although not to any extent greater than under rate-of-return regulation.

It is more complicated for the regulator to design the whole menu effectively than the X factor alone. If the menu is structured inappropriately, the utilities will simply choose the regulator's minimum X factor, even when some other menu structure might have persuaded the utility to choose a higher X factor.

It remains important to use good information and benchmark inputs (where available) to setting the X factor and for assessing a reasonable range of attainable X factor reductions. This is consistent with the menu-driven asymmetric information results of Laffont and Tirole.[38] This scheme should motivate the utility to assess its own possibilities for providing an increased share of dividends to consumers and for still making profits. The approach should also reduce the costs associated with the regulatory processes and might even increase regulatory commitment, because the regulator has taken ownership of the menu and allowed the utility to make the final choice.

Design of the menu

The design of the menu is more critical to its success than the setting of the X factor. There are two main objectives:

- The firm must be provided with a strong incentive to achieve and exceed its performance targets
- It must be attractive for firms with a higher incremental performance potential to choose the incentive options with more demanding performance targets

To achieve this, the regulator should offer rewards that increase with performance against targets. Firms that choose higher performance targets should realize higher rewards if they achieve the target. The expected rewards should be maximized for the option that corresponds to the firm's own expectation of its performance. This will create an incentive for the firm to assess its performance capability accurately and to reveal its unbiased estimate of expected performance. However, if the firm is risk-averse, it may be nervous about taking on more risk, despite the inherent incentive to choose a higher performance target, especially if it is forced to use a new and riskier technology. However, appropriate risk premiums or capping possible rewards and penalties to avoid extreme outcomes could be built in.[39]

Practical experience of menu regulation in the electricity industry

The academic enthusiasm for menu regulation is proving infectious, and a number of regulators have been tempted to try it. The approach has been used by the Federal Communications Commission in the U.S. for the price cap regulation of local access charges. The Australian Competition and Consumer Commission has evaluated a menu of three price cap options with different productivity targets in its issues paper on the regulation of transmission service providers.[40]

Ontario distribution utilities

The Ontario Energy Board (OEB) attempted to introduce menu regulation in the electricity distribution sector in 1999 and met with much hostility. The menu was set with relatively little data collection or analysis, because the OEB had never regulated before in this sector. The approach made some sense from the point of view of regulatory burden and asymmetry of information. The OEB was about to regulate more than 250 municipal electricity utilities (MEU) for the first time, and the data gathering and analysis would have taken several years. However, the MEUs, having never been regulated before in this way, and having never had their performance measured or compared, had no information available to them on which to base their choices.

The total factor productivity varied widely from MEU to MEU. It was impossible to know whether the variation was due to superior management performance or external factors, such as consumer density or environmental differences. The MEUs argued that the menu amounted to a "one size fits all" approach that was inappropriate, given their diversity. They would have had to achieve significant productivity gain in order to meet the minimum return on equity ceiling, well in excess of historic performance levels ascertained by the OEB under a sampling process.

In any event, the MEUs had been operating under an effective price freeze for some time, which had forced cost savings to already occur. They argued that the linear relationship between the productivity factor and the return on equity was arbitrary. There were significantly higher risks in selecting a higher productivity target, but the return on equity ceiling limited only the upside and not the downside.

The OEB was forced to drop the menu of options and determined that for the first generation of performance-based regulation of the distribution sector, a single productivity factor should be used for all utilities, combined with an earnings-sharing mechanism.

NGC and the balancing services activity price cap restriction

The successor to National Grid Company's (NGC) system operations transmission services (TSS) incentive scheme (described in Chapter 3), which was called the Balancing Services Activity Price Cap Restriction, was initially put in place as a result of the application of a menu approach. The regulator felt that NGC had been "unduly pessimistic" in its forecast of system balancing costs under the new electricity trading arrangements (NETA) and was concerned that there was a significant gap between its estimates and those of NGC as to the balancing costs, which made it difficult to negotiate the appropriate cost target. In order to bridge the gap, the regulator proposed four options for NGC's incentive arrangements under NETA. The four options included performance targets, sharing arrangements as well as caps and collars, to provide a range of risk and reward performance options that were designed to mitigate the initial uncertainty faced by NGC under the NETA.[41] This incentive scheme is discussed in more detail in the supplement to this chapter.

A useful regulatory technique

The menu approach has its basis in economic and regulatory literature, but there is increasing experience of it in procurement contracts and, more recently, in the electricity industry. It has given regulators the ability to obtain information (including estimates of capital expenditure) that is less biased than it might otherwise be under rate-of-return regulation and information as to expected productivity gains for use in setting price caps. It has certainly encouraged firms, on a voluntary basis, to opt for more aggressive performance targets that, although they may be ambitious, are achievable. In short, it looks to be a useful regulatory technique to achieve mutual gains for the utility and the consumer.

The Discounted Cost Pass-Through

The concept of the discounted cost pass-through

The discounted cost pass-through is a very simple technique designed to incentivize economic purchasing and competitive procurement. Where a tariff involves the pass-through of costs to the consumer, a discount is applied as a percentage deduction from the amount passed through. The discount operates to create an incentive to reduce the costs that are passed through. For example, a regulator could apply a discount in an effort to incentivize the procurement of necessary security machines as competitively and economically as possible and allow the costs to be passed through to the consumer less the amount of the discount. If the discount is too high, it will be a disincentive to invest in the security machines. If it is set too low, the utility will achieve an easy gain at the expense of the consumer.

Experience of the discounted cost pass-through

In Norway, Statnett is allowed to pass through only 50% of the costs that it necessarily incurs to accommodate incremental load growth. Clearly, it is unrealistic to expect it to be able to beat a target set at a level that assumes that all equipment and services can be procured at half price. The discount can only be set realistically at a very small percentage, or by reference to standard industry discounts.

Northern Ireland Electricity, which was restructured using the so-called "single buyer model" has a discount factor applied to the power procurement costs that it is entitled to charge under its bulk supply tariff. This is designed to create an incentive on it as the single wholesale buyer in the Province to procure power as cheaply as possible.[42]

This technique has limited application outside the procurement context. It can operate unfairly if the discount is not set realistically to create a cost-saving target that the firm is actually capable of beating. It can really only work in an environment where the achievable discounts can be predicted with some certainty.

Yardstick or Benchmark Regulation

The concept of yardstick or benchmark regulation

Yardstick regulation involves comparing the performance of utilities in the same class in particular areas and benchmarking them against each other or against some external, efficient benchmark. The prices the utility can charge will depend upon the benchmarks created as a result of this comparison. The approach can be formal or informal. Informal yardstick regulation involves performance targets set by reference to the best-observed performance or best practice of a peer group. The formal approach involves setting a specific performance or cost target by reference to the performance of another utility in the peer group. For example, if the firm can beat the costs of a specified utility or peer group, it is allowed to keep the difference.

Homogeneity and sample size are important

While this approach has the advantage of helping to overcome the asymmetry of information on the part of the regulator, it does require a sample of a reasonable size that is homogeneous. If the sample were heterogeneous and sufficiently large, it may be possible to set up cohorts within the sample to correct for any differences in the firm's operating characteristics and its external circumstances.

Yardstick regulation creates risk for the utility to the extent that its costs are different from the benchmark or yardstick for reasons that are outside its control, such as geology, load density, climate, local overhead costs, and local taxes. Benchmarks or yardsticks do provide incentives and can be used successfully where cost data are not readily available.

Benchmarking in Latin America

One development has been to base prices on efficient long-run costs, derived from engineering models of the utility. This approach began in Chile in the early 1980s (and was followed in a number of Latin American countries), where prices were based on the long-run marginal costs of "an efficient operator." The approach was utilized in the U.S. telecommunications industry in the 1990s, using detailed cost models.

In various Latin American countries, certain elements of tariffs are benchmarked against international best practice. With respect to these elements, utilities are not allowed to charge more than would be charged by an efficient international utility operating in the same or similar circumstances. In some cases, this seemed to involve a good degree of judgment that was not very transparent and did not appear to have been entirely objective. It has proved to be a controversial approach in some cases, and has sometimes been counterproductive in attracting investors.

Input into setting the X factor

Benchmarking is chiefly used in the U.K. for the purpose of regulating electricity distribution and water companies, and has been a strong input into the setting of the X factor, as we have seen. The performance of the utilities is compared in a variety of different performance areas, particularly in terms of the standards of service to consumers (*e.g.,* connections, meter reading, billings, revenue realization, efficiency gain), and the pressure the comparison makes tends to act as an incentive to improve performance generally. In practice, it has proved difficult to isolate genuine efficiency differences from cost differences due to different circumstances.

In principle, the trade-off between resetting prices in line with costs and incentives for cost reduction might be overcome if an external or exogenous benchmark could be used. This has been attempted in the U.K., for example, by comparing the performance of the independent water companies through an informal variant of "yardstick competition" described earlier.

Productivity benchmarking using input: output ratio comparisons may be misleading. It will be dependent upon the chosen parameters and the peer group members. Performance comparisons with other domestic and international peers tend to provide the most comprehensive platform for analysis of productivity. The electricity distribution sector has been the subject of many benchmarking studies and yardstick regulation. Tooraj Jamasb and Michael Pollitt have considered benchmarking and yardstick regulation with respect to transmission, using both domestic and international comparators.[43] However, international comparisons to create benchmarks can become quite complicated (*e.g.*, comparing service in London, New York, and Tokyo), because the circumstances and needs that transmission companies serve are different, and the transmission systems represent something of an accident of the history of their development. Also, the expectations of the quality of service and data collection methods may be different. Variations in input prices, such as taxes, wage rates, and cost of capital, that are outside the control of the utility will need to be taken into account.

Jamasb and Pollitt conclude that regulators in many countries are embracing incentive regulation and benchmarking, especially in jurisdictions that have deregulated. They identify a number of problems, including the comparability and quality of data for international benchmarking, that may only improve over time and require cooperation between regulators. The European regulators have agreed to standardize their data collection in an effort to cooperate. The data collection should focus on high level costs and rates of change.

Quantitative techniques for comparative efficiency measures

There are various analytical tools for calculating the relative operating efficiencies of firms within a peer group relative to the observed best practice and for measuring the recent efficiency growth rates of the utilities in the peer group. This gives the regulator the ability to increase the power of the incentives by using information relative to those of the best performing utilities. It is particularly helpful in price cap regulation in setting the X factors for a group of comparable utilities. RPI-X regulation embodies forecasts of what productivity improvements can be achieved in the future. In practice, regulators seldom use one technique to the exclusion of any other. There are three main quantitative techniques:

- data envelopment analysis (DEA)
- multilateral total factor productivity (TFP) measures
- cost functions

DEA is an extension of simple ratio analysis rigorously generalized to handle multiple outputs and inputs. It shows an "efficiency frontier" for an industry that produces a single output from a single input. The efficiency frontier is defined by those observations that envelope all of the other output/input relationships. Those firms not on the frontier are ranked relative to the frontier according to their distance from it. Under this approach, any difference between the firm's actual performance and the frontier is ascribed to inefficiency. As a result, the variables outside the control of the utility need to be identified. In practice, there are many outputs, many inputs, and linear programming techniques that are used to maximize an objective function containing all of these variables.

DEA identifies an "efficient peer group" that is used to make up the hypothetical reference point on the frontier. It can be used to examine technical efficiency without the need for price data, although to derive cost efficiency estimates, the same data are needed for TFP—price and quantity data for each input and output. However, DEA is sensitive to the way outputs and inputs are specified, and measurement errors have a potentially large impact. Sam Lovick concludes that DEA needs a relatively large sample size and that its technicalities and methodology can be difficult to communicate to nontechnical audiences.[44]

TFP measures have enabled comparisons to be made of rates of change of productivity between organizations, but have not enabled comparisons to be made of differences in the absolute level of productivity in combined time-series, cross-section data. This is because conventional TFP measures do not satisfy the important technical property of transitivity (that states direct comparisons between two observations, *m* and *n*, should be the same as indirect comparisons of the same observations via the intermediate observation *k*).

Caves, Christensen, and Diewert created the multilateral TFP index in order to enable reliable comparisons to be made between absolute levels, as well as growth rates of productivity.[45] It satisfies the technical properties that are required to compare TFP levels accurately within panel data. This index has some advantages over DEA. It is a robust technique that is relatively insensitive to data errors, does not require a large number of observations, provides information on growth rates as well as levels, and can be readily understood by non-technical audiences. However, it provides information on cost efficiency and not component parts of technical and allocated efficiency, and it requires price and quantity information from all outputs and inputs.

Cost functions are estimated using ordinary least squares econometric techniques. This approach is less arbitrary, has more robust statistical properties, and allows for the inclusion of operating environment variables. The disadvantage is that it does not explicitly estimate a production frontier.[46] Firms that have an actual cost below that predicted from the estimated cost function are said to be more efficient than average, while those with an actual cost above that predicted are said to be less efficient than average.

Jamasb and Pollitt conclude that the choice of methodology cannot be settled on theoretical grounds.[47] Therefore, benchmarking should not use one of these techniques to the exclusion of all others.

Benchmarking can be applied to transmission

On the face of it, yardstick or benchmark regulation could be used to regulate transmission in countries where there is more than one transmission owner, or, if there is only one, using international comparators. However, to work well, the benchmarks must be credible, and utilities will always be concerned if the differences between their performance and the benchmark involve factors that are outside their control. Jamasb and Pollitt conclude that: "Benchmarking exercises should be viewed as just a transitional regime until competition can be introduced into the sector or international best practice arrives." They take the view that the primary function of benchmarking is to aid decision-making. Regulators should still make informed judgments in setting prices and performance targets.

Performance Target Schemes

The concept of performance target schemes

A performance target scheme is based on the simple idea that if a firm manages to reduce the relevant costs below a specified figure, it is allowed to keep a percentage of further cost savings it is able to make, up to a financial limit or ceiling. Similarly, it must absorb a percentage of any shortfall in cost savings, down to a specified floor. The same approach can be applied to other performance indicators, and the firm can be rewarded appropriately for exceeding the target and be penalized for failing to meet it.

This approach has proved to be extremely effective in the case of NGC's TSS incentive arrangement, which we discussed in chapter 3 and earlier in this chapter. As we have observed, because of the difficulties of setting the targets, the percentage of cost savings to be shared, and the shortfall to be absorbed as well as the floor and ceiling, the schemes are set for short periods (typically two years).

General principles to be applied

It is useful to bear in mind two general principles to be applied in setting performance target schemes:

- The incentive properties will be greater, and the effects more satisfactory from the point of view of the consumer, if the firm is

exposed to the whole of the downside risk of not meeting the performance target. It should be protected, if the potential downside risk is considerable, by a floor or collar, which would be negotiated between the firm and the regulator[48]
- The performance or cost targets should not be so narrowly focused that the firm concentrates its efforts purely in those areas and, as a result, lets its performance standards in other areas fall[49]

The Joskow scheme

Joskow has given the following as an example of a scheme involving performance targets:[50]

"$R_t = C_t + g(C_t - C_t^*) + d(M_t - M_t^*)$

where:

R_t = the revenues the regulated firm is allowed to recover in the prices it charges
C_t^* = a cost target established by regulators for providing service efficiently, including its cost of capital (and return on investment)
C_t = the firm's realized costs of providing service
M_t^*= a target level for a service quality index (*e.g.*, network availability) established by regulator. The higher is M, the better is service quality
M_t = the regulated firm's realized value for the service quality index

$0 < g < 1$

$0 < d < 1$

In simple terms, the regulated firm is rewarded if it can reduce its costs below the target C_t^*. Because g is the factor that shares part of the savings with the consumer, the smaller it is, the greater the incentive to reduce costs. If C_t^* were determined each year by adjusting the current level of costs by an inflation index and a productivity index, and g and d were set equal to zero, the result would be the standard RPI-X or price cap formula. However, if d is zero, it will provide poor incentives to improve the quality of service. It also rewards the regulated firm if it can achieve service quality levels that exceed the target M*. The larger is d, the greater the incentive to improve service quality.

At regular intervals and at the end of every review period, cost and service quality performance would be measured, and the base revenue level and incentive parameters of the regulatory mechanism would be adjusted in accordance with a pre-agreed formula to reflect realized performance. The adjustment can be referred to as a "ratchet." Ratchets necessarily soften incentives for cost reduction and service quality enhancement, but they are generally necessary to take into account both the firm viability constraint (*e.g.*, if the cost target turns out to be too low) and the rent extraction goal (*e.g.*, if the cost target turns out to be far too high) that we discussed in Chapter 6. If the ratchet achieves the wrong balance, it can be renegotiated. Floors and ceilings on profits or the retention of cost savings and efficiency gain can also be included, for the same reasons. As with all approaches to regulation, it is a question of getting the balance right between the interests of the investors and the interests of the consumer.

However, Joskow comments "no incentive regulatory mechanism could fully replace continuing regulatory oversight."[51]

Incentives for efficient connections and interconnections

In many countries, where attracting new investment is important, the issue of creating incentives for the timely connection of new generators and consumers to the transmission system should not be overlooked. A foreign investor may well be deterred by a statement from the transmission utility that it cannot expect its proposed new industrial plant or generator to be connected to the transmission system for several years or, even worse, that there is a queue that the investor will have to join. In fact, queues for the connection or interconnection of new generation are not uncommon.

As we shall discuss in chapter 8, some utilities are subject to regulatory rules that require them to provide information for system studies or, indeed, to carry out system studies within a certain period of time in order to enable generators and loads to assess the cost of connecting at various points on the system. They are also required to publish statements of opportunities to connect to the system such as NGC's Seven Year Statement, which is a useful planning tool and is described in more detail in chapter 8. Many transmission licenses require the transmission owner to offer terms for connection within a relatively short period of time (*e.g.,* three months).

In some jurisdictions, improvements in generation connection or interconnection are needed from the point of view of speed and certainty of outcome, as well as price. Performance targets could be set, both in terms of time and cost. The transmission owner could be rewarded if it exceeded the targets, and be exposed to the cost consequences, subject to a floor and a ceiling if the amounts involved were very substantial, if it failed to meet the targets. Interestingly, NGC has a separate regulatory regime relating to new connections under which it is allowed to cover its costs and a reasonable rate of return. A new incentive scheme is proposed to allow it to agree to higher rates of return for connections.[52]

One of the reasons that generator interconnection queues have built up in the U.S. is the result of the interaction of FERC Orders 888 and 889.[53] The generator starts by filing a request for physical interconnection to the transmission system with a transmission owner under Order 888. There may be a first-come-first-served queue for this, but once the request has been granted, the generator must file a request under Order 889 with the system operator for the capacity to transport a certain amount of energy to each of its potential customers under the open access same-time information system (OASIS).

This is based on the fiction of a "contract path." The generator may find itself at the end of a queue consisting not only of generators, but also traders and utilities needing to secure a path to enable them to do business with customers. The focus of the OASIS filing is typically on transmission constraints. Paths may be constrained because of notional contract paths or actual power flows.

This creates considerable uncertainty, partly because of the need to deal with two different entities that may apply different interpretations of the rules, and partly because of the inherent uncertainties to do with the nature of transmission constraints and the intentions of other market participants. It can result in applicants filing for more contract paths and capacity than they really need, to cover all eventualities in order to deal with these uncertainties. This only exacerbates the queuing and the uncertainty. The problems created by Order 889 would not arise under the kind of locational marginal pricing scheme that we discussed in chapter 4.[54]

FTR incentive schemes

As we shall discuss in chapter 8, setting appropriate performance standards on a comprehensive basis for transmission is not a straightforward exercise, particularly in view of the fact that certain technical standards may not provide the transmission service the market needs. Where a system of FTRs is in place, it should be possible to put into effect the kind of performance target incentive scheme we discussed in chapter 4. Rather than focusing on technical standards such as the number of hours of availability of equipment, the incentive scheme would focus on the outputs—the cost of power actually delivered under changing demand conditions. In this case, a performance target incentive scheme would define the outputs in terms of maintaining FTR funding or revenue adequacy for the system operator that would reflect market participant and consumer interests.

Clearly, there are circumstances that fall outside the control of the transmission owner. This means that the design of the incentive scheme, particularly the floor or limit on the exposure to any revenue shortfalls that would arise, would have to take these circumstances into account. Moreover, it is difficult to design a scheme with a potential upside for the transmission owner, because maintaining the system to create additional capacity may have no value if there is no congestion. However, as discussed in chapter 4, the transmission owner could be rewarded with additional, probably short-term FTRs if it could sustain increased availability for the relevant duration of the FTRs.

Rate Freezes and the Rate Case Moratorium

The concept of rate freezes and rate case moratoria

Under the title of "performance-based regulation" (PBR), several states in the U.S. have implemented incentive schemes consisting of rate freezes, rate case moratoria, and earnings-sharing plans. Under a rate freeze, the company cannot change its prices during a defined period. Under the rate case moratorium, rate cases designed to increase or decrease rates are not permitted, but some elements may be changed. These arrangements have been coupled with earnings-sharing plans that resemble the sliding-scale regulation described earlier, where there is an explicit sharing of realized earnings between the utility and its customers. The customers receive, either through lower prices or direct payments, a share of the utility's achieved earnings in excess of a predetermined threshold return. The schemes may be utilized in combination with price and revenue caps. These approaches essentially grew out of and are consistent with rate-of-return regulation.

Risky for the utility

A rate case moratorium can be viewed as a special case of a price caps where the Y factor is zero, and the X factor equals the rate of inflation. It may have strong cost-cutting incentive properties and be popular with consumers, but it puts the inflation risk on the utility. It can threaten its financial viability or increase its cost of capital, because the shareholders will be taking a risk they cannot manage. Tenega

Nasional Berhad (TNB) in Malaysia has pointed out that it has no incentive to over-invest because of its fixed tariff. Equally, it has little financial incentive to invest in the first place.

Used in combination with other regulatory techniques

Clearly, not being able to increase prices should have the incentive effect of encouraging utilities to cut costs. The popularity of this technique stems, in part, from experience with targeted incentive schemes that have appeared to encourage the utility to focus on cutting cost. However, the focus is often at the expense of other aspects of its performance, notably maintaining quality of service. Several states, where a restructuring has already taken place, have combined price freezes on retail rates with PBR schemes for unbundled distribution services. In this way, the retail customers can be protected during the transition to retail competition. The schemes may also involve the pass-through of uncontrollable costs (such as costs associated with natural disasters and changes in environmental regulations or taxes).

The schemes may also explicitly incorporate multiple incentives such as cost reduction and service quality improvements, so that cost cutting is not undertaken at the expense of maintaining the service the consumer requires.

Evaluation of the Existing Incentive Models

Framework for evaluating the incentive models against objectives

We shall now evaluate the various incentive schemes that have been described against the regulatory objectives that were set out toward the beginning of this chapter.

We shall limit the evaluation to those incentive or performance-based models that are applicable in the context of transmission expansion. Hence our principal goals are to:

- attract investment (long-term capital attraction)
- achieve the most efficient use of existing assets (utilization efficiency)
- create incentives for operational efficiency (supply-side efficiency)

We recognize that the regulator must protect the consumer as to prices (rent extraction for consumers). The long-term viability of industry participants (utility viability) is equally important. Some of the models we have looked at—the revenue cap, sliding scale, discounted cost pass-through, rate freeze, and rate case moratorium—are too limited in their application, have identified problems, or need to be used in conjunction with other primary mechanisms. In order to focus our thinking and create the right recipe for regulated expansion, we shall look at only four techniques that appear to have the properties we are looking for and have the best chance of achieving our goal, either alone or in combination.

Figure 7-1 compares the models (by scoring them low, medium, or high) in terms of their effectiveness against each of those objectives.

Regulatory approach	Supply side efficiency	Long-term capital attraction	Utility viability	Rent extraction for consumers	Utilization efficiency
Price cap	High	Medium	Medium	Medium	Medium
Menu regulation	High	Low-medium	Medium	Medium	Medium
Yardstick	Medium	Low	Medium	Medium	Medium
Performance target	High	Low-medium	High	High	Medium

Fig. 7-1 Comparison of regulatory approaches against objectives

The table assumes that:

- The guidelines prescribed by Joskow and Schmalensee described earlier in this chapter[55] have been followed in establishing the incentives under each of the approaches
- The incentives are equally high-powered in their incentive properties

- None of them creates any perverse incentives in all the circumstances.

Figure 7-2 compares the degrees of information needed by the regulator under each approach

Regulatory approach	Regulator	Utility
Price cap	Medium	High
Menu regulation	Medium	High
Yardstick	Medium	High
Performance target	High	High

Fig. 7-2 Information requirements and regulatory burden

The immediate conclusions, for our purposes are:

- The *price cap model* creates excellent supply-side efficiency incentives, and experience has shown that it has reasonably good investment incentive properties if a negative X or Y factor is used. However, the real difficulty is that the short-term of the review period will mean that investors will primarily be looking for a return within that period. It cannot be relied upon to incentivize investors to be there for the long haul
- The *menu approach* is a means of putting in place one or more of the incentive schemes that we have examined. It is not really a model in itself. If it is well done, it can create superior supply-side efficiency incentives, but it may not be able to focus on attracting investment as well as the price cap, unless it is utilized to set a negative X factor in a price or revenue cap. Much will depend upon how the menu is constructed, and it could contain an explicit capital investment element. Unfortunately, there is very little relevant operational expe-

rience with menu regulation, at this writing. It is attractive, because it creates an incentive within an incentive for the utility to choose the most challenging incentive that it is capable of achieving, and this will benefit the consumer

- The *yardstick approach* has lower long-term capital attraction and supply-side efficiency incentive properties, partly because of the difficulties of establishing the comparators and potential productivity gains and partly because of the differing investment needs. It can only realistically be used in jurisdictions where there is a sufficient number of transmission owners in similar situations to make comparisons meaningful. For example, it could be implementable in the U.S. Utilities will always argue that there are good reasons they are different to their comparators. However, it is a very useful tool for setting the X factor in price cap and revenue cap regulation. It will not be easy to use yardstick regulation to attract long-term capital because no two systems will have exactly the same investment needs or cost structures. It has never been used to regulate transmission on its own
- *Performance target schemes* have excellent supply-side efficiency properties, but can only be set for a relatively short term. They will not attract long-term capital, because the utility will want to be assured of a return within the review period. They can be used alongside another mechanism, such as the price cap or as part of a menu approach. They probably require more information than the other three models in order to set the targets correctly

Although all four models score equally on utilization efficiency, they are not the primary tool to achieve this goal. As we have seen in chapter 4, the best method of achieving efficient use of transmission is through its pricing. All four models could achieve the necessary pricing flexibility.

As far as the balance between rent extraction for consumers and the firm viability constraint is concerned, all four models have good properties. If the incentives are well set, even though they will be short term in nature, they should achieve mutual gains for the consumer and the regulated firm.

We shall create a recipe from these models at the end of the chapter.

Transitional Issues

Regulatory commitments may be difficult to change

It is easier to put in place incentive regulation starting with a blank sheet of paper, than in a situation where private sector utilities have been subject to rate-of-return regulation for years, as in the U.S. The issue of how to move from a traditional rate of return approach to incentive regulation needs to be analyzed from the point of view of the regulatory commitments that are in place, such as those to be found under regulatory rules, approvals, processes, and the legal framework of the regulator.

The regulatory commitments may be difficult for a regulator to change unilaterally, if they are established in licenses, concessions, rules, or approvals, unless the legal framework allows. If new legislation is not possible, the regulator must find some means within the existing legal framework that allows it to design formulae for incentive schemes. Some regulators maintain that they are not able to enter into binding commitments with the firms they regulate, because they may be required to change their policies by the government or legislature or because of their broader consumer protection role. In some countries, the government has an explicit or implicit policy-setting role, but it is not clear whether this can be used to enable the regulator to depart from its commitments (particularly in licenses or concessions).

Acceptance of changes by the regulated utility is important

If the regulated firms do not oppose the move to performance-based approaches, then the transition will be much easier. They may welcome a performance-based approach if the schemes are well designed and give them an opportunity to earn more in return for taking more risk that is within their core competencies and management control. They will react well to the opportunity to commercialize transmission and run it as a business, rather than a public service. However, if they do not welcome the new approach, perhaps because it is seen to be unfair or to increase risk, consensus building and mediation would be very helpful in arriving at the best answer for both the regulator and the transmission owners.[56] We will discuss the use of consensus building in utility regulation in chapter 10.

If negotiations fail, the only answer is for the government to legislate to empower the regulator appropriately. In extreme cases, transmission owners may raise constitutional issues of whether existing rights have been somehow adversely affected or even expropriated. They might complain of expropriation without compensation, which is outlawed under many constitutions.

It is possible that the current regulatory processes inhibit the transition to performance-based techniques, in which case a radical overhaul, not only of the procedures but also of the regulatory mind-set, will be necessary. Performance-based techniques involve a lighter regulatory touch that may not come naturally to experienced rate of return regulators. There will also need to be different data collection, modeling, and analysis.

Benchmarking as a transitional mechanism

One path to transition from rate-of-return regulation to incentive regulation is through benchmarking or yardstick competition. The regulated firms can be asked simply to give the regulator information annually on their performance in key areas (which usually has the effect of improving performance in those areas in the first year, because the utility is curious as to why the data have been requested). The regulator will analyze the data and make further data requests until it is satisfied that it can establish credible benchmarks and comparators. From there, it will be able to establish a range within which all utilities may set their charges, rewarding the most efficient with a higher rate of return. This would be consistent with the U.S. concept of a "just and reasonable rate."[57]

Steps from rate of return to incentive regulation

Vogelsang, with Crew and Kleindorfer, have suggested some steps to transition from rate-of-return regulation (ROR) to performance-based, incentive regulation (PBR) that involve announcing a clear transition plan at the outset to reduce investor uncertainty. The steps involve the following:

- A rate case is held to set the initial regulated prices on an ROR basis using a three-part pricing approach (as discussed toward the end of chapter 4)
- Initial price caps are benchmarked using cost data previously developed under the ROR approach
- The transmission owner is given flexibility to develop a three-part pricing structure (consistent with the ROR approach) efficiently
- A PBR-based profit-sharing scheme to help ease the transmission owner into price caps is established. The initial sharing bands (to mitigate downside and upside profit risks) can provide large cost and profit pass-throughs to mimic the effect under the ROR approach
- The profit-sharing bands are gradually relaxed until they disappear altogether, leaving only the price cap under which the transmission owner assumes the entire cost risk[58]

Operating costs represent variable costs and are normally based on existing operating cost levels. They are rolled forward into the price cap. The prices that are calculated are designed to recover the anticipated costs for the year in question. The properties of the price cap will create an incentive to reduce these variable charges.[59]

Transition must be well handled

Whatever regulatory approach is chosen, it is critical that both the design and the transition are handled extremely well and with the utmost goodwill and transparency. It will be so much easier if the regulated utilities take ownership of the changes, because they will then respond well to the incentives. It is worth paying a small price to align the objectives of the new approaches to their interests. The benefits to the consumer will far outweigh the costs.

Conclusions

Incentive regulation is superior to ROR

There is no doubt that incentive or performance-based regulation has distinct advantages over ROR. It encourages the kind of efficiency, investment, and innovation that we are looking for. It achieves the pricing flexibility we need in order to utilize the key ingredients we discussed in chapter 4. It will create consumer and market-oriented service that will result in the transmission system being utilized as fully and efficiently as possible. Incentive regulation will result in the effective use of the core competencies and develop the management skills of transmission owners to achieve cost savings, increase profits, and reduce prices to consumers.[60] Regulatory incentives can, to an extent, simulate those incentives that are presented to firms in competitive markets. The empirical evidence from the U.S. telecommunications industry suggests that incentive regulation can result in lower prices, higher earnings, increased network modernization, and with no noticeable reduction in service quality. The evidence from the U.K. electricity industry is similar. According to Chunrong Ai and David E.M. Sappington: "Network modernization is more pronounced under incentive regulation than under rate-of-return regulation."[61]

Incentive regulation achieves short-term efficiency gain

However, the major success of incentive regulation has been in achieving only relatively short-term efficiency gains, and, as we have observed, it is not clear that it motivates investment or efficiency gain that has a payback beyond the interval between the price reviews. Our main concern is how incentives can be created to encourage long-term investment that will result in needed transmission expansion. The NGC experience is interesting in that its price cap formulae and performance target incentive schemes have been supplemented by a rate of return approach on new connection (interconnection) assets. It is still subject to regulatory rules relating to the approval of investment plans and capital expenditure budgets. The performance target (TSS) incentive arrangements do accelerate needed upgrades and investment, particularly if they help NGC to beat the targets, but there is a very narrow window of opportunity for it to achieve a return before the next review. The early investment then feeds into the regulated asset base and the base price control (price cap).

Improving the price cap

The concern that the price cap is not as well suited to encouraging long-term investment in capacity as it is to creating short-term efficiency incentives argues for some reform of the way in which the price cap formula is applied. For example:

- There should be clarity and consistency of the accounting treatment that should not vary from one review period to the next, particularly with respect to the valuation of the regulated asset base
- The cost of capital is typically set at the time of the periodic review for the following five years. Regulated investors will raise capital throughout this period, when very different market conditions might apply. The regulator could make allowances for these conditions at the time the capital is raised or, retrospectively, at the next review (if a reliable and transparent method of carrying forward regulatory commitments can be found through the system of regulatory accounts)
- The time profile of the investment could be changed. Financing costs of investment are typically front-loaded, and the returns are back-loaded over a very long payback period. The regulatory uncertainty over rates of return from one period to the next means that the delayed profile of the return is much more likely to deter investment. Currie suggests applying a nominal return to the regulatory asset base and modifying the formula for updating it to ensure the same overall rate of return, but with more front-loading of the payback. This would reduce some of the regulatory risk[62]

Under-investment under price cap regulation

In the context of telecommunications, Crew and Kleindorfer argued that many price cap proposals do not provide appropriate incentives for investment in product and processing innovations.[63] The resulting under-investment can lead to significant welfare losses from lower efficiency, higher prices, and less product innovation. They regarded the central difficulty in designing responsive proposals to promote investment innovation as the linking of the utility's benefits to those delivered to the consumer, while protecting the consumer and potential new entrants against the monopoly power of the incumbent, in the form of over-investment and not executing the planned investment in a manner that minimizes the cost. In other words, the trade-offs between the regulatory goals we have utilized in our evaluation of the options are the major source of difficulty in designing a regulatory regime to attract investment in transmission.

The Best Recipe

Design is driven by regulatory objectives

The key step in the recipe to design a good regulatory regime for transmission expansion is for the regulator to decide exactly what its objectives are and, more importantly, what its priorities are. It is seldom possible to achieve all objectives at once. This exercise involves identifying what the most pressing needs of the consumer are. If transmission expansion is clearly a priority, society could afford to create significant incentives for investment, because the benefits would far outweigh the costs.

In proposing their menu-based proposal outlined earlier, Crew and Kleindorfer provided a basic recipe:

- Any proposed regulatory scheme should provide some protection against exploitation of monopoly power
- If the utility is to have a limit placed at the upper end of its earnings, it will require a floor to its rate of return

- Where price cap regulation is used, the level of the price cap must enable changes in the existing price structure to achieve efficient pricing of services
- Capital recovery and service quality should be addressed explicitly to ensure that incentives are created for efficient product and process innovation. Consumers should be offered continuing high-quality service, especially with respect to those services for which competition is absent
- Continuing regulatory oversight is necessary to ensure that the utility is not making excess profits, the quality of service is maintained, and, more generally, the regulator is protecting the public interest
- The regulatory processes should minimize transaction costs[64]

The key ingredients

The list of the key ingredients for regulated transmission expansion is:

- a price cap formula with a negative X or Y element to attract investment (if necessary)
- an incentive scheme to make sure that the existing transmission facilities are utilized to their maximum and most flexible potential
- if the transmission owner is also the system operator, a performance target scheme (like the NGC TSS arrangements) to achieve savings in the indirect costs of transmission
- flexibility to create efficient transmission prices with respect to the location and use of the transmission system
- yardsticks for use in setting the X factor, if relevant comparators can be found
- a review period of 5-10 years with consistency and clarity of accounting treatment from one review period to the next
- incentives that are set on a comprehensive basis under a single scheme, in order to ensure that average performance levels in all areas are maintained or improved
- capital expenditure and operating budgets set on a rolling basis

- allowances made for unforeseen investment needs and permitting, undergrounding, and re-routing costs outside the control of the investor
- the calculation of the cost of capital that reflects the real risks of transmission
- accelerated depreciation for assets with very long lives
- the calculation of the cost of capital that gives more front-loading to the payback
- a mechanism to adjust the cost of capital in the light of prevailing market conditions
- the ability to invest to meet unforeseeable investment needs and to recover certain costs over which the investor has no control
- clear rules relating to the downward revaluation of existing assets, which only take place if it can be proved the investment was never likely to have been justified
- a significant degree of regulatory commitment seasoned with a mechanism that achieves the necessary degree of flexibility, but on a basis that is *predictable*. If the regulator has to change its priorities and goals, it should give as much advance notice as possible
- benefits of efficiency or productivity gain that are enjoyed for a pre-determined period and not lost if they are achieved late in a particular review period
- real potential upside for the transmission owner—if the downside is significant or risky, the scheme will be seen as unfair, and it will deter investment (because carrots always work better than sticks)

Even if the rate of return is implicit in the price cap formula, regulators should pay attention to it. They should also recognize that transmission business risk has increased. They should allow accelerated depreciation over shorter asset lives. Transmission investors will be deterred by a combination of their continued low beta (the measure of the risk of their business described in chapter 2), a low rate of return, and straight-line depreciation over long asset lives. The cost of capital is still a key ingredient, even under incentive or performance-based regulation. The regulatory asset (rate) base is another.

The way the market typically values the business is to use the regulatory asset base as a benchmark, adjusting it to reflect the level of cost savings that can be achieved over the above those expected by the regulator. It will also assess the scope for the cost savings to continue to exceed the expected levels and the difference between the cost of capital allowed by the regulator and the actual cost of capital, and, if there is a difference, whether this will continue beyond the current review period.

The menu-based regulatory approach has many attractions, because it deals well with the asymmetry of information the regulator suffers. In theory, the utility should know to what extent it can improve its performance and achieve cost savings or efficiency gain. However, the approach is risky for the regulator if it gets the menu wrong. If it is done well, it will create an incentive within an incentive for the utility to choose the highest-powered incentive that it can realistically expect to deliver. It creates a mutual gain for the utility and the consumer.

Regulatory scrutiny of investment plans

There is probably no guarantee that if the utility is rewarded with a higher rate of return resulting from all these incentives, it will use the higher earnings to invest in needed transmission facilities. The incentives are in place for short periods, and the investments have very long lives. As we discussed in chapter 1, there are many reasons that transmission owners have been reluctant to invest.

Investment decisions must be planned and approved in an active process. Regulatory rules are needed requiring transmission owners and developers to present investment plans for regulated expansion to the regulator for approval. Regulating investment plans is never easy when the transmission owner has greatly superior information relating to the performance of its assets. There is scope for a further incentive scheme to encourage implementation of transmission projects with minimum delay and for cost-savings in carrying out the implementation to be achieved.

Although it constitutes conduct regulation which we would prefer to avoid, oversight of transmission investment plans and capital expenditure budgets for regulated expansion will probably always be necessary. The regulator will want to know what is being built and why. For example, it may want the transmission owner to focus on relieving a constraint that will unlock significant generating capacity or reduce local market power, or to focus on building some redundancy in a weak radial system to improve reliability. When investment funds are scarce, the regulator is bound to have a view on what should be built from the consumer and market perspective. It will certainly want to check that the expansion does not increase, rather than decrease, congestion rents.

The regulator will need to anticipate the effects of additions to the regulated asset base on the costs that users of the transmission system are required to bear. The inherent lumpiness of transmission investments as well as the short-term responses to incentive schemes can cause charges to jump from one level to another within a short period. Equally, if the investment has the effect of reducing congestion, the regulator will need to ensure that consumers benefit, if locational marginal pricing is not in place.

An incentive for the long term

The greatest contribution that a regulator can make is not to abdicate responsibility for the long-term viability of the system to transmission owners and system operators by placing them under aggressive incentives to achieve short-term cost savings and efficiency gain, otherwise the regulator will preside over a deterioration in the transmission system. The fatal accidents and lack of availability of railway tracks in the U.K. are a prime example of this effect. Companies can be expected to react to short-term incentives to return profits to their shareholders. What they need is a real incentive to be there for the long term. In this way, they will be able to fulfill two duties—the classic fiduciary duty to the shareholders and also a duty to provide the changing transmission service that the market and the consumers will need over time.

The visible hand of regulation

Although we have tried to create the best recipe for regulated transmission expansion, there is no real certainty that it will create the kinds of incentives that transmission investors really need to be there for the long haul. Regulation is inherently imperfect, but we can try to make it better. Markets, even if they are imperfect, may work better.

As Alfred E. Kahn explains:

> *The visible hand of regulation is not a sufficient answer ... the role of the government is essentially negative – setting* maximum *prices ... Specifying* minimum *standards of service ... In these circumstances regulation cannot supply the same assurances as competition that performance will be positively good-efficient, progressive, risk-taking, innovative.*[65]

CHAPTER 7 APPENDIX

The Design of Incentive Schemes

In this appendix, we shall look at two different approaches on either side of the Atlantic.

FERC's efforts to encourage investment in transmission

The concerns over the lack of transmission investment in U.S. have led to FERC rethinking its approach to transmission pricing and its traditional regulatory approach. It has been looking at performance-based regulation (PBR) and at modifying rate-of-return regulation (ROR) for transmission in an effort to encourage investment.

FERC had no reason to develop performance-based regulation

In the past, it is probably fair to say that FERC had no particular reason to try to develop a system of performance-based ratemaking for transmission. The reason was that a utility would have built transmission to serve its native load customers. The "need" for the investment would have been tested under state siting laws and regulatory prudence reviews by reference to customer needs. Once "need" had been determined and the transmission lines had been constructed, the capital expenditure was added to the asset or rate base, for which native load customers were obligated to pay. The bulk of the revenues collected for transmission services arose from the rate-based treatment of the investments by the state regulators, and not as a result of rates set by FERC.

If any non-native load use of the line under FERC jurisdictional transmission rates had produced revenues for the utility, that revenue would have been returned to native load customers by the state regulators, who would reduce the utility's retail revenue requirements by the amount of the revenues received for this additional use. In this way, the native load customer became the virtual guarantor of the utility's revenue requirements. In the past, when transmission access was voluntary, the perverse result would arise that efficient owners of transmission facilities, which maximized the throughput of their facilities, received no reward for these efforts. By the same token, transmission owners that were inefficient or denied access to their systems, suffered no penalty for their behavior. The efficient utility would suffer the loss of its additional revenues that would be credited back to its native load customers, while the inefficient utility would be able to recover the costs created by its inefficiency fully from its native load customers. This interplay between state and federal regulators in the U.S. has made the introduction of incentive regulation for transmission very difficult.

Lack of a bright line between the jurisdiction of FERC and the states

It was not always easy to answer the question of where the boundary between the respective jurisdictions of FERC and the state regulators actually lay. Transactions involving wholesale market use of the transmission systems are FERC jurisdictional, but, unfortunately, retail transmission transactions do not present as clear a picture. Some lawyers believed that the state regulators had jurisdiction over retail transmission transactions only when retail rates were fully bundled (to aggregate energy, transmission, and distribution charges). On this line of reasoning, it would follow that once the rates were separated or unbundled, the jurisdiction would pass to FERC. This is consistent with some past practice and the language of the Federal Power Act. The state of Texas is not within FERC's jurisdiction.

However, there was a countervailing view that FERC's only transmission pricing authority pertained to wholesale use of the grid and that the states had jurisdiction as to how to price transmission services for all retail uses. There was a third view that all transmission pricing, whether retail or wholesale, would be subject to the jurisdiction of FERC, if it chose to exercise its authority. However, on March 4, 2002, the U.S. Supreme Court affirmed the FERC's 1996 decision in Order No. 888[66] to assert jurisdiction over "unbundled" retail transmission.[67] The Court evaded the question of whether FERC regulation could reach the transmission component of "bundled" retail transactions.[68]

FERC Order 2000

Performance-based regulation or modified rate-of-return regulation?

There are proposals in FERC Order 2000 in relation to performance-based regulation (PBR) and modified rate-of-return regulation (ROR) for transmission.[69] FERC did not propose to impose PBR, but preferred to achieve a voluntary, incentive-based regulatory regime. Performance-based schemes could be made to work for all transmission owners and regional transmission organizations (RTO), including public power utilities and cooperatives.

Because ROR tends to allow a rate of return that is lower than that sought by competitive enterprises or those assuming a greater degree of risk, FERC suggested modifying ROR with higher rates of return to reflect greater risk, levelized rates, incremental pricing, and accelerated depreciation.

FERC was attracted by performance-based regulation,[70] but put down certain markers to the effect that:

- It would need to include both rewards and penalties for inadequate performance
- It could not be used in part or on a piecemeal basis
- It must not compromise reliability
- There should be an equitable sharing of benefits between investors and customers
- It must create incentives for both efficient operation and investment

- To the extent possible, the rewards and penalties should be prescribed in advance, based on known and measurable benchmarks

FERC indicated that it was prepared to consider a number of approaches such as:

- a higher rate of return on equity on the transmission assets
- allowing the transmission owner to retain the benefits of cost savings attributable to RTO formation
- acceleration of transmission cost recovery in rates
- non-traditional valuation of transmission assets such as an estimate of replacement costs for assets
- liberalized allowance of levelized or non-levelized rate methods (designed to recover capital costs through a uniform payment over the life of the asset)

The transmission owners would have to meet all of the requirements to become an RTO before a performance-based pricing proposal would be accepted.

In Order 2000, FERC recognized the need to create incentives to invest in new transmission facilities. Many of the comments that were made to FERC concentrated on the need for improved rates of return, accelerated depreciation, and asset revaluation. FERC did not get into the detail of the design of performance-based rates and incentives, but invited the industry to submit proposals that it would consider on a case by case basis. However, the Order did give assistance in outlining the filing procedure for what it called "innovative rate" proposals.

Despite the genuine attempt on FERC's part to incentivize new investment, it has been disappointed with the response. FERC Chairman, Pat Wood, said at the RTO Workshop on October 16, 2001[71] that FERC had "written the book", but "nobody had made the movie" of the proposed PBR schemes. FERC referred to the British approach and suggested that RTOs should examine its strengths and weaknesses.

Analysis of "innovative rate" filings

David Clement has analyzed seven applications that have been made to FERC under the invitation given in Order 2000 for "innovative rate" proposals. The application of the International Transmission Company that related to the recovery of capital gains tax from the sale of the transmission assets to a new company (where the existing owners no longer wished to remain in the business) was the only one to be accepted. The others were denied (or FERC did not rule upon them), because they did not meet procedural requirements for RTO policy development, mainly involvement of stakeholders in developing the proposed incentives. In addition, an analysis of proposals indicated that many of the proposals lacked other requirements outlined in Order 2000 and subsequent FERC decisions. For example, they lacked:

- a cost benefit analysis
- a demonstration that the proposed rate treatment was appropriate for the proposed RTO (leading to efficient use of and investment in the transmission system and improved reliability)
- an explanation as to the reason the rate proposal was just, reasonable, and not unduly discriminatory
- a demonstration that the proposal would not impede the RTO from meeting the minimum characteristics and functions required under Order 2000

Accelerating depreciation creates the strongest investment incentive

David Clement has demonstrated that it is not clear that the applications would have stimulated transmission system investment. He looked at their impact on a modest $50 million transmission investment. Using a hypothetical transmission company, he analyzed the net present value (NPV) of operating cash flow as a determining factor in deciding whether the proposal would have a significant impact on new investment. He applied each of the seven proposals to the hypothetical model and concluded that, for the most part, the impact of the proposals on new investment was fairly modest. As Figure 7-3 shows, accelerating straight-line depreciation, assuming a 15-year asset life would have had by far the largest impact, followed by an assumed 20-year asset life for depreciation purposes. Increasing the return on equity to 13% would have had less impact and some of the applications proposed would have an even more modest increase. Several of the applicants were subject to a price moratorium that was a significant deterrent to investment. In short, the applications were modest, and, in some cases, scarcely created incentives at all. Moreover, some elements of the proposals had complex interactions with other aspects and had not been fully evaluated.

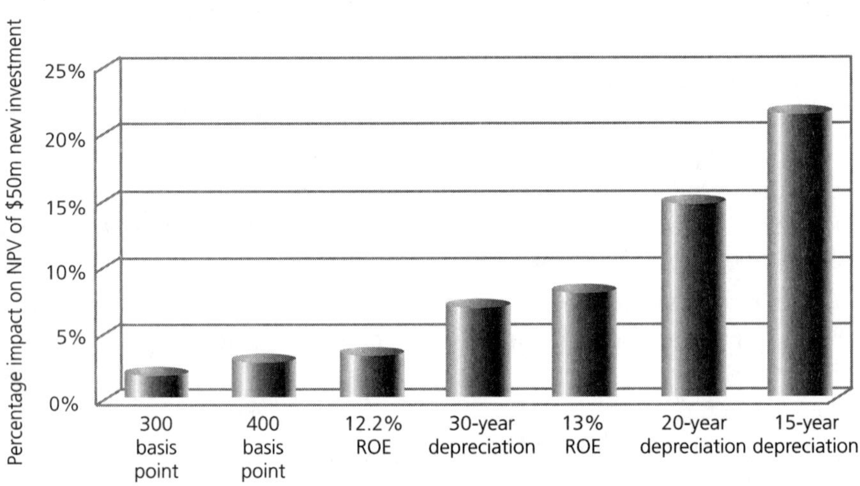

Fig. 7-3 Investment properties of "innovative rate" filings (Source: David Clement, CERA)

David Clement concluded that FERC placed a higher priority on the status of the RTO designs for the applicants' regions than with the design of the proposed incentives.[72] FERC seems to have ignored Joskow's warning (which he stressed in italics):

Transmission regulatory reform should not be viewed primarily as a "carrot" that is desirable only as a way to entice reluctant utilities to form and participate in RTOs meeting the Commission's criteria.[73]

New England incentive pilot scheme

National Grid USA, on behalf of its subsidiary, New England Power Company (NEP) submitted an application to FERC proposing a transmission incentive pilot scheme along the lines of the TSS scheme. The proposal was designed to reduce increased congestion costs in New England by minimizing the amount of time that a 345 kV transmission line would be taken out of service while it was being refurbished. The work was originally intended to be completed using conventional methods that involved taking the line out of service. However, system conditions were such that there would be considerable savings in congestion costs if live-line work could be carried out. The increased cost of the live-line work was expected to be more than offset by significantly reduced energy and congestion charges, providing net savings to consumers. The transmission incentive pilot scheme was thoroughly reviewed by ISO New England and was endorsed through its governance process, achieving a vote of 80% of its members. The anticipated savings in congestion were estimated by the ISO to be potentially up to $11.8 million. National Grid USA would have been exposed to a potential loss if the work had been completed late, when a daily penalty payment would have applied.

FERC turned down the application "because it is inconsistent with the principles of performance-based regulation."[74] This statement was not explained. The reason that was given for the rejection was that under the existing cost-recovery methods, NEP would fully recover its costs of performing live-line maintenance, which NEP had not claimed as being innovative. FERC said that there was "no reason to pay an incentive premium for performing such noninnovative procedures."[75]

Voluntary basis

The proposed voluntary and case-by-case approach could help with the problem of the transition from ROR to PBR, but it calls into question whether FERC will be in a position to ensure that a particular proposal will actually result in the performance that FERC would like to incentivize. In other countries, the regulator has usually taken quite an active role in imposing the same PBR approach on all regulated utilities of the same class, even though the precise numbers contained in the incentive formulae may have been different.

Return on equity, levelized rates, and accelerated depreciation

FERC did provide greater specificity on the issues of return on equity, levelized rates, accelerated depreciation, and incremental pricing for new transmission investments. It agreed that the risk profile of the transmission business has changed as a result of industry restructuring and that the historical data typically used to evaluate the return on equity may not be reliable since they reflect a different industry structure. It indicated a willingness to consider a moratorium that was tied to the rates the transmission provider charges on transmission assets with respect to bundled retail power sales. The moratorium option might be tied to the existing transmission rate level, or to the existing return on equity.

On the issue of levelized rates, which are rates designed to recover all capital costs through a uniform payment over the life of the asset, FERC indicated that levelized rates were preferable in an RTO environment, because all customers, regardless of when they take service, face the same price. The levelized rates with respect to existing investments would be higher than normal levelized rates and would address concerns that the mere fact of restructuring will decrease revenues.

FERC was prepared to make a distinction between accelerated depreciation for existing transmission assets and accelerated depreciation for new transmission facilities. There was clearly a willingness to consider accelerated depreciation in the case of new transmission facilities. However, FERC was not willing to adopt a suggestion that it should allow assets to be revalued upwards. This is known as an acquisition premium.

Confusion created by the regulatory regime

It is clear, that despite Order 2000, there is confusion among planners in the U.S. as to the regulatory regime at both the federal and state level. As Kiah Harris has observed:

> *The confusion amongst the planners could disappear if the state and federal regulators could define a consistent framework to plan within and allow transmission business structures that give some clarity as to how the investment is returned. The regulatory confusion at the state and federal level is the primary impediment to generation and transmission expansion.*[76]

Apart from the confusion as to the jurisdiction of FERC and state regulators over transmission that we have discussed earlier, there are regulatory rules at the state level that will inhibit investment, particularly in interconnectors between regions. For example, if a new line is built by a utility that owns transmission assets in one area, and the benefits accrue in part to users or consumers served by utility in a neighboring area, the regulatory regime can be such that the whole of the costs must be borne by the customers of the investing utility, which will deter investment. To solve the problem, the costs and benefits could be split, and part of the cost could be recovered under a rule-based system through the other transmission owner that could pass the cost on to those customers who benefit. The rule could be designed as a backstop in case the transmission owners cannot agree among themselves, and a transfer payment between them could be codified. The FERC notice of intended rulemaking or standard market design is helpful in this respect. It contains the principle that the beneficiaries of transmission expansion should pay.[77]

U.S. legal framework

In theory, none of this confusion should exist, because there are several well-known cases that provide the legal framework for capital attraction in regulated infrastructure projects in the U.S. U.S. regulators have an obligation to establish equity returns of a standard that will attract capital. As David Clement points out:

> *as long as electric utilities are vertically integrated with large businesses other than transmission, it may be argued that they are meeting this standard, but if one looks at electric transmission as a separate business, the performance in this regard is highly suspect.*[78]

The two leading cases in question are:

- Bluefield Waterworks and Improvement Co. v. Public Service Commission.

> *A public utility is entitled to such rates as will permit it to earn a return on the fair value of the property which it employs for the convenience of the public equal to that generally being made at the same time and in the same general part of the country on investments in other business undertakings which are attended by corresponding risks and uncertainties; but it has no constitutional right to profits such as are realized and anticipated in highly profitable enterprises or speculative ventures ... A rate of return may be reasonable at one time and become too high or too low by changes affecting oppor-*

> *tunities for investment, the money market and business conditions generally.*[79]

- FPC v. Hope Natural Gas Co.

> *The rate-making process under the Act [The Federal Power Act], i.e., the fixing of "just and reasonable" rates, involves a balancing of the investor and the consumer interests ... The investor interest has a legitimate concern with the financial integrity of the company whose rates are being regulated ... By that standard the return to the equity owner should be commensurate with returns on investments in other enterprises having corresponding risks. That return, moreover, should be sufficient to assure confidence in the financial integrity of the enterprises, so as to maintain its credit and attract capital.*[80]

What are the implications for transmission expansion?

The provisions of Order 2000 with respect to innovative rate proposals appear to be less concerned with creating incentives to expand transmission and more to do with incentives to join or form an RTO. We conclude this from the fact that the provisions are written in the language of ROR and not incentive regulation as we have discussed in this chapter. We also reach this conclusion in the light of the innovative rate proposals made under it that were turned down on procedural grounds, in large part related to RTO formation. Whatever the intention behind these provisions, they have not been effective.

The NGC Model

The performance-based model under which NGC is regulated is well designed and has many admirers. Not only has there been good practical experience with this model, both the price cap and the TSS incentive arrangements also have satisfied strong economic analysis as to their theoretical justification.[81]

The Transmission License

NGC's transmission license is a regulatory instrument (enforceable only by the regulator) in which the detailed regulatory rules are written, including the price cap and incentive scheme formulae. It requires NGC to offer terms for connection to and use of its system to anyone that applies. It is not permitted to decline to provide service on grounds of inadequate capacity, unless the application relates to use of its interconnectors with neighboring systems. NGC has a monopoly duty to expand its system to meet the needs of market participants. In order to carry out this duty, it prepares detailed investment plans and a capital expenditure budget. The plans and the budget are reviewed and approved by the regulator (with the aid of consultants) at the time of each price cap review. An allowance is then included in the calculations for the resetting of the price cap. NGC is free to set its own budget.

NGC uses an investment cost-related pricing (ICRP) transport model. The ICRP model calculates the increase in MW/km flows, assuming a simplified network and ignoring how power actually flows and any security arrangements.

Planning and operating standards

NGC is subject to planning and operating standards that are given regulatory force through the license. The standards cover the level of transmission capacity that NGC should maintain, the capacity of transformers and other equipment, and interactions with distribution systems. All of the standards are now combined into a single standard, the components of which include generator connections, the main interconnected system, and connections to demand groups. Operational and planning criteria are covered within the one standard. NGC is required to build sufficient transmission capacity so that, at the time of system peak, the relevant parts of the system can accommodate power flows even without two circuits. In addition to the usual system frequency and voltage limits that are imposed, there is an obligation to ensure that there are sufficient circuits in operation so that the system can withstand a single "security event," which typically comprises the loss of a double circuit overhead line, transformer, or cable, or in bad weather, certain two circuit pairs.

It is sometimes argued that these standards act as a proxy for an economic assessment of the costs and benefits of planning operating the system under different conditions. Sally Hunt and Graham Shuttleworth have argued that these standards should be replaced with well-designed economic standards.[82] There have been complaints that the standards are too high and that adequate security could be achieved at less cost. If the standards were to be relaxed, NGC's revenue and capital expenditure would reduce, but there could be other costs to market participants and consumers from the potential insecurity and lack of reliability.

NGC is a transco

In England and Wales, NGC is both the transmission owner and the system operator. Both activities are regulated, but the revenues associated with each activity are subject to separate incentive schemes. The role of the system operator changed materially on the introduction of new electricity trading arrangements (NETA) in March 2001. It is now responsible for system balancing and acts as a principal in the market to minimize all the indirect costs of transmission and balancing under an incentive scheme. Part of the logic for it to trade as principal is to enable it to deal more effectively with the potential market power and strategic behavior of other market participants.

Price cap formula

The transmission network use of system charges are regulated under an RPI-X mechanism for a five-year period. The X factor for 2001-2006 was set at 1.5%. If other users undertake work on the system (primarily on the existing connection assets) that falls within the scope of the regulated business, the permitted revenue is correspondingly reduced. This is limited to the maintenance of certain connection assets. The RPI-X formula covers the main infrastructure and pre-vesting (pre-restructuring) connection assets. Post-vesting connection assets are subject to ROR. There is a correction factor to adjust for over- or under-recovery of revenue covered by the price control in the previous year. There is also an allowance to reflect any unexpected changes in the capital expenditure required in order to accommodate new generator connections or interconnections to the system on a timely basis. NGC can increase its charges to cover the increased depreciation, and is permitted a 6.25% return on the additional spending. Another adjustment allows NGC to pass the costs of U.K. local taxes directly to customers. The formula is as follows:

$$M_t = \left[1 + \frac{RPI_t - X_g}{100}\right] P_{t-1} - D_t - K_t + G_t + U_t$$

where:

M_t means the maximum revenue in relevant year t

RPI_t means the percentage change (whether of a positive or a negative value) in the arithmetic average of the retail price index (RPI) published or determined with respect to each of the six months from May to October (both inclusive) in relevant year t-1 and that are published or determined with respect to the same months in relevant year t-2

X_g has the value one and a half (1.5)[83]

P_{t-1} means the amount derived from the following formula:

$$M_t = \left[1 + \frac{RPI_t - X_g}{100}\right] P_{t-1} - D_t - K_t + G_t + U_t$$

save that:

(i) in relation to the relevant year commencing on April 1, 2001, P_{t-1} shall have a value equal to U.S. $1,178,100,000

(ii) in relation to the relevant year commencing on April 1, 2002, P_{t-2} shall have that value; and

(iii) in relation to the relevant year commencing on April 1, 2003, P_{t-2} shall have the value derived from the following formula:

$$P_{t-2} = P_{t-3}\left[1 + \frac{RPI_{t-2} - X_g}{100}\right]$$

where:

P_{t-3} shall have the value derived from the following formula:

$$P_{t-3} = \$1{,}178{,}100{,}000\left[\frac{W}{Y}\right]$$

W is the arithmetic average of the RPI published or determined with respect to each month of the relevant year commencing on April 1, 2001 on the assumption that the RPI for January 1987 equals 100

Y is 175.17 (being the forecast of the RPI prepared by Business Strategies Limited in December 2000 in respect of the relevant year commencing on April 1, 2001 on the assumption that the RPI for January 1987 equals 100)

D_t means a correction factor to be applied to transmission network revenue and is equal to the value of user maintenance in relevant year t

K_t means the correction factor (whether of a positive or negative value) that is derived from a formula forth in the license

G_t means a revenue adjustment factor derived from a formula set forth in the license

U_t means a revenue adjustment factor reflecting changes in non-domestic rates (local taxes) and the license fee, and is derived from the formula set forth in the license

Regulated asset base

The regulatory value of the assets is assumed to be equal to the value based on a generic life equal to the remaining average life for all the pre-privatization assets. This has resulted in an assumed 20-year life on £4 bn of asset value depreciated on a straight-line basis over 20 years between 1990 and 2010. Ideally, the profile should track individual assets to derive a regulatory value.

Post-vesting connections

Because the level of connections was unknown in 1990, a different approach was adopted to deal with what were described as "post-vesting connections." NGC charges only the shallow entry costs to generators (discussed in chapter 4) and is allowed to earn a 6% real, pretax rate of return on these investments (based on RPI indexation of asset values). A proposed inceentive scheme would allow it to agree a higher rate of return.[84] The risks involved are relatively low and the approach serves to incentivize NGC to avoid delays in connecting generators. Remote reinforcement (deep entry) costs are regulated under the price cap formula. A "single till" approach is adopted. The regulator looks at the total assets involved to derive NGC's revenue requirements, but deducts the post-vesting connection asset income from the total amount.

Review process

The prices, the X factor, and the formula are reset at the end of each 5-year period. The regulator typically undertakes an extensive analysis of operating costs, the capital expenditure budget, and the cost of capital. The level of revenue and its rise and fall during the following period are set so that the net present value of future revenues less operating costs, discounted at the assumed cost of capital, is equal to the "regulatory asset value." This value is determined by historic investment depreciation and the valuation placed on the assets at the time of restructuring and privatization. The review process is essentially an elaborate negotiation between the regulator and the company and takes about 14-18 months. It is also open to other interested parties to make their views known to the regulator.

It has been observed that, over time, the assessment of costs by the regulator and regulated utilities tends to converge, and the incentive qualities of the price cap formula may become less important.[85]

NGC has a monopoly on information, and the regulator is at the usual disadvantage in terms of asymmetry of information upon which to base its analysis and assessment of NGC's capital and maintenance expenditure and the technical content of its plans. The concept of a reference network described in chapter 8 could be useful in this respect.[86]

The transmission system in the U.K. is relatively robust, and the main concern is with maintaining the facilities and replacing the aging assets. The planning standards are technical in nature, and it might be difficult for the regulator to assess the implications of moving to cost-based standards. However, the approach has been very successful. By 2005, NGC will have cut its prices by 41% in real terms, or an average of 3.5% per annum.[87] Since 1989, more than 25 gigawatts (GW) of new plant has been connected to the system in England and Wales, and 22 GW has been disconnected. NGC cut its controllable operating costs by 57% at an annual rate of 6% (achieved over the previous 4-year regulatory cycle).

The initial uplift management incentive scheme

Even in 1990, when the new market opened, it was clear that NGC had no financial incentive to minimize the costs of operating the system, the England and Wales Pool that the wholesale market (although it did have a license and statutory duty to manage the system efficiently). The "uplift" in the Pool rose dramatically. The uplift included the cost of ancillary services and constraints, but not reactive power, energy uplift (a feature of the now defunct Pool), and transmission losses. When the incentive scheme was put in place, NGC made a considerable effort to bring these costs under control. Since April 1994, when the first incentive scheme to manage the indirect costs of transmission was put in place, the uplift costs have fallen dramatically, as shown in Figure 7-4.

Cost	1993/4	1994/5	1995/6	1996/7	1997/8	1998/9	1999/00	2000/01
Incentivized uplift cost	570	478	356	335	328	267	207	203
Payment to NCG (£m)	0	25	23	9.1	10.2	11	27.3	16.8

Fig. 7-4 System operations costs since TSS incentive schemes were introduced in 1994 (Source: NGC)

An incentive scheme that is distinct from the price cap

The original NGC Uplift Management Incentive Scheme (UMIS) proved to be very effective in achieving performance that managed the uplift costs, particularly the costs of constraints. It was originally known as UMIS, but as it developed, it was renamed the Transmission Services Scheme or TSS. This is how we have referred to it, even though it has now been replaced by a new scheme reflecting NGC's additional role in balancing the system under the new electricity trading arrangements (NETA). It is now called the "Balancing Services Activity Revenue Restriction" in NGC's license.[88] OFGEM's consultation paper also referred to it as the "NGC system operator incentive scheme."[89] NGC was given a target of indirect, uplift costs to beat. The costs included the costs of congestion, losses, ancillary services, and "must run" or "constrained on" generation. A floor and a ceiling were put in place to protect NGC from extreme circumstances and also to protect the consumer from windfall gains. The incentive was reviewed and revised every two years in the light of experience. Ignoring the indirect costs of transmission service would have created an incentive for NGC to increase the congestion costs and not to invest (although all substantial investment was driven by NGC's planning standards).

Development of TSS incentive schemes

These incentive arrangements have increased in sophistication over time.[90]

Before the Balancing Services Activity Revenue Restriction (the latest TSS scheme) was introduced, the incentive scheme had four separate elements:

- a transmission services cost target, associated with the procurement of ancillary services (except for reactive power) and other costs including congestion management
- a reactive power uplift cost target for the procurement of reactive power
- a cost target reflecting transmission losses and the way in which the costs of the losses were passed on to wholesale buyers (which was designed to give NGC responsibility for optimizing losses)
- an energy uplift cost target for costs arising from generator availability changes, generator shortfalls, and demand forecast errors

The sharing factors were asymmetric, and, by 1999, the elements of the scheme were as shown in Figure 7-5.

Scheme	Target	Incentivized range	Asymmetric sharing factors	Maximum gain/loss
Transmission services uplift	£219m	£176m-£304m	50% below target 25% above target	+/- £21m
Reactive power uplift	£46.5	£41.5m-£56.5m	50% below target 25% above target	+/- £2.5m
Energy uplift	£25m	£0-65m	12% below target 5% above target	+£3m/-£2m
Transmission losses	5.05-5.25TWh	4.73-5.57TWh	50% below target 25% above target	+£4m/-£2m

Fig. 7-5 Incentive schemes 1999/2000 summary (1999/2000 prices) (Source: Lehman Brothers Global Energy Research Report on NGC, October 15, 2001)

Balancing services activity revenue restriction scheme

Changing the scheme to accommodate the introduction of the new electricity trading arrangements (NETA)[91] involved extensive studies that indicated actual costs could differ from expectations by several hundred million dollars. Much of the reason for this uncertainty reflected the lack of any means of knowing how market participants would behave under NETA and how the costs of managing constraints and losses might change as a result of NETA. As we discussed in chapter 7, the regulator offered NGC a menu of options (when negotiations over the cost targets failed to produce an agreement), and NGC made its choice from the menu. The new Balancing Services Activity Revenue Restriction Scheme was only put in place for an initial period of one year, but has since been rolled over on new terms for a further year (to March 31, 2003). However, the regulator intends to set a longer-term incentive scheme that will have greater incentive properties.

A single, comprehensive incentive scheme

The new scheme bundles all of the costs that NGC may incur in its role as system operator and also in balancing the system. Unlike the previous scheme, it incentivizes NGC to bring down the total cost, rather than individual elements. This is intended to allow NGC greater flexibility to make appropriate trade-offs when making decisions that affect the procurement of different services, particularly as many of the services (*e.g.*, energy-related ancillary services, congestion management, and balancing) all use the same energy. Therefore, NGC was incentivized to beat a preset target of total costs amounting to £460 m. The initial sharing factor on the upside was 60% and on the downside was 50%, with a cap of £60 million and a collar of £45 million.[92]

Upside for NGC

These incentive schemes associated with system and market operations have been a source of additional profit for NGC. Although the schemes may have accelerated some capital expenditure in order to enable NGC to make the most of the opportunity to increase its profits under these schemes, the capital costs associated with this type of investment have been relatively low. It has been NGC's skill in managing the system and in maintaining the assets that account for its success.

Are two schemes better than one?

The question arises whether the twin incentive schemes that apply to NGC in its roles as transmission owner and system operator create opportunities for NGC to arbitrage the two schemes to its advantage. However, the periods of the TSS schemes have been so short that there is very little time within which NGC can invest and capture a benefit. In theory, if the incentives are properly set and there is a cap, which allows the consumer to share in some of the upside, it should not matter if the two incentive schemes can be arbitraged. The risk of the consumer paying too much will be small.

Theoretical analysis of the TSS incentive schemes

We have looked at the incentive schemes that have been applied very successfully with respect to NGC's roles as system operator and market administrator. Not only have they been justified by practical experience, but they have also been justified by strong theoretical analysis.

Thomas-Olivier Léautier has shown that an uplift management rule, similar to the TSS schemes, induces optimal expansion of the transmission network, provided that the transmission company is responsible for the *full* cost of the outturn plus the transmission losses, valued at the system marginal price. In his proof, he used the Laffont-Tirole menu of revenue sharing rules described earlier in this chapter that trade-off cost minimization against consumer rent extraction.[93] This meant that the regulator could implement this trade-off independently of the choice of the optimal transmission capacity.

Where Léautier differed from the TSS model was with respect to the extent of the exposure to the cost of the outturn. NGC was not responsible for the full cost of the outturn. This lowered the power of the incentive. He argued that it made both economic and regulatory sense to expose the transmission owner to the full cost of the outturn, but to mitigate the risks associated with it through the negotiation of an appropriate floor and ceiling (collar and cap). In this way, the regulator could minimize the risk of criticism for paying a substantial amount of money to the transmission owner for having beaten the cost targets, and the transmission owner could protect itself from the risk of insolvency as a result of an unlimited exposure to the cost of the outturn.[94]

For the most part, NGC was exposed to only 30% of the outturn, and a floor and ceiling were also in place. The marginal value of the transmission capacity for NGC was, therefore, only 30% of the marginal social value. As a result, NGC would underexpand. Léautier showed, in a fully worked example, that the full marginal value of transmission exceeded the marginal cost, but NGC's share, which was only a third of the value, was lower than the marginal cost. He concluded that when the incentives were periodically reset, the regulator should have concentrated all of its bargaining power on further increasing NGC's share, rather than on renegotiating the cost target.

What does the future hold?

NGC thinks there may be a move away from the TSS schemes that apply to system operations and a more detailed approach to the planning process may be adopted. It is difficult to plan investment of assets with 40 or 50 year lives from current, short-term price signals.

It might be that the price cap formula will decouple operating expenditure and capital expenditure, and that operating expenditure will be calculated on some moving average formula. Ultimately, the regulator may have no alterative but to fall back on the mechanistic approval of capital expenditure plans and budgets, which was a technique successfully employed in the water industry in England and Wales. In the water industry, it was possible to segregate investment against measurable output. In transmission, there is no simple measure of the output of a particular reinforcement.

Conclusions with respect to the NGC model

The model that regulates NGC has clearly been successful. It is a product of much thought, ingenuity, and experience. Its success in practice is not its only justification. It has both academic support and theoretical justification. As Thomas-Olivier Léautier observed:

> *The work's conclusions are optimistic: first, the complex laws of physics can be married with the principles of economic regulation. Second, the regulation contract we propose is relatively simple and built on the existing regulation contract in England and Wales. This should encourage policymakers to create for-profit transmission companies: they can be confident that transmission companies can be effectively regulated and induced to expand the transmission grid, so as to support the establishment of efficient power markets.*[95]

NOTES

1. Ingo Vogelsang, "A 20 Year Perspective on Incentive Regulation for Public Utilities," Boston University (2000)

2. Michael A. Crew and Paul R. Kleindorfer, "Incentive Regulation in the United Kingdom and the United States: Some Lessons," *Journal of Regulatory Economics*, vol. 9, no. 3 (May 1996)

3. Stephen C. Littlechild, "Regulation of British Telecommunications' Profitability," Report to the Secretary of State, Department of Industry (February 1983)

4. William J. Baumol, "Reasonable Rules For Rate Regulation: Plausible Policies For An Imperfect World" (1967), pp. 108-123. Fred E. Sudit, "Automatic Rate Adjustments Based on Total Factor Productivity Performance in Public Utility Regulation," in Michael A. Crew, ed., *Problems in Public Utility Economics and Regulation* (Lexington: Lexington Books, 1979), pp. 55-71. Michael A. Crew and Paul R. Kleindorfer, "Incentives for Efficiency in the Nationalised Industries: Beyond the 1978 White Paper," *The Journal of Industrial Affairs*, vol. 7, no. 1 (Autumn 1979), pp. 11-15. Ingo Vogelsang and Jorg Finsinger, "A Regulatory Adjustment Process for Optimal Pricing by Multiproduct Monopoly Firms," *The Bell Journal of Economics*, vol. 10 (1979), pp. 157-171

5. Paul L. Joskow and Richard M. Schmalensee, "Incentive Regulation for Electric Utilities," *Yale Journal on Regulation*, vol. 4: I (1986) p. 1,049

6. See David E.M. Sappington, "Strategic Firm Behavior Under a Dynamic Regulatory Adjustment Process," *The Bell Journal of Economics*, 360 (1980). Ingo Vogelsang & Jorg Finsinger, "A Regulatory Adjustment Process for Optimal Pricing by Multiproduct Firms," *The Bell Journal of Economics*, Econ. 10 (1979), pp. 157-171

7. Paul L. Joskow and Richard M. Schmalensee, "Incentive Regulation for Electric Utilities," *Yale Journal on Regulation*, vol. 4: I (1986), pp. 1,049

8. This was supported by an empirical study of the determinants and impacts of incentive regulation with respect to generation heat rates and plant utilizations in the late 1970s and early 1980s. Sanford V. Berg and Jinook Jeong, "An Evaluation of Incentive Regulation for Electric Utilities," *Journal of Regulatory*

Economics 3 (1991), pp. 45-55. They found that the introduction of specific cost-component incentives improved overall operating cost performance

9. Jean-Jacques Laffont and Jean Tirole suggest that optimal incentive mechanisms may generally yield expected returns on investment that are, in fact, somewhat higher than the firm's cost of capital. Jean-Jacques Laffont and Jean Tirole, "Using Lost Observations to Regulate Firms," 94 *Journal of Political Economy*, 614 (1986). The basic argument is that the regulatory firm viability constraint (or goal) limits the use of penalties to provide incentives, so that if strong incentives are desirable, generous bonuses may be required. It is not clear how seriously this result should be taken in practice, since it may be driven by the interpretation of the firm viability constraint

10. Jean-Jacques Laffont and Jean Tirole, *Competition in Telecommunications* (Cambridge, Massachusetts and London England: The MIT Press, 2000), pp. 39-60

11. In some incentive schemes such as NGC's 1999-2000 schemes, the downside risk may be greater than the upside benefit in beating the cost-reduction target. The asymmetric sharing factors for the uplift costs relating to transmission services, reactive power and losses were 50% below target and 25% above target, for example

12. Stephen C. Littlechild and Michael E. Beesley, "The Regulation of Privatized Monopolies in the United Kingdom," *RAND Journal of Economics*, vol. 20, no. 3 (Autumn 1989), pp. 454-472

13. Frank Ramsey, "A Contribution to the Theory of Taxation," *Economic Journal* (March 1927). Marcel Boiteux, "On the Management of Public Monopolies Subject to Budgetary Constraints," *Journal of Economics Theory*, 3 (1971), pp. 219-240

14. Bradley and C. Price, "The Economic Regulation of Private Industries by Price Constraints," *Journal of Industrial Economics,* vol. 37 (1988), pp. 99-106. T.J. Brennan, "Regulating by Capping Prices," *Journal of Regulatory Economics*, vol. I (1989), pp. 133-147. W. Neu, "A Theoretical Look At Price Capping Mechanisms: Some Clarifications and One Recommendation," W.I.F.K. Diskussionsbeiträge zu Telekommunikationsforschung, no. 47. Bad Honnef. Germany (December 1988). Ingo Vogelsang, "Price-Cap Regulation of Telecommunications Services: A Long-Run Approach," in M.A. Crew, ed., *Deregulation and Diversification of Utilities* (Boston/Dordrecht/London: Kluwer Academic Publishers, 1988)

15. Michael E. Beesley and Stephen C. Littlechild, "The Regulation of Privatised Monopolies in the United Kingdom," in Cento Veljanovsky, ed., *Regulators and the Market*, IEA Readings No. 35 (London: Institute of Economic Affairs, 1991), p. 35

16. José Gómez-Ibáñez, forthcoming book "Commitment and Flexibility," chapter 9, on setting the efficiency factor in the U.K. water industry

17. The failure to do so led two major disputes in the mid-1990s in the U.S. in the telecommunications sector with respect to interstate access charges

18. Stephen C. Littlechild and Michael E. Beesley, "The Regulation of Privatised Monopolies in the United Kingdom," *RAND Journal of Economics*, vol. 20, no. 3, (Autumn 1989), pp. 454-472

19. Ibid.

20. Ibid.

21. In a sense, the review period simulates a long-term fixed-price contract. See William J. Baumol, "Reasonable Rules for Rate Regulation: Plausible Policies for the Imperfect World," in A. Phillips and O. E. Williamson, eds., *Prices: Issues in Theory, in Practice and in Public Policy* (Philadelphia: University of Pennsylvania Press, 1967). It is well-known, however, that fixed price contracts only work for short periods after which the potential for breach and renegotiation arises

22. It would not be fair to say that he had reneged on a commitment, as the price caps were not scheduled to take effect until April 1, 1996, but he had announced the "final price caps." See the comparison of Ian Byatt and Stephen Littlechild's 1994 price cap reviews in Dieter Helm, "Discussants Comments," pp. 31-35 in Michael E. Beesley, ed., "Utility Regulation: Challenge and Response," *London Institution of Economic Affairs* (1995)

23. Shimon Awerbuch, Leonard S. Hyman, and Andrew Vesey, *Unlocking The Benefits of Restructuring: A Blueprint for Transmission* (Vienna, Virginia: Public Utilities Reports Inc., November 1999), chapter 7

24. Ivo Vogelsang, "A 20-Year Perspective on Incentive Regulation for Public Utilities," Boston University (2001)

25. Office of Water Regulation (OFWAT), "Setting Price Limits for Water and Sewerage Services: A Framework and Approach to the 1994 Periodic Review" (November 1993), p. 6

26. For example, in 1994, the operating cost per unit of water supply was a key performance indicator, but in 1999 the analysis changed to the costs of a unit of water delivered, to reflect the increased attention to leakage since the 1995-96 drought

27. The water regulator, Ian Byatt, was given a knighthood as recognition of his services

28. Stephen C. Littlechild, "Regulation of British Telecommunications' Profitability," report to the Secretary of State, Department of Industry (February 1983). Richard L. Schmalensee, "Good Regulatory Regimes," *RAND Journal of Economics*, vol. 20 (1989), pp. 417-346

29. T. Langset, F. Trengereid, K. Samdal, J. Heggset, "Quality Dependent Revenue Caps-A Model for Quality of Supply Regulation," The Norwegian Water Resources and Energy Directorate and SINTEF Energy Research, Norway, n.d.

30. Michael A.Crew and Paul R. Kleindorfer, "Incentive Regulation in the United Kingdom and the United States: Some Lessons," *Journal of Regulatory Economics*, vol. 9, no. 3 (May 1996)

31. There are various descriptions of the history of sliding scale regulation, including Irvin Bussing, "Public Utility Regulation and the "So-Called" Sliding Scale" (New York: Columbia University Press, 1936) and Ralph Turvey, "The Sliding Scale: Price and Dividend Regulation in the 19th Century Gas Industry," Topics Note No. 16, National Economic Research Associates, London, England (1995)

32. David E.M. Sappington, Johannes P. Pfeifenberger, Philip Hanser, and Gregory N. Basheda, "The State of Performance-Based Regulation in the U.S. Electricity Industry," *The Electricity Journal* (October 2001). In this study, schemes relating to 28 electric utilities in 16 states were analyzed

33. David Currie, "The Labour Party's Approach to Utility Regulation: An Appraisal," in Michael Beesley, ed., "Regulating Utilities: Broadening the Debate," *IEA Publication* (1997). Stephen Littlechild, the former England and Wales electricity regulator, said in the Utility Week Supplement on 24 Seven Utilities Limited, n.d.: "I've always resisted an explicit profit-sharing mechanism in the past because companies might be able to use accounting manipulations to avoid returning a proper share of the savings to customers"

34. David E.M. Sappington and Dennis L. Weisman, "Designing Options in Incentive Regulation Plans", chapter 6, in *Designing Incentive Regulation for the Telecommunications Industry* (Cambridge, Massachusetts: The MIT Press, 1996). The firm's choice of options may also include traditional cost of service regulation in addition to specific incentive plans. For an overview of incentive regulation in the U.S., see David E.M. Sappington, Johannes P. Pfeifenberger, Philip Hanser, and Gregory N. Basheda, "The State of Performance-Based Regulation in the US Electric Utility Industry." *The Electricity Journal* (October 2001), pp. 71-79

35. Johannes P. Pfeifenberger, Paul R. Carpenter and Paul C.Liu, "REx Incentives: PBR Choices That Reflect Firms' Performance Expectations," *The Electricity Journal* (November 2001), pp. 44-51

36. Jean-Jacques Laffont and Jean Tirole, "Using Cost Observations to Regulate Firms," 94 *Journal of Political Economy* 614 (1986)

37. Michael A. Crew and, Paul R. Kleindorfer, "Incentive Regulation, Capital Recovery and Technological Change in Public Utilities," in Michael A. Crew, ed., *Economic Innovations in Public Utility Regulation* (Boston: Kluwer Academic Publisher, 1992), chapter 4

38. Jean-Jacques Laffont and Jean Tirole, "A Theory of Incentives in Procurement and Regulation" (Cambridge, Massachusetts: The MIT Press, 1993)

39. Johannes P. Pfeifenberger, Paul R. Carpenter and, Paul C.Liu, "REx Incentives: PBR Choices That Reflect Firms' Performance Expectations," *The Electricity Journal* (November 2001), pp. 44-51

40. Australian Competition and Consumer Commission, "National Electricity Market: Statement of Regulatory Intent for the Regulation of Transmission Revenues," Issues Paper (May 1998)

41. Office of Gas and Electricity Markets (OFGEM), "NGC System Operator Price Cap Control and Incentive Schemes under NETA: Final Proposals" (December 2000)

42. The Department of Economic Development, Belfast, Northern Ireland, Transmission and Public Electricity Supply License Document granted to Northern Ireland Electricity PLC, Schedule 5A (license modification dated December 6, 2000)

43. Tooraj Jamasb and Michael Pollitt, "Benchmarking and Regulation of Electricity Transmission and Distribution Utilities: Lessons from International Experience," DAE Working Paper 0101, University of Cambridge, Department of Applied Economics (December 2000). See also by the same authors, "International Benchmarking and Yardstick Regulation: An Application to European Electricity Utilities," DAE Working Paper 0115, University of Cambridge, Department of Applied Economics (October 2001) and "Benchmarking and Regulation: International Electricity Experience," *Utilities Policy* 9 (2001), pp. 107-130

44. I am grateful to Sam Lovick for his explanations and for sharing a paper with me on this subject produced by him for clients of London Economics LLC

45. D. W. Caves, L. R. Christensen and W. E. Diewert, "Multilateral Comparisons of Output, Input, and Productivity Using Superlative Index Numbers," *The Economic Journal,* 92 (1982), pp. 73-86

46. An empirical cost function framework has been developed. See W. E. Diewert and A. O. Nakamura, "A Survey of Empirical Methods of Productivity And Measurement," mimeo (1998)

47. Tooraj Jamasb and Michael Pollitt, "Benchmarking and Regulation of Transmission and Distribution Utilities: Lessons from International Experience," DAE Working Paper 0101, University of Cambridge, Department of Applied Economics (December 2000), pp. 29-30

48. Thomas-Olivier Léautier, "Regulation of an Electric Power Transmission Company," *The Energy Journal*, vol. 21, no. 4 (2000)

49. Sanford V. Berg and Jinook Jeong, "An Evaluation of Incentive Regulation for Electric Utilities," *Journal of Regulatory Economics*, 3 (1991), pp. 45-55

50. Regional Transmission Organizations, Notice of Proposed Rulemaking: Comments of Professor Paul L. Joskow, Docket No. RM99-2-000 (August 13, 1999)

51. Regional Transmission Organizations, Notice of Proposed Rulemaking: Comments of Professor Paul L. Joskow, Docket No. RM99-2-000 (August 13, 1999), p. 8

52. Office of Gas and Electricity Markets (OFGEM), NGC System operator incentive scheme, 2003/04—2005/06, initial consultation document, October 2002

53. Promoting Wholesale Competition Through Open Access Non-discriminatory Transmission Services by Public Utilities and Recovery of Stranded Costs by Public Utilities and Transmitting Utilities, Order No. 888, 61 FR 21,540 (May 10, 1996), FERC Stats. & Regs. ¶ 31,036 (1996). Open Access Same-Time Information System and Standards of Conduct, Final Rule, Order No. 889, 61 FR 21,737, FERC Stats. & Regs. ¶ 31,037 (1996)

54. Kiah Harris, "Impediments to System Expansion," presentation to the Sixth Electricity Conference, International Bar Association, Section on Energy and Natural Resources Law, Berkeley, California (February 26, 2002)

55. Paul L. Joskow and Richard M. Schmalensee, "Incentive Regulation for Electric Utilities," *Yale Journal on Regulation*, vol. 4:I (1986), pp. 1-49

56. Jonathan Raab, *Using Consensus Building To Improve Utility Regulation* (Washington, D.C. and Berkeley, California: American Council for an Energy-Efficient Economy, 1994)

57. Promoting Wholesale Competition Through Open Access Non-Discriminatory Transmission Services by Public Utilities and Recovery of Stranded Costs by Public Utilities and Transmitting Utilities, Order No. 888, 61 FR 21, 540 (May 10, 1996), FERC Stats. & Regs. ¶ 31,036 at 31,767 (1996)

58. Shimon Awerbuch, Leonard S. Hyman, and Andrew Vesey, *Unlocking The Benefits Of Restructuring: A Blueprint For Transmission* (Vienna, Virginia: Public Utilities Reports, Inc., November 1999), chapter 6 supplement

59. Shimon Awerbuch, Leonard S. Hyman, and Andrew Vesey, *Unlocking The Benefits Of Restructuring: A Blueprint For Transmission* (Vienna, Virginia: Public Utilities Reports, Inc., November 1999), chapter 6

60. Ibid.

61. Kent Lassman, "A Primer on Price Cap Regulation," Issue Analysis No. 85, Citizens for a Sound Economy Foundation, Washington, D.C. (February 25, 1999)

62. David Currie, "Mutualisation of Debt-Only Vehicles: Which Way for RPI-X Regulation," the Beesley Lectures on Regulation, Series XI, London Business School (November 13, 2001)

63. Michael A. Crew and Paul R. Kleindorfer, "Incentive Regulation, Capital Recovery and Technological Change in Public Utilities," in Michael A. Crew, ed., *Economic Innovations in Public Utility Regulation* (Boston: Kluwer Academic Publishers, 1992), chapter 4

64. Michael A. Crew and Paul R. Kleindorfer, " Incentive Regulation, Capital Recovery and Technological Change in Public Utilities," in Michael A. Crew, ed., *Economic Innovations in Public Utility Regulation* (Boston: Kluwer Academic Publishers, 1992), chapter 4

65. Alfred E. Kahn, *The Economics of Regulation: Principles and Institutions* (New York: John Wiley & Sons, 1970), vol. 1, p. 18

66. Promoting Wholesale Competition Through Open Access Non-Discriminatory Transmission Services by Public Utilities and Recovery of Stranded Costs by Public Utilities and Transmitting Utilities, Order No. 888, 61 F. R. 21,540 (May 10, 1996), FERC Stat. & Regs.¶ 31,036 (1996), aff'd, Transmission Access *Policy Study Group v. FERC*, 225 F.3d 667 (DC Cir. 2000)

67. *New York v. FERC*, 535, U.S., Docket No. 00-658 (March 4, 2002)

68. For a fuller discussion see CMS Cameron McKenna, "Focus on Energy," vol. III no. 1 (March 2002)

69. FERC Order 2000, Docket No. RM 99-2-00 (issued December 20, 1999)

70. In her "LEAP Letter" of May/June 2000, Janet Gail Besser points out that there are a number of challenges to implementing PBR for transmission. "Properly setting the cast-off rates may be difficult, and the anticipation of a traditional rate case as the starting point may be a disincentive for companies to propose PBR. Indirect costs are difficult to forecast. Ensuring revenue adequacy requires balancing [congestion] and price signals-accounting for embedded and incremental costs, and addressing the dilemma that indirect cost reduction can lead to lower revenues." Janet Gail Besser, "The Importance of Pricing Transmission Properly," LEAP Letter, Lexecon Inc. (May/June 2000)

71. FERC Docket No. RM01-12-000

72. David Clement, "The FERC's Innovative Rates: A New Tool for Transmission," presentation to CERA Electric Transmission Executive Retreat, "Creating a

Competitive Grid: Opportunities and Challenges," Chicago, Illinois (October 23-25, 2001)

73. Regional Transmission Organizations, Notice of Proposed Rulemaking: Comments of Professor Paul L. Joskow, Docket No. RM99-2-000 (August 13, 1999), p. 6

74. FERC Order On Transmission Incentive Pilot, Docket No. ER01-2922-000, New England Power Pool, 97 FERC P 61,093 (2001) at 61,480

75. Ibid, at 61,480

76. Kiah Harris, "Impediments to System Expansion," presentation to the Sixth Electricity Conference, International Bar Association, Section on Energy and Natural Resources Law, Berkeley, California (February 26, 2002)

77. FERC Notice of Proposed Rulemaking, 18 CFR Part 35, Docket No. RM01-12-000 (July 31, 2002)

78. E-mail to the author

79. *Bluefield Waterworks and Improvement Co. v. Public Service Commission*, 262 U.S. 679 (1923), pp. 692-693

80. *FPC v. Hope Natural Gas Co.*, 320 U.S. 591 (1944), p. 603

81. It is advocated by Professor Paul L. Joskow in Regional Transmission Organizations, Notice of Proposed Rulemaking: Comments of Professor Paul L. Joskow, Docket No. RM 99-2-000 (August 13, 1999). Thomas Olivier Léautier, "Regulation of an Electric Power Transmission Company," *The Energy Journal*, vol. 21, no. 4 (2000)

82. Sally Hunt and Graham Shuttleworth, *Competition and Choice in Electricity* (London: John Wiley & Sons, 1996)

83. This is the efficiency factor for 2002-2006

84. Office of Gas and Electricity Markets (OFGEM), NGC System operator incentive scheme, 2003/04—2005/06, initial consultation document, October 2002

85. Lehman Brothers Global Equity Research Report on NGC, October 2001

86. Goran Strbac, "Concept of Reference Network: A Tool for Transmission Regulation and Expansion," presentation to Harvard Electricity Policy Group, San Diego, California (January 24, 2002)

87. For an account of the England and Wales regulatory regime since 1990, see Steve Thomas, "National Grid Co–too lightly regulated?" *Power in Europe* 336 (October 27, 2000), pp. 27-29

88. Department of Trade and Industry, Electricity Transmission License Special Condition AA5 (September 28, 2001), pp. 25-57. It is beyond the scope of this text to provide a detailed description as the provisions with respect to this restriction and the price cap formula extend more than 30 pages

89. Office of Gas and Electricity Markets (OFGEM), "NGC System Operator Incentive Scheme from April 2002: Final Proposals" (February 2002)

90. The real difficulty with an incentive scheme of this type is that there is considerable uncertainty as to the level of the costs in question. In the early days of the schemes, even drawing on the experience gained with them over the previous review periods, the uncertainty was such that NGC was happy not to let an agreed scheme run beyond a two-year period

91. Jack Neushloss and Fiona Woolf, "Review of the England and Wales Trading Arrangements: The Proposal to Cure The Ills by Euthanasia of the Pool," *The Electricity Journal* (December 1999), pp. 27-38

92. Office of Gas and Electricity Markets (OFGEM), "NGC System Operator Incentive Scheme from April 2002: Final Proposals" (February 2002)

93. Jean-Jacques Laffont and Jean Tirole, *A Theory of Incentives In Procurement and Regulation* (Cambridge Massachusetts: The MIT Press, 1993)

94. Thomas Olivier Léautier, "Regulation of an Electric Power Transmission Company," *The Energy Journal*, vol. 21, no.4 (2000), pp. 61-92

95. Thomas-Olivier Léautier, "Regulation of an Electric Power Transmission Company," *The Energy Journal*, vol. 21, no. 4 (2000), p. 81

CHAPTER 8

Preparation: Standards and Planning Coordination

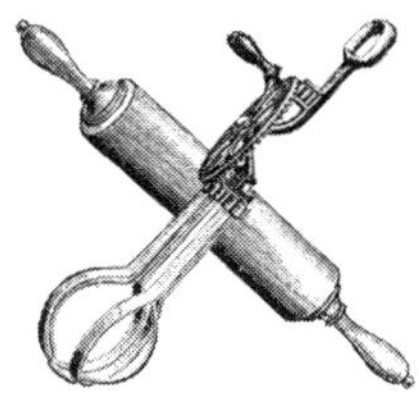

Ingredients of this Chapter

Once the key ingredients—the raising and the setting agents—have been assembled, the next task is to embark on the actual preparation. In this chapter, we shall begin by looking at planning and performance standards as an ingredient and starting point for a planning coordination process, why they are necessary, and the reason they are usually regulated. We shall then discuss a number of issues relating to the setting of standards—the objectives, the trade-offs, the need for standards to be measurable and realistic, the differences between economic and technical standards, as well as the expertise and information needed to set the standards. We shall then consider who should set and enforce the standards and some issues related to implementation and enforcement. Because the traditional planning and operating standards are highly interrelated with the performance targets, indicators, and standards that are used in incentive or performance-based regulation, we shall look at them together.

Having discussed standards as a key ingredient in the planning coordination process, we shall move to the process itself. We shall consider the need for central coordination of transmission planning and who should carry it out. We shall look at a number of different planning tools that are available to identify the optimal expansion that is needed. After a discussion of the kind of plan that can be produced, we will look at regulatory approval of expansion plans and the role of the regulator. Finally, we shall consider who should build the projects identified in the planning process.

The Need for Planning and Performance Standards

Standards, rules, and procedures

Many jurisdictions impose—either through legislation, regulation, or market rules—performance, planning, interconnection, connection, operating, and reliability (security) standards with respect to transmission. Standards are a key ingredient for the purposes of protecting consumers as to quality of service, adequacy and reliability.

There are also standards that relate to operations between transmission owners and system operators in different control areas that are fundamental to the ability of the systems or markets to operate on a regional basis and for interregional trading to take place. There are standards where harmonization will greatly facilitate operations and trading, where there is no justification for local differences. There are also a number of other performance standards that need to be set and enforced or, at least, measured for the purposes of regulatory performance-based incentive schemes.

Standards can be contained in laws and regulations. They can be cast as technical rules or operating procedures in market rules and protocols. The distinctions between the provisions of standards, rules, codes, protocols, and procedures are now quite indistinct. In this chapter, we shall refer to any requirement as to the planning or performance of a transmission system or transmission owner—wherever it is to be found—as a standard. Although we will not be focusing on the standards that apply purely to system operations, they are highly interrelated.

Planning Standards

Planning standards under the spotlight

Most jurisdictions have well-established planning standards in place, the most common planning standard being n-1. In other words, the system is planned to withstand the loss of a single component. A single component could be a line, transformer, or generator, for example. It does not have to be the largest infeed. The standard gives a list of consequences that are unacceptable, such as loss of supply. Transmission planning standards are different from generation planning standards that focus, for example, on the margin between demand and available generation. An n-2 standard is to be found in England and Wales and in New Zealand.

In recent years, performance, planning, interconnection, and operational standards have been put under the spotlight. The question of relaxing standards has arisen because of:

- pressure from generators and loads that do not need (or wish to pay for) such a high degree of security
- technological advances that allow more accurate monitoring of system conditions (*e.g.*, temperature sensors)

They are often codified into Grid Codes, market rules, or technical rules during the course of deregulation, in order to create some transparency as to the standards to which the transmission system is to be planned, built, and operated. This facilitates new entry and provides a mechanism for keeping these standards under review. It enables users and regulators to have some input into how they are set.

Planning standards are useful to create some certainty for system operators, transmission owners, and market participants as to what they should expect and the minimum entry requirements. If there were no planning standards, there would be no benchmark by which to assess whether expansion was needed to maintain reliability and security of supply.

It is clear that in attempting to reach agreement with a user over the standard of connection assets and level of reinforcement to be provided, without some system of planning standards that the transmission owner and all users must adhere to, the position of the incumbent transmission owner could be extremely difficult. It could be put in a position where it would be called upon to reduce standards in individual cases. System integrity and reliability could be impaired through no fault of the transmission owner or the system operator, simply by the demands placed upon it by users to make special arrangements in order to reduce the costs of connections and reinforcement of the system. Enforceable standards minimize the scope for dispute and litigation.

The relationship with operating standards

Reliability can usually be maintained provided the operational standards of security are high enough. If there has been only modest investment in the system, then to maintain reliability by keeping to the operational standards, it may be necessary to prevent certain generation from running or not supply some demand (or both). However, if the transmission access arrangements and market rules are well designed, it should be possible to create a regime where it is not in the interests of generators and loads to skimp on the quality of the transmission system service or investment they are prepared to pay for.

Regulatory protection

In New Zealand, where there is no regulator, planning standards have been a source of controversy and dispute. In the U.K., where the standards are considered to be technically high (most of the system is built to an n-2 standard), the question did arise as to whether generators and loads should be free to negotiate different standards with NGC. This gave rise to concerns about the relative bargaining powers of different types of generators and loads and also the potential for discriminatory practices between users on the part of NGC. As a result, both NGC and users enjoy some regulatory protection[1] through NGC's license and the Grid Code. By contrast, Transpower in New Zealand is not able to impose certain standards as a minimum.

There is a strong argument in favor of minimum performance and planning standards being imposed (rather than agreed) in order to achieve the right balance between cost and quality of service, rather than leaving it to the market participants to agree among themselves. The right balance is always difficult to achieve, and market participants will, inevitably, look at the issue from their own perspective. Without understanding the consequences, users will want to negotiate a relaxation of standards if it saves them money, and will not be willing to accept arguments that the standard is necessary to maintain system reliability and the quality of service to other users of the transmission system. As experience in New Zealand has indicated, negotiations can be lengthy and without any satisfactory outcome.

Planning standards, however, are usually technical standards that remain in place for years without being reviewed. There is an argument that it is not easy to revise them. Once one part of the system is built to a particular standard, it may be extremely difficult to build other parts, particularly those that are adjacent, to another standard and to provide service tailored to an individual user. Regulators have reviewed them on the basis of the changing objectives for the industry and the consumers that it serves. However, they are increasingly focused on replacing technical standards with economic (cost-saving or cost-related) standards.

The reason for this is that standards do impose costs, and there is an inevitable trade-off between cost and quality of service. If the standards are set too high, they will raise prices, and if they are set too low, they could reduce the quality of service, compromising system reliability. This is the primary reason that they are usually part of the regulatory regime.

Performance Standards and Targets

Objectives

The setting of standards in the past may have been driven largely by engineering and reliability considerations, as well as the historical development, of the system that was designed with the primary objectives of connecting new generation and loads and reinforcing the system to meet demand growth. Now, the focus is on performance and output on a forward-looking basis to meet very specific regulatory and market-based performance objectives. Regulators will look at the need to maintain or improve performance or control costs in areas that are critical to the needs of consumers and market participants, and the proper functioning of the market. In the transmission context, the standards could be designed to incentivize:

- better control of the direct costs of transmission (*e.g.*, maintenance and construction costs)
- better control of the indirect costs of transmission (*i.e.*, congestion, certain ancillary services, and electrical losses)
- maintaining or improving the technical quality or operational characteristics of various elements of the transmission system
- maintaining or improving the availability or transfer capability of the transmission system (or key parts of it)
- investment and other action to minimize forced outages of the transmission system
- more rapid restoration times after forced outages
- facilitation of the transmission rights market by maintaining capacity at a level that supports and maximizes the value of the rights

Performance must be capable of measurement

It is paradoxical that the electricity industry, which has a long tradition of setting and performing according to standards, does not have a long tradition of rewarding utilities by reference to performance measurements against the standards. The electricity industry does keep fairly detailed records of incidents that result in interruptions in service to consumers. These statistics tend to be a measure of how well a system works. However, they are extremely difficult to interpret in the transmission context, because the number of serious problems (blackouts) is very small. Some of the incidents that occur are the result of factors such as the weather, which are outside the control of the transmission owner. The statistics can exhibit variations and volatility that are also difficult to interpret.

Perhaps a better indicator of transmission system performance is the extent to which generation despatch is constrained. If the answer is "never", the system has probably been overbuilt. Although records of when consumers have been cut off and when there are constraints are usually available (and hence monitoring can be carried out), there may not be comprehensive records available when the system has been operated outside the operational standards. Provided a critical contingency did not occur, nothing noticeable will have happened and the operator will have "got away with taking the risk."

Performance-based regulation involves setting key performance indicators, performance or cost targets, benchmarks of performance, and standards of service and construction. We will refer to them all as standards. It goes without saying that performance must be measured at periodic intervals and at the end of the regulatory review periods. Typically, performance will be measured at the time the charges are submitted for regulatory approval, or when an incentive formula is reviewed and reset. Therefore, in order to avoid disputes, the standards must be capable of accurate measurement and not open to interpretation.

Standards should be realistic

The reason it is challenging to set standards for transmission (rather than distribution) is that the standard measures of performance that are critical to a consumer, *e.g.*, minutes lost or the number of outages (*i.e.*, lack of service) can be attributable to many factors outside the control of the transmission owner and the system operator (*e.g.*, lack of generation or a distribution fault). Clearly, for a transmission standard or performance target, the only customer minutes lost that matter are those due to a transmission system deficiency.

The main difficulties of setting standards or targets for transmission are the low probability and high impact of interruptions to customers as a direct result of transmission performance, the difficulty of setting a target level of constraint costs (which are affected by market participants' behavior), and the near impossibility of an outsider monitoring whether operational standards of security are being complied with. This combination of low probability and high impact of lack of transmission performance is not useful as the basis for performance indicators or incentives. The low probability of the event occurring would weaken the power of the incentive, and, if the event did occur, the appropriate level of penalty would probably have to be set high enough to put the transmission owner out of business.

Moreover, standards and cost targets need to be carefully crafted to work together, otherwise they can create perverse incentives. For example, a standard that is designed to minimize interruptions in service for more than a target period of time may lead to behavior that results in a series of interruptions, each lasting less than the specified period. Penalties or an obligation to pay compensation for lack of service may, if severely set, result in a reluctance to take circuits out for maintenance when it is needed. If the maintenance is justified, then a reluctance to take transmission facilities out of service would result in poor reliability and potential failures that might be penalized. All justified maintenance outages should be taken at the right time. The key objective is to avoid giving the transmission owner the incentive to take only a short-term view.

The obvious performance measures would be in terms of circuit availability or transfer capability, restoration time after system incidents (subject to the actions of the system operator and certain *force majeure* events), and the delays in returning from planned maintenance, if the delay had a significant effect on reliability or congestion.

The allocation of functions and responsibilities between the transmission owner and the system operator will need to be taken into account, partly because their performance depends upon the same assets to some extent (and in some cases, they may even use the same staff). Their interaction is extremely complex and varies from system to system. Both of them have a role in maximizing the availability of the transmission facilities. Although much may depend on the industry structure and the market rules, the primary role of the transmission owner is to maximize and maintain transmission capacity and transfer capability, whereas the system operator manages the indirect costs of transmission with a view to minimizing them (*e.g.*, through efficient congestion management and procurement and use of ancillary services).

As we discussed in chapter 7, benchmarking between comparable transmission systems can be useful to set performance standards and targets. However, international comparisons may be problematic. Because the needs and circumstances vary, the transmission systems represent an accident of the history of their development and local load growth, and data collection methods are often different.

Economic standards and cost targets

As we have observed, there has been a move away from technical standards to economic performance targets. Standards or targets that relate to the control of costs may be more easily measured, but they may be difficult to set at the right level, unless there is good information available as to anticipated costs and the potential for cost savings and productivity gain. Similar issues arise in setting the X factor in the price cap we discussed in chapter 7. Because guesswork is usually involved in setting the targets, floors and ceilings may be set to protect both the transmission owner and the consumer. The significance of the impact that the incentive may have on creating shareholder value for the transmission owner is often a guiding factor in the floor and ceiling negotiations. For this reason, the cost-based or economic standards will typically be reviewed more often than technical standards. An incentive scheme based on any performance target, as we have seen in the case of NGC, involves investment that will need to be appropriately remunerated. This involves designing all of the aspects of the regulatory regime for transmission so that they work together to incentivize the achievement of the desired outcome.

If the standards are economically based (*e.g.*, valuing customer interruptions at a $/KWh figure and calculating constraint costs), then they should work seamlessly in all areas. It is only when standards are expressed deterministically (which they often are) that they may justifiably vary from area to area.

Technical standards and performance indicators

The technical standards are usually set by reference to current performance indicators (but not necessarily at current performance levels, although they are often used as the initial indicators), if only to monitor any deterioration in performance. They are usually relatively easy to measure. Standards may be easier to set where new technology is used, where the time needed for maintenance can be established with more certainty. Very old assets are inevitably more unpredictable, and historic performance may be no indicator as to future forced outages and maintenance and repair requirements. As Figure 8-1 shows, outage rates for transmission equipment usually follow a "bathtub" curve with a progressive drop at the beginning of the life of the asset when manufacturing problems might surface, followed by a long period with relatively few problems and then a sharp increase as the asset reaches the end of its useful life.

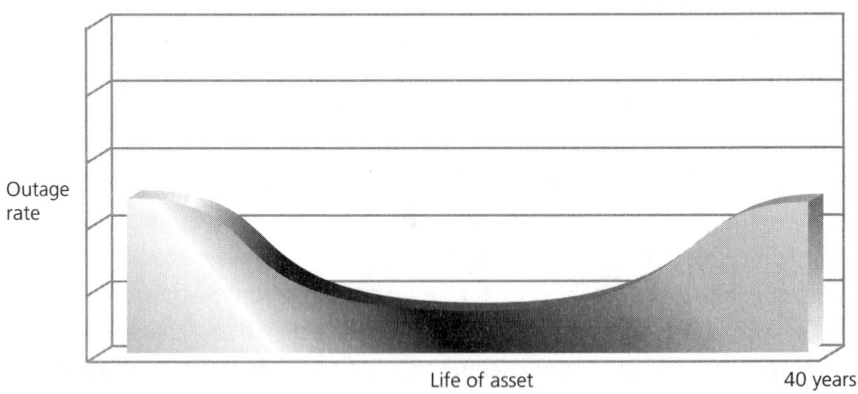

Fig. 8-1 Likelihood of failure over life of asset

However, there may be little value in creating incentives to improve performance in certain technical areas, because there might only be marginal beneficial effects on the quality of service that the consumer receives, and the costs may not be warranted. Improving performance in technical areas may be helpful in dealing with certain reliability problems. Technical standards will not be a useful tool to deal with congestion that primarily causes cost. Economic targets will produce more effective performance to reduce those costs than technical standards. The linkage between reliability and congestion is such that a cost-based scheme designed to reduce congestion may also improve reliability. Congestion can also be mitigated by reducing the security margin and reducing the amount of maintenance being carried out. These two approaches reduce reliability of the system in the short- and long-term, respectively. Investment to improve reliability may also reduce congestion.

Examples of technical performance standards

The types of technical performance measures that relate purely to the transmission system asset ownership, investment, management, and maintenance (as opposed to system and market operations) that would be suitable for use in regulatory schemes will depend upon the objectives, the existing standards, and the ability to measure. Examples might include the:

- availability of the transmission system as a whole (or key sections)
- availability of interconnectors with neighboring systems
- availability of associated protection systems
- sustained forced outage rate for all (or the key) transmission lines, interconnectors, and transformers (number of incidents), drawing a distinction as to whether the cause was within the control of the transmission company, *e.g.*, operating or maintenance error, storms, or sabotage
- mean duration of forced outages using the same distinction
- percentage of successful auto-reclosure achieved following transient faults on transmission lines

- percentage of incorrect protection system responses
- percentage of time the system operates below the planning (n-1 or n-2) standards

In theory, it would be possible to measure the availability of parts of the transmission system (*e.g.*, circuits, transformers, static VAR compensators, synchronous compensators, capacitor banks, protection systems), but this runs counter to the principle that it is better to incentivize transmission owners to achieve the required performance with respect to the entire transmission system on a comprehensive basis.[2] Otherwise, a perverse incentive may be created to not carry out maintenance, because it involves taking out part of the system on which availability will be measured. Maintenance should be encouraged when a facility is not needed to be in service. The performance target should be focused on the overall quality of transmission service to the market and the consumer, not on whether one line or transformer was available for a specified number of hours.

It is important to not take the possibility of improving the ability of the system to withstand storms out of the equation entirely, because, although it will involve additional cost, a transmission owner can build to different standards to withstand storms. There needs to be some agreed standard, otherwise the transmission owner will obviously minimize its investment and maintain that it is not at fault if it takes a long time to repair the system after a storm. Bad weather events are foreseeable to some extent and certainly need to be taken into account in setting standards.

Knowledge and expertise required to set standards

The expertise to set technical standards resides in the industry. The number of transmission owners and system operators involved in a single jurisdiction adds complexity, because all systems have a different history of their development that impacts the way in which the planning standards have been set. If there is more than one transmission system, there will be a need for greater information and coordination in order to set standards covering regional coordination and interoperability, as well as standards for individual transmission systems that do not merely reflect the lowest common denominator.

Interaction with market rules

All standards and transmission performance indicators, whether they are cost-based or technical, will be highly interactive with the market rules and their implementation. The standard setting and enforcement will also depend upon the degree of complexity and detail of the market rules. Harmonization of the existing standards and the best practices to be reflected in the market rules will need to be undertaken.

Review of standards

Setting the standards initially is one thing, but they also need to be kept under regular review. Not only are the industry and the technologies that it uses highly dynamic, but so are market forces. Unexpected operational outturn can result in the need to look at existing standards with an open mind as to whether they are still justified or whether the standards should be strengthened or relaxed. It may be decided, with the benefit of experience, that certain traditional standards can no longer be justified from the point of view of cost-benefit to the consumer. If they are economically set, this problem should not arise, except to the extent that the valuation of an interruption to supply may change with time, for example.

Who Should Set and Enforce the Standards?

The options

The question arises as to who should set and enforce the performance standards and how they should be kept under review.

The choices are:

- the regulator or the entity that sets or approves tariffs, if there is no regulator
- a self-regulating technical organization like the North American Electricity Reliability Council (NERC)
- a hybrid model like a Grid Code review panel (used in all three U.K. jurisdictions)

Arguments in favor of the regulator setting and enforcing the standards

There are a number of arguments in favor of the regulator (or the agency that sets or approves tariffs, if there is no regulator) taking the major role in standard setting and enforcement. We have already noted that all standards involve costs for market participants and, ultimately, the consumer. It would be the role of the regulator to decide if the standards were justified after having reviewed the necessary cost-benefit analyses. Inevitably, it will have to decide whether the costs of meeting the standards are justified, because they will be included in the permitted cost or revenues of the regulated transmission owners. For this purpose, the regulator will need to take into account and rule on the differing viewpoints put forward by the various stakeholders. It will need to consider three key issues:

- Barriers to new entry in all sectors of the industry (particularly generation and transmission) will be created if the standards are set too high, and the cost allocation rules are arbitrary and inefficient. Standards can create the potential for the exercise of monopoly abuse and market power by transmission owners that is very much the concern of the regulator
- Given that the standards will be highly interactive with the trading arrangements or market rules, the role of the regulator in overseeing the market and the changes to the market rules will have an important effect on the way in which the standards are reviewed

- Monitoring and enforcing compliance with the standards will also be a key ingredient in the role of the regulator in monitoring how well the market is functioning and whether the market rules need to be reviewed or revised

These arguments support a single regulatory authority having full responsibility for setting and enforcing standards. However, the regulator is unlikely to have all the technical expertise and industry knowledge necessary to enable it to envisage every eventuality and to understand the day-to-day operational needs of each transmission system within its jurisdiction. For this reason, in some countries, the regulator asks the transmission owners to propose standards and to keep them under review. The standards are scrutinized by the regulator, often with the help of consultants (as in the U.K. and Australia), and then approved. The role of the consultants can be critical, especially where the standards represent a contentious issue. The consultants must be neutral, independent, and credible. They must not have any particular industry bias.

An industry body or self-regulating organization

The regulator could also use an industry body or self-regulating organization (SRO), such as the NERC, to provide the technical expertise it lacks. It could give the industry body an advisory role or delegated powers to carry out certain functions associated with standard setting, monitoring, and enforcement on its behalf. However, given the extent of its duties, particularly in relation to tariff setting, market oversight, market power monitoring, and monopoly regulation, it is difficult to justify the creation of a second regulatory agency purely for the purposes of standard setting, monitoring, and enforcement. In any event, it would be entirely inappropriate for any organization other than the regulator to set regulated performance standards or targets that were based on the cost performance of an individual transmission owner.

Advantages of an SRO. A specialized, industry-based, self-regulating organization is an attractive model, of which NERC is a well-known and well-respected example. An industry body should be in touch with the relevant issues and be able to draw on the best technical expertise. Much will depend upon the governance mechanism as to whether the organization is able to act independently and neutrally, especially where there are controversial issues and complex inter-area coordination problems that need to be addressed. NERC recently adopted an independent board structure to overcome arguments that it was capable of being captured by certain industry interests. However, the industry has a strong advisory role.

Disadvantages of an SRO. The model suffers from the obvious disadvantage that it may be criticized for adopting the lowest common denominator that the industry is prepared to live with. If the governance structure includes an independent board to overcome conflicts of interest, it is important that the board should not become completely isolated from the concerns and advice from the industry that must have broad representation. It is also open to criticism for being an organization unlikely to be motivated to look after the interests of the consumer. It will tend to focus purely on technical standards, without taking into account the commercial and market implications. For example, NERC standards have not traditionally taken into account the market needs or congestion costs. The regulator would still need to approve all standards that are set, because of their impact on prices and quality of service, for which it may need to hire the necessary unbiased, consulting expertise.

The role of the system operator

The system operator clearly has a key role to perform for two reasons. The system operator will be the front-line monitor of performance and market rule compliance. It will also be the primary source of information for standard enforcement purposes. Secondly, as custodian of the market rules in many jurisdictions, it will be in a position to advise on the need to change standards in order to meet the needs of the market and to enable the market to operate more effectively. There is a strong argument in favor of giving the system operator the primary monitoring and enforcement role in relation to technical standards, if it is an entity that is independent from the transmission owners and market participants.

However, there are two important issues that must be addressed in relation to system operators—the complexities of the independent governance mechanism and the ability of stakeholders (that may have conflicts of interest) to influence the monitoring and enforcement that is carried out. The system operator must be in a position to enforce standards on a basis that is entirely non-discriminatory within clearly defined rules and regulatory objectives.

Who polices the system operator?

Another important issue that needs to be addressed is the system operator's compliance with the rules, and which entity will carry out the monitoring and enforcement of the system operator's performance. Given that these organizations will have been set up as independent bodies with the sole focus of complying with and implementing standards and rules, there is a strong argument that any non-compliance should be dealt with by the regulator, because it will create serious concerns from a market and public interest perspective. However, the regulator may well be in difficulty in obtaining the necessary information, because it will not be able to undertake daily or real-time monitoring for itself.

There is an argument for setting up a separate monitoring regime that could be both proactive and also provide a mechanism for market participants to give information or to register complaints as to the performance of the system operator. This could perhaps be delegated to a self-regulating organization of the type described earlier. These complaints could then be investigated by the organization, which would then report to the regulator. While this has its attractions, it also could have a significant cost if elaborate software is to be developed for day-to-day or real-time monitoring. In a transparent market, every market participant is an active policeman of the performance of the system operator, well able and motivated to lodge complaints against the system operator with the regulator on its own. However, market participants cannot be relied upon to protect the consumer interest.

Grid Code review panel: A hybrid model

The final model is a hybrid model such as the Grid Code review panel. This is a technique that was incorporated into the original England and Wales Grid Code to ensure that the technical standards and rules it contained did not become ossified or inflexible in light of the needs of the market. Usually, the market rules will have their own mechanism for review and change, but the regulatory regime can also impose review and revision obligations. The panel is comprised of market participants, transmission and distribution owners, and the system operator.[3] Therefore, it is self-regulating, but the model is hybrid in nature, because there is a strong role for the regulator in approving changes. A regulator also has the ability to compel changes to be made if it does not agree with the panel's proposal. The regulator may even appoint a representative to sit on the panel, either as an observer or as a full member. In other words, it is hybrid, because it represents a combination of the regulator and industry participants.

The panel has no role in monitoring and enforcement. The Grid Code is enforced contractually through the Connection and Use of System Code and by the regulator through the industry licenses.

Implementation and Enforcement of Standards

A single agency to set and enforce standards

The monitoring and enforcement of standards are often looked at separately from the setting and imposing of standards. However, there is an argument that if a single agency is empowered to carry out all of these activities, it is likely to produce a better answer. This is because its experience in monitoring and enforcement will add value to the knowledge needed for setting and overseeing the implementation of the standards in the first place. NERC has found that there was considerable scope for interpretation and dispute in the implementation of the standards it had set, even where they were thought to be common sense and fundamental to the operation of any transmission system. The difficulties that it encountered may, in part, have been drafting problems, where the rules were written as guidelines rather than detailed obligations. Over the years, as NERC tightened up the drafting, the scope for dispute over interpretation and implementation may have diminished, but it still remained.

Technical criteria may not be strictly enforced

It is sometimes the case that although technical standards are imposed on system operators and transmission companies, they are not strictly enforced through contractual or regulatory mechanisms. This may be because there are limited means to monitor compliance (as we have observed with respect to operational standards) or because the system operator or transmission owner is trusted to comply with them. In other cases, it may be that the standards are deliberately relaxed in order to reduce costs in suitable cases or to enable the system operator to avoid an emergency or system incident. Well-drafted standards may allow for both of these possibilities.

Transpower, in its submission to the post-winter (2000) review of the New Zealand electricity system[4] expressed the view that investing in expensive and long-lived assets such as new transmission lines to relieve constraints, especially those that occur under abnormal operating conditions, such as a dry-year events on a mainly hydro system, will not always be economically justifiable. A temporary relaxation of local security criteria or standards was likely to represent a more efficient means of managing dry-year events that occur only infrequently. As a transco (discussed in chapter 3), it was able to make trade-offs between asset ownership and system operations. While Transpower had its own contingency plans and had the ability to act unilaterally in response to market conditions, it was looking for wider industry support to be given to the temporary relaxation of security standards, given the lack of a regulator.

Compensation and penalties

The non-compliance may also have created a right on the part of a market participant to compensation under the market rules. It would normally be regarded as unfair for the enforcement agency to be able to penalize an organization for the same non-compliance a second time. In any event, the question arises as to what the enforcement agency should do with the fines or penalties. Compensation for economic loss and penalties by way of punishment are separate concepts. This argues for the market rules, the regulatory regime, and the enforcement mechanism to be well coordinated.

The Key Ingredients in Planning

Traditional approach to planning

Using the planning standards or criteria as a starting point, transmission system planners have traditionally undertaken the planning process as an exercise in anticipating the transmission needs of generation and loads using techniques that assumed a degree of vertical integration. Planning on this basis, therefore, tended to exhibit some of the characteristics of the command-and-control style of planning.

At its very simplest level, it involved preparing a forecast of load levels in total and at each node on the system and a system model. In addition to the load forecast, anticipated patterns of generation would be factored into the model with a likely generation expansion and closure (retirement) plan. The planners would take into account the concerns and needs of the system operators from a reliability perspective, and the planning standards or criteria would be applied. This would involve contingency analyses covering loss of in-feeds, system peaks, potential voltage collapse, synchronous resonance, and other issues. The main objective was to be able to supply load with the necessary reliability. The planners would look at a variety of different circumstances within the planning criteria and contingency analyses to derive a series of options available to cure the perceived transmission needs and problems. The plan would typically cover five to seven years ahead.

There would be a thorough cost-benefit analysis, and a series of net present value (NPV) calculations would be carried out. The cost of capital and the present value of funds over depreciation of the assets that might be constructed would be calculated. Economies of scope and scale would be assessed, and new technologies would be evaluated. The cheapest solution might not be chosen if the system-wide benefits of a more expensive solution were thought to be greater.

Undertaking the planning in an unbundled industry

Clearly, the planning process becomes more complex in unbundled industry structures and de-regulated markets, where different functions are carried out by different corporate entities.

The advent of new generation and plant retirement make planning more difficult for a transmission owner and a system operator in a competitive wholesale market, where generation planning is no longer centralized, but is left to the market to elicit opportunistic behavior in response to energy price signals that need to be locational if they are to be effective. New technologies and construction techniques have accelerated generation plant development and construction times, and volatile energy prices can result in the retirement of older, less flexible, or uneconomic plants at short notice.

Demand (load) forecasting becomes more difficult, because the relevant data have to be collected from a number of different sources. Demand should become more elastic in response to price signals, particularly where retail competition is introduced. The system operator may retain the best data with respect to demand (although not necessarily with respect to potential growth). Although distributors and retailers can help to build demand forecasts, obtaining meaningful and informed input from consumers is difficult. Ashok Manglick, the planning manager of Transgrid explained the main reason: "One of the difficulties, at least in Australia, is that electricity is not an 'issue' to end-users until the lights go out and that doesn't happen frequently enough to keep their attention."[5]

In addition, power flows may change as a result of deregulation. Variations by time of day or season may become even more noticeable. All this will be in response to the new signals that the trading systems and markets create.

If the ingredients that we discussed in chapters 4 and 5 are in place for market-based transmission expansion projects to flourish, transmission projects could be proposed and implemented relatively quickly by comparison to their larger regulated counterparts. As we shall discuss, they could affect the need for or viability of other planned transmission projects that are in the course of development.

The combined effects of Order 888 and Order 889

The combined effects of Orders 888 and Order 889 in the U.S. make planning extremely difficult.[6] As we discussed in chapter 7, under Order 888, a developer wishing to construct a new generation plant must apply to the local transmission owner for interconnection, and the request would be placed in a queue of applications that are studied in order of submission. There is another queue for the open access same-time information system (OASIS) under Order 889, for which the developer must apply to the system operator to use as many (fictional) contract paths as it thinks it may need in order to reach its customers. The OASIS filing is focused on only transmission constraints where paths may be constrained because of applications for notional contract paths or actual power flows.

The queue consists of generators and other market participants needing to secure contract paths to enable them to do business with their customers. Obviously, it is extremely difficult in a dynamic marketplace for market participants to anticipate their needs. Typically, generating plant developers and market participants will file applications for more capacity than they need to cover all eventualities, and it is clear that not all the plant for which applications are made is ultimately built at the planned capacity (or at all).

The combination of Order 888 and 889 processes have left great uncertainty with generation developers, transmission owners, and other market participants. The planning process involves enough inherent uncertainties without more being added by these rules.[7] As we have observed, the problems created by Order 889 would not arise under the locational marginal pricing system we discussed in chapter 4, which FERC proposes to adopt under its mandatory standard market design.[8]

The need for central coordination of the planning process

In the deregulated world of competitive markets, the notion of central planning from the old command-and-control environment is anathema. The vision of the Task Force on Electric System Reliability in a report to the U.S. Department of Energy was:

> *[I]n the future, decisions regarding whether to build transmission or generation or both, or to dispatch customer load reductions will be made by multiple market participants, with decisions about one approach or another being informed by but not necessarily integrated with decisions about other approaches.*[9]

Arguably, the philosophy of no central planning lies behind the design of the Argentine expansion system (discussed in chapter 5), in which no coordinated planning is formally undertaken, even though some of the transmission companies do issue some elements of indicative planning statements. People may agree that there were benefits of vertical integration when it came to planning, but there are greater benefits to vertical separation. The question is whether, after this separation, it is necessary to attempt to capture the benefits of vertical integration and, if so, how.

There are several arguments in support of a centrally coordinated planning process, particularly from the perspective of regulated expansion, but also merchant expansion:

- It will throw up coordinated information and an overview of all committed investment plans that will be useful to the regulator, market participants, and potential investors that will help to minimize the risk of stranded or imprudent investment
- Because one project could conflict with another or destroy the capacity of existing or proposed facilities, it would be useful if potential investors and transmission owners were to understand each other's plans, so they can conduct the necessary negotiations and create coalitions among themselves
- A regulated transmission expansion project can create costs and benefits that the proponents of the project do not pay or receive. A project can change the value of energy at many locations on the system. If there is an FTR scheme in place, the proponents should capture most of the pecuniary external benefits and pay most of the pecuniary external costs. Some benefits may not be taken into account, because they appear as externalities (*e.g.*, emergency support or improved reliability and security of supply). It is necessary for the regulator, at least, to look at the system and the market as a whole in order to assess which other participants will benefit and what, fairly, they should be asked to pay. A planning coordination process will assist the regulator to do this
- Because a project can create both costs and benefits to many parties, some of which are not easy to quantify (*e.g.*, reliability benefits), it is useful to undertake a reasonably comprehensive and coordinated analysis of whether the total benefits are likely to outweigh the total costs and whether the most cost-effective solutions have been chosen. (As there are always winners and losers, some market participants might be prepared to pay for an expansion to proceed, and others might be prepared for it to not proceed)

- A project that is profitable for its sponsors may not be optimal from the point of view of the system as a whole, because of economies of scale, lumpiness, and integer problems where the benefits may take the form of costs avoided rather than revenues achieved. An integer problem arises because several projects are possible in different combinations, with the private profitability of each project being dependent upon which of the other projects are undertaken and in what order
- Expansion projects that have lower negotiation and transaction costs are more likely to be implemented than projects that might be better from an economic, reliability, or social standpoint, but have higher negotiation or transaction costs

Central coordination will assist both merchant and regulated transmission developers as well as regulators to make better informed and more efficient decisions with respect to transmission expansion.

Central coordination perceived to favor regulated expansion

Merchant investors in transmission sometimes express concern that centralized coordination of a planning process may give advantages to the position of the incumbent, regulated transmission owners. The planning process could take on a life of its own, looking for large-scale transmission projects as solutions to all needs, eclipsing the possibility of market-based transmission and generation projects to succeed and demand-side options to be taken into account. Because merchant projects rely for their revenues on transmission property rights and, if permitted, the exploitation of differences in energy prices in different regions, merchant developers will feel they do not compete on a level playing field with the larger regulated project developers, where the capture of other benefits (such as the reduction in ancillary services costs, market power, or price volatility hedging costs) is taken into account by the regulator. The process might identify or even design a regulated project that could undermine the economics of a more risky merchant-based project, because of the socialization of the investment costs by the regulator.

Arguments against central coordination

There are other arguments against central coordination of planning that are made particularly by merchant developers of transmission projects. If efficient pricing creates signals to market participants as to when and where it would be economic to invest in transmission, it is necessary to give the market the necessary freedoms and time to respond. A centralized planning process, in whatever form it takes, could interfere with those freedoms. They argue that there should be no need for new regulated transmission projects that respond to the economics of the energy market and the costs of congestion and losses. They could produce an expensive solution to a transient problem.[10] The only planning process that would be required would be for reliability projects or transmission expansion needs in respect of which there had been a market failure (*e.g.*, to facilitate imports to improve competition or alleviate local market power).

Market failure is difficult to assess

There is no clear definition of market failure in the context of transmission expansion. It is difficult to know whether the market has failed to respond to an opportunity because of defects in market design (*e.g.*, the absence of property rights) or because it was not economic. If it is clear that the defects in market design might have deterred investment, then corrections in the design should be made. If small projects have not been built in response to signals in the energy market created through efficient pricing and good market design, the chances are that they were not economic (and therefore not needed) in the first place. The investment would not have made a material impact on prices. In this case, the market would have reacted appropriately, rather than failing to respond to the opportunity.

The question of whether the market will respond to need for transmission investment will tend to turn upon whether prices in the transmission rights market the developer can achieve are sufficient to promote these investments. These prices relate to the future value that the market will place on the reduction of congestion, hedging risk, and the ability to arbitrage energy prices in the future.

There is an argument that if energy prices in the spot market are not sufficiently high or if the prices between locations are not sufficiently different, projects driven by the need to improve reliability are not likely to be economic for merchant developers. Equally, spot prices at a level that does make reliability projects economic may not be politically acceptable. It is the expectation of high spot prices that funds investment. If it is believed that the high spot prices will not be allowed to stand politically, then there will indeed be no investment, as it is cheaper to complain to a politician than to invest in some wires. Ultimately, the question of market failure may have to be assessed by regulators and politicians. The simple question is how long can society tolerate high congestion costs before politicians and regulators are forced into the conclusion that the market has failed to respond and a regulated solution is needed, for which planning is required?

A matter of policy

Even the most ardent believer in market economics admits that there could be situations in which the market fails to respond, particularly where there are very large economies of scale that could and should be captured, or where there is a significant free rider problem. As a matter of policy, a government could decide to encourage small, market-based transmission projects that do not capture economies of scale, to the exclusion of larger projects, if it is clear that the corresponding impact on the environment and on local communities justifies the premium that the loss of economies of scale might entail. An alternative policy would be to regard the economies to be of more benefit to a greater part of society and that mechanisms should be put in place to ensure that they are captured.

It would be good to believe that the market will solve all the transmission expansion problems that society will face and that the need for central coordination and regulated projects will be rarely encountered. Although Argentina has adopted a policy of decentralized transmission expansion with no central coordination, no other countries have followed this path, even though they embrace market-based expansion.

Central coordination of planning is implied in deregulated industries

Most deregulated power markets utilize some degree of a centrally coordinated planning process. In its Order Providing Guidance on Continued Processing of RTO Filings, FERC required all regional transmission organizations (RTO) to carry out regional planning.[11] It has placed a similar requirement in its standard market design.[12] A central view of the market as a whole is implied in the test that many regulators apply in approving the need for a project or the recovery of its cost from consumers. It is usually an economic test that looks at whether the overall benefits of the project exceed its cost from the perspective of the market or society as a whole, both present and future. For example, the Australian Regulatory Test will be met if the project "maximizes the net present value of the market benefit having regard to a number of alternative projects, timings and market development scenarios..."[13]

This may not require command-and-control-style central planning, but it does imply collecting information to give a centrally coordinated overview. The distinction is a little difficult to define in the deregulated world. Perhaps central planning now implies the ability to order utilities to build or not to build, whereas coordination implies a consensus-building process in which information is gathered together and shared. That will help merchant and regulated investors plan and develop projects. The industry participants discuss their committed plans openly and together seek solutions to unmet needs for transmission. The challenge is to design a planning process that does not inhibit market-based expansion, but, rather, reconciles market-based and regulated transmission projects. If the process can be designed in a way that is transparent, inclusive, and that builds consensus, transmission expansion will be facilitated, the market will be better served, and the consumer will get a better deal.

Who Should Carry Out the Coordination?

The question of who should plan is a simple one, but has resulted in a number of different answers. The choices are the:

- system operator (ISO or RTO)
- transmission owner
- regulator
- government
- central agency
- independent, system "architect", or outsourced contractor given the role of managing the assets and preparing investment plans[14]
- independent transmission administrator

Different objectives of the parties

It is important to remember that there is unlikely to be a community of interest. There are clearly tensions between the different objectives of the ISO, transmission owners, market participants, regulators, governments, and other stakeholders throughout the planning coordination process that will need to be reconciled. At its simplest, for example:

- The ISO or RTO will be interested in expansion that eases its ability to maintain system reliability and to administer the market
- The transmission owners will focus on earning a reasonable return on an increasing asset base and on keeping congestion rents if the market design permits them to do so[15]
- Market participants will be concerned with access to customers, cheaper sources of power, and the ability to arbitrage energy prices and hedge risk

- Regulators will have an interest in minimizing the cost to the consumer, maintaining quality of service, and an overall desire to promote competition
- Governments will have the interests of the public at heart, but may have social, economic, environmental, and political agendas that create policy conflicts
- A central agency may be driven by the desire for greater involvement and control
- An independent transmission system "architect" should have no particular agenda apart from maintaining its reputation, maximizing its revenues under its appointment contract, and positioning itself for renewal of that contract. If the contract is well designed, it will create incentives to manage the assets efficiently, maximize availability, and plan and implement needed expansion projects as cost-effectively as possible
- An independent transmission administrator will also be motivated by the terms of its contract

It is helpful if all these seemingly irreconcilable interests can be put on the table and taken into account in the design of the process. Indeed, it would be useful to define an efficient and transparent planning coordination process that does not introduce unreasonable delays and also sets out the order in which the various issues need to be dealt with and resolved. Otherwise, there is a danger of unproductive effort for everyone.

Most candidates are at a disadvantage

Most of the candidates, if they were to attempt to assume the role of system planner on their own, would suffer from the obvious disadvantages of:

- lack of knowledge of the assets that only the transmission owner (or its asset manager) can have from monitoring and maintaining them and that the ISO will have from watching the performance of the system
- being susceptible to lobbying by the various interested parties

- finding it difficult to resolve adversarial positions
- not having sufficient information to carry out the coordination
- in the case of a transmission owner (and probably also the system operator), a conflict of interest with the role of planning coordinator that implies lack of bias in favor of any solution other than one that is best for the market and the consumer

Close to the assets

If the coordination process is well managed, transparent, and inclusive of as many of the interested parties as possible, the logical coordination role should rest with those who are closest to the best sources of information as to the existing assets, notably the transmission owners or the system operator, who observe the performance of the transmission system on an ongoing basis. After all, there are plenty of "policemen" in the form of other stakeholders and the regulators to keep the coordination team from favoring their own interests. However, they will only be effective if there is transparency and good information available to everyone. It is a fact of life that transmission owners have the best knowledge of their own assets. For example, nobody but the transmission owner is likely to know if it is possible to string a 345 kV line where a 69 kV line currently exists. System operators have good knowledge of the asset performance. The ideal answer is to engage both of them.

The choice between the system operator and the transmission owner

The transmission owner could plan and propose projects and submit them to the system operator. Together, they could discuss the system operators needs and look for solutions without preempting projects that are proposed by non-incumbent transmission developers (on a merchant or regulated basis).

However, joint appointments tend to blur responsibility. We need to appoint one entity as planning coordinator in order to make the responsibilities entirely clear. There are three factors that drive the choice between the system operator and a transmission owner for appointment as planning coordinator:

- If, as a result of its governance structure, the system operator is truly independent and commands universal credibility, then it is more likely to be accepted as an unbiased transmission planning coordinator than a for-profit transmission owner, which will be expected to look for profit-enhancing projects that only it can implement
- The fact that a credible independent system operator has proposed or approved a project should make it easier for the developer of the project (especially if it is an incumbent transmission owner) to convince the regulator, siting authority, and other permitting agencies establishing the need for the project
- The independence of the system operator should also enable it to obtain information more easily, particularly if the information is commercially sensitive and needs to be kept confidential. For example, a merchant or regulated developer of a transmission project may be reluctant to reveal any information to an incumbent transmission owner, if it feels that the transmission owner may use the information to achieve a competitive advantage or the ability to propose a competing project. This argument can be undermined by the fact that if the project is to be interconnected with the system of the incumbent transmission owner, both parties will need to consider the system reinforcement. The information requirements probably come at a time when the project is likely still to be commercially sensitive

If there is a transco and a single transmission system, the transco is obviously the only possible candidate. However, where there is a multiple ownership of transmission, and generation and demand-side solutions are a real possibility in order to meet transmission expansion needs, the system operator would be a better option. This is because it will be viewed as more impartial as to the choice of solutions to implement. The system operator may have a bias towards transmission solutions, because it may want to minimize the costs of ancillary services and "must run" generation that it passes on to market participants.

Even though there are strong arguments that transcos can be regulated to avoid favoring their own transmission solutions, and that the system operator, market participants, and other transmission owners will closely police a transco, the debate in the U.S. has swung against transcos in the role of planning coordinator, at least where the potential for competition in transmission expansion exists. As FERC put it, transcos have "the incentive and ability to bias the [needs assessment plan] in favor of their competitive interest."[16] If there are no regulatory incentives to build new transmission capacity, the question of bias will not arise, because they are unlikely to invest.

An independent planning coordinator (such as the transmission administrator in Alberta) looks like an attractive model, but it would not be the ideal candidate, because it would not have constantly updated information as to the performance of the existing assets that was as good as the information available to the transmission owners and the system operator.[17] The transaction costs would be higher, and the transmission administrator would be unlikely to have the same incentives as the transmission owners to provide efficient services, because its remuneration is difficult to link to system costs.

Who oversees?

Even if the planning coordination process is inclusive of all interested parties, the more important issue is who has ultimate authority to oversee the process to ensure that it is fair, that all interests are fully taken into account, and that all options are evaluated. On the face of it, this role should fall to the regulator (or the entity that approves tariffs, if there is no regulator), because of its market oversight role. In any event, it will probably have to approve the transmission expansion plan for regulated projects at some stage, in order to determine how the costs are to be allocated.

Creating Transparency

Planning rules

Many jurisdictions have codified their planning rules in order to create transparency and enforceability. A planning code is a key ingredient of the Grid Codes that are used around the world. In some cases, planning rules are to be found in the market rules.

The objectives of the planning rules are to:

- promote interaction between the transmission owners, planners, and users of the system with respect to any proposed developments by users that may impact the performance of the transmission system
- provide for the supply of information required by the planners
- facilitate connections and interconnections
- provide for the supply of information by the planners and the transmission owners to users
- specify the standards that will be used in the planning and development of the existing transmission systems

The rules cover planning procedures, data, and standards. They are usually driven by the technical needs of a conventional planning process conducted by planning engineers. Much of the detail of the planning rules relates to the planning data that are needed. The rules could be enhanced by an overall statement of objectives and planning policies.

Statement of opportunities

In several countries, transmission owners publish long-term statements of the circuit capacity, forecast power flows and loading on each part of their systems, and the fault levels for each transmission node. They are either required to do so by the regulator, or they publish the statements voluntarily. They have become known as Statements of Opportunities that look forward over a time horizon that is between 5 to

10 years. In England and Wales, NGC's transmission license requires it to include in its statement:

> (a) *such further information as shall be reasonably necessary to enable any person seeking use of system to identify and evaluate the opportunities available when connecting to and making use of such system; and*
> (b) *a commentary prepared by the licensee indicating the licensee's views as to those parts of the licensee's transmission system most suited to new connections and transport of further quantities of electricity.*[18]

These statements have become an extremely useful tool for both generation and transmission planning and are widely respected by market participants, new entrants in generation, and regulators. If they are well prepared on a very comprehensive basis, they could form the starting point for the planning coordination process.

The Five Year Statement that National Grid USA and Niagara Mohawk voluntarily prepare provides a wealth of information excellently presented in an easily accessible form, covering demand projections, the generation and reserve margin, the capacity position, transmission system topology, and the transmission system performance and capability. The statements include opportunities for new demand and new generation to connect to the system and opportunities to relieve congestion. A power flow model has been developed to give a probabilistic view of certain levels of power flow. This can be used to provide a view of the probability that one project would solve more problems than another. There are maps showing areas where new generation could worsen or alleviate congestion. In some markets, it will be extremely important to draw the attention of market participants and loads to the current and anticipated levels of congestion costs. In Argentina, Transener publishes a similar statement. Figure 8-2 is taken from the Five Year Statement of National Grid USA for 2002-2008, and shows the interface transfers and capability for power transfers from New England to New York, which is used to forecast the need for transmission reinforcement or constrained operation.

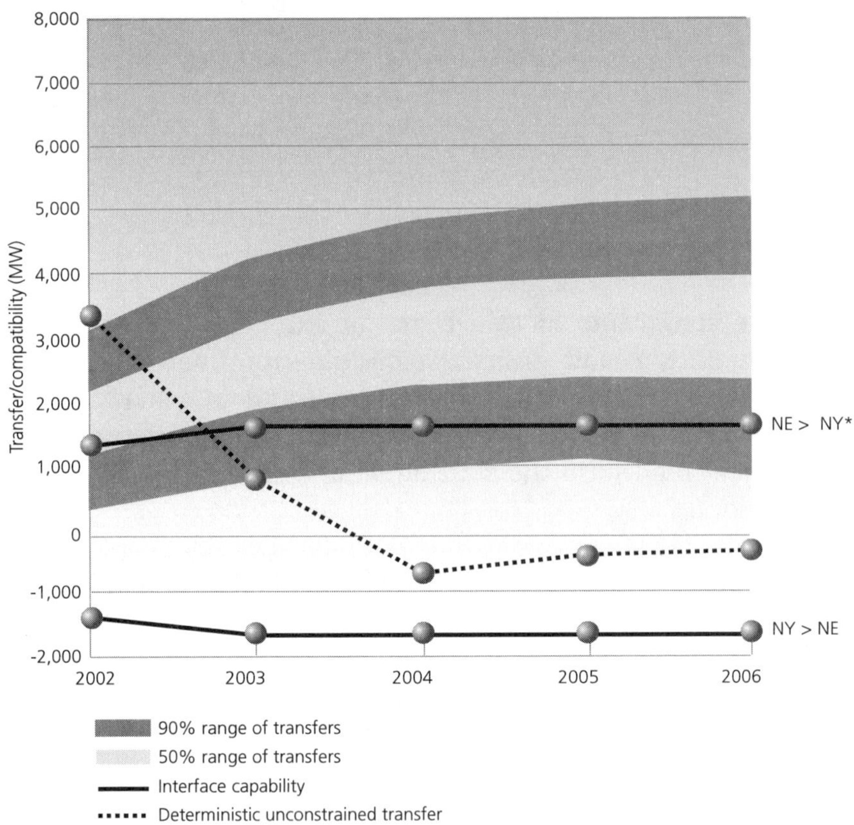

Fig. 8-2 New England to New York (Source: National Grid USA)

The statements could be put together as part of an annual process involving the collection of data from generators, transmission owners, distributors, consumers, and new entrants contemplating new generation projects or demand-side solutions. It would include committed merchant transmission projects, as well as projects contemplated by incumbent transmission owners. This approach could enable the proponents of market-based projects to find out in advance what the committed plans of generators and incumbent transmission owners were, without working in the dark.

Rules or incentives could be used to ensure that the necessary data are made available for planning coordination and system studies. FERC has cited as a best practice that:

> *Transmission Providers will post on their web site what, in their view, are the optimal and non-optimal sites on their system for locating prospective generating facilities. Transmission Providers need to identify areas where, for example, due to load growth, siting would require minimal network upgrades. Also, the Transmission Provider should identify areas where, for example, due to transmission constraints, significant network upgrades would be required, and the expected delay before such upgrades will be made.*[19]

Some stakeholders in Alberta and New England have specifically asked that accurate and up-to-date databases are available to everyone, and that third parties be able to perform studies, particularly if the ISO or transmission owners cannot perform the studies in a timely fashion. What is important is that there should be an accurate and up-to-date database available to everyone, good transparency, an open and inclusive process, and a common planning model that commands confidence.

Merchant transmission developers and independent transmission companies may be reluctant to provide information about their transmission assets and their development plans to the planning coordinator, if there is a chance that another transmission owner or a potential investor may gain a competitive advantage if the information remains public. This issue is similar to the one that transmission planners face with respect to planned generation projects. The rules can require the planning coordinator to keep the information confidential until such time as the project is fully committed or it begins the regulatory and permitting approval process.

System studies

All planning processes involve the carrying out of system studies to assess the impact of a new generation, demand, or transmission project on the existing system. The issue arises as to who should carry them out, because there is often a fear of conflicts of interest or bias if the transmission owner has the monopoly right to carry out system studies on its own assets. The transmission owners can be required under the regulatory or market rules to provide the necessary data to enable third parties to carry out studies (providing the models remain consistent).

The Coordination Process

Avoiding a two-stage process

In chapters 3 and 5, we discussed:

- the possibility of a regulatory backstop process for needed expansion, in the event of the failure of the market to build the necessary transmission project

- the desirability of taking into account generation and demand-side alternatives to transmission expansion that might be more cost-effective than a transmission project

In designing a coordination process, we should try to avoid duplication of effort by incorporating all projects and solutions in parallel, rather than in series. If a process were designed that looked at all solicited demand-side and generation options against projects proposed by transmission developers followed by a second step that, in essence, looked like a traditional least-cost planning exercise in the light of market failure, transmission owners and developers would be bound to complain. We should avoid creating two processes where one will do.

The steps

The coordination process could be carried out on an annual or biennial cycle, looking over a five- to seven-year time horizon. It would, inevitably, involve a number of steps:

- The planning coordinator assesses the requirements of the relevant planning and operating standards, reliability criteria, and market and technical rules
- A planning database is established and maintained by collecting data on existing transmission assets from transmission asset owners, the necessary data relating to generators and loads directly connected to the transmission system, and on distribution and neighboring systems interconnected with the transmission system
- The planning coordinator collects information for demand forecasts and inputs from transmission owners, market participants, large consumers, and the owners and operators of the interconnected systems as to their planned investments and transmission needs (including information that it might be required to keep confidential as to planned projects)

- The system operator or RTO publishes its assessment (baseline needs) of the operation and performance of the transmission system and its adequacy and reliability. It provides input as to expansion or performance enhancements required to maintain or improve reliability. It may also add its views as to the need to increase the competitiveness of the market to alleviate market power, to facilitate imports, to reduce ancillary services costs, and to reduce price volatility
- The planning coordinator runs a consultation or indicative solicitation process to assess the degree of likely generation and demand-side responses available to improve reliability or reduce congestion
- The system operator or market administrator publishes information regarding congestion costs and where locational marginal prices are in place, a summary of nodal energy prices for the relevant period, to assist the market (in which regulated transmission developers may also participate) in identifying economic projects
- The planning coordinator evaluates applications for connection or interconnection of committed generation, demand and transmission projects, and long-term transmission service requests by generation and transmission developers and loads, and assesses transmission capacity adequacy
- The planning coordinator obtains wide stakeholder input, particularly as to needs, costs, and local community and environmental sensitivities, on a regular and on-going basis
- The planning coordinator publishes its methodology for assessing and determining need (including scenario planning and details of the relative probabilities of various future outcomes, *e.g.*, fuel prices, generation closure, and demand growth or decline)
- The planning coordinator publishes its assessment of the transmission performance and capacity enhancement needed that may relate to reliability, congestion, increasing the competitiveness of the market, facilitating imports, the costs of ancillary services, and price volatility, based on an assumption that the

committed generation, demand, and transmission projects will be built. It will not design specific projects

- The assessment is then discussed by the transmission owners, market participants, and other stakeholders (and may also be discussed with the regulators), and their input is taken into account in a process that endeavors to reach consensus or acceptance by stakeholders
- Once the assessment is accepted by a substantial majority of substantial stakeholders (after they have all had a reasonable opportunity to have their views taken into account), the planning coordinator will publish its final assessment
- The planning coordinator then invites (or, if there is an obligation to build, instructs) the relevant transmission owner or owners to develop projects on a regulated basis (or on a merchant basis, but it is likely that the market would, by now, have failed to respond) in order to meet the needed capacity and performance enhancements that have been identified in the assessment, and to file the necessary applications with the relevant permitting, siting, and environmental agencies within a specified period
- If the projects are not developed and complete applications are not submitted to the relevant agencies within the specified period, the planning coordinator prepares and issues a request for proposals (open to regulated and merchant developers) for projects to meet or assist in meeting the needed performance and capacity enhancements described in the assessment, inviting generation, transmission, and demand-side solutions. The planning coordinator would not attempt to redesign the projects or evaluate their costs. This exercise would be left to the normal regulatory, siting, and permitting processes
- The competitive bidding evaluation process must be conducted under transparent rules and objective evaluation criteria. The winning projects will be those that score highest on the published evaluation criteria, for example, applying a specified methodology to calculate a cost-benefit ratio, perhaps setting a target threshold

The entire process is summarized in Figure 8-3.

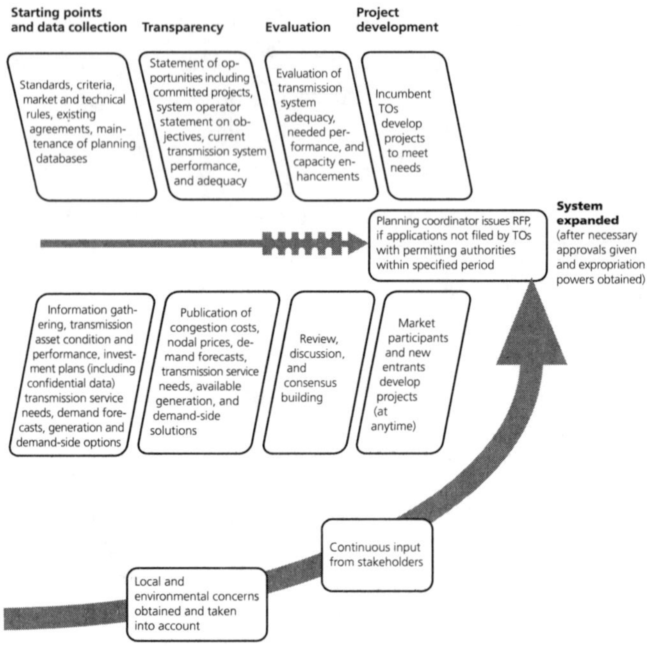

Fig. 8-3 The planning coordination process

Is a competitive procurement process necessary?

On the face of it, a competitive solicitation process has some of the characteristics of central planning in what is intended to be a consensus building, coordination paradigm. It is proposed as a backstop in the event of lack of market response to create an incentive for transmission owners to develop the needed projects expeditiously and to file their proposals with the relevant agencies. This proposal is a compromise between running a solicitation process automatically at this backstop stage and leaving the development of projects entirely in the hands of the incumbent transmission owner or owners, in the event of market failure.

Regulated transmission owners might well argue that, at this stage in the process, solicitation on a competitive basis is cumbersome, expensive, slow, and unnecessary. They may feel that they are being pushed into an expensive and lengthy procurement process that may well involve outsourcing the construction of a project in cases where it might be more cost-effective for them to build it themselves. Some of them will go out to competitive tender for construction services, in any event. The regulator will scrutinize their costs before they are allowed to recover them in their charges and conduct prudency reviews. The planning coordinator does not need to "first" or "second" guess the regulator. The process could add another layer of least cost planning.

Another argument against the competitive solicitation of transmission lines is the sheer difficulty of bidding a single or nonvariable annual charge or fee for the project. The uncertainties of the siting and permitting processes, the negotiations with landowners and objectors, and the likelihood that rerouting will be necessary mean that it is virtually impossible to calculate a price that does not leave the developer with risks that are outside its control and that would make the project unbankable. If the risks were very considerable, it might be possible to construct a mechanism to allow the developer to recover additional costs incurred for reasons entirely outside its control (*i.e.*, a cost-pass through) and at the same time to create a cost and performance target incentive to manage and minimize the cost and delay risks. However, the experience in Argentina indicates that developers of transmission projects have been able to put together bids that involve a single annual charge or fee, and, indeed, they have been prepared to do so very competitively. Transener, as we saw in chapter 5, did suffer cost at the hands of a landowner, which it was not in a position to manage.

Advantages of a solicitation process

As we saw in chapter 5, the experience of competitive bidding in Argentina shows that it is a powerful mechanism to put a downward pressure on expansion costs. VenCorp, in Australia, has used it to procure some small projects. It would be useful to give merchant developers of transmission and generation more information and an opportunity to put forward proposals. There is no reason that regulated transmission owners or any other project developer should not put forward proposals as well (as long as they do not ask for a regulated subsidy, as we shall discuss later). Developers would be able to put proposals forward on a merchant or regulated basis.

The other advantages of a competitive solicitation process from the point of view of the consumer are that it should result in the developer bearing more of the risks, particularly with respect to timetable, cost, and the achievement and maintenance of target or design availability. The solicitation should result in lower prices for the consumer as it should encourage technical, financing, and contracting innovation. FERC has said "[all] projects in the [regional transmission] plan should be built following a competitive solicitation."[20]

A requirement to this effect has been included in its standard market design.[21]

Figure 8-4 shows the New England solicitation.

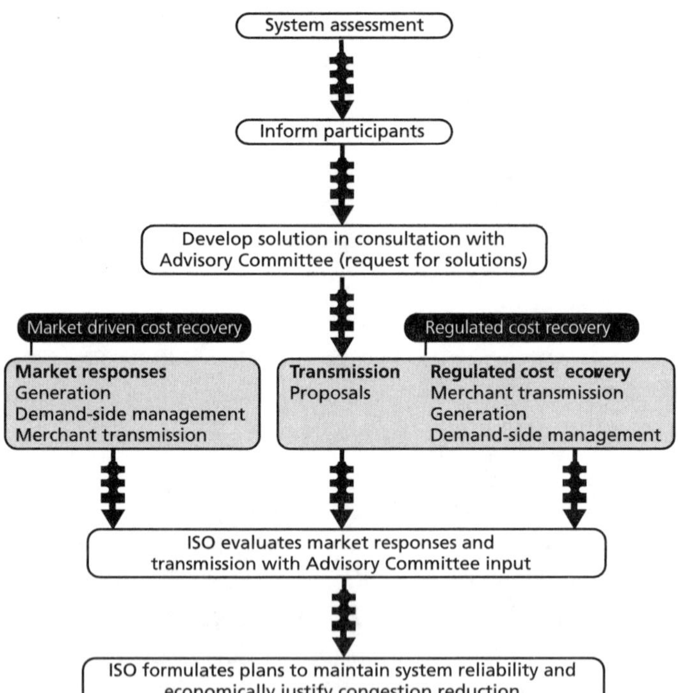

Fig. 8-4 Solicitation process-New England example (Source: TransEnergie US)

There are significant difficulties in designing a solicitation process for both generation and transmission projects

A competitive solicitation process is always cumbersome. It takes time to organize and carry out, and the cost to bidders can be very significant.

It is obviously difficult to carry out a solicitation process that will attempt to evaluate some very different solutions in an environment where market participants are free at any time to propose and pursue market-based generation and transmission projects within or outside the coordination process that we have just discussed.

In the context of a competitive solicitation process with published, objective evaluation criteria, it is never easy to establish fully objective criteria to compare the costs and benefits, particularly the non-economic benefits of generation, transmission, and demand-side solutions (the exercise of comparing apples with pears or chalk and cheese). On the face of it, the differences between development of generation and transmission projects are enormous. As Eric Hirst and Brendan Kirby point out:

> *The competitive business environment of generation pushes investors to faster planning, shorter deployment times, and less sharing of commercially sensitive information. The regulated business environment of transmission pushes it to slower planning and longer deployment times (to accommodate an inclusive public process) and the sharing of information.*[22]

Some sort of evaluation can be carried out, but it is likely to involve ad hoc judgments and subjectivity. The obvious difficulties in comparing transmission, and generation and demand-side alternatives are:

- The asset lives are different—generation assets typically have shorter asset lives, and load management schemes may have a very short duration[23]

- Transmission facilities usually have very high availability factors, much higher than those that could be achieved by the generation or demand-side options
- The capital costs of transmission are typically high, and the operating costs are very low, whereas the operating costs for generating plants are typically high and depend critically upon fuel prices that are not always predictable
- The returns on most transmission investments are currently regulated, whereas, in a deregulated industry, the returns on generation investments are subject to market forces and may command higher rates of return to compensate for the different risks
- Distributed generation and load resources cannot always solve the problems on the transmission system, such as transient stability or the need to replace aging or obsolete facilities. The connection of distributed generation to the transmission system may impose new costs in terms of both local and remote system reinforcements

If generation and demand-side solutions are to be taken into account in the coordination of the transmission planning process, it will be necessary to establish an energy price that reflects the value of electric power as transmitted, in order to derive a value for a transmission project that will enable comparisons leading to an efficient choice to be made among the alternatives.

It may also be difficult to compare one transmission project with another, whether merchant or regulated. The cost reduction benefits may be easier to compare than non-quantifiable benefits, such as an overall reduction in market power and price volatility. The planning coordinator would need some policy guidance from the regulator as to the relative importance of these benefits in certain circumstances.

The California ISO experience with solicitation of transmission alternatives

The California ISO issued a request for proposals for cost-effective and reliable alternatives from generation and/or load alternatives to the Tri-Valley project proposed by Pacific Gas & Electric in Northern California. The ISO rejected all four bids, because they failed one or more of the evaluation criteria that included the commencement date, the ability to provide the proposed services, cost, safety, impacts on the market (in particular, generator market power), environmental implications, the reliability criteria adopted by the ISO, and operating characteristics. The main reasons that the bids were rejected were because they were substantially more expensive than the transmission project and provided less capacity to the system.

A year later, the ISO decided not to repeat the process, but accepted the proposals for the construction of the Valley-Rainbow transmission project by San Diego Gas & Electric, partly because of the high California electricity prices and partly because the ISO decided the project should be considered part of "the broad strategy by the state of California to put in place a robust transmission system to support reliable service to consumers."[24] The project would have increased the transfer capability between neighboring states as well as creating a variety of local reliability benefits that otherwise would have had to be purchased through reliability must-run contracts.[25]

Solicitation process as a backstop

Because of all these difficulties, a competitive solicitation process is suggested as a backstop, if transmission owners do not develop projects within a specified period that will be acceptable to the regulatory, permitting, and siting agencies as a starting point in the permitting process. The planning coordinator would not need to evaluate the projects, as that could be left to the regulatory and permitting processes. It would simply act as the timekeeper for the development of the projects. However, if the deadline had passed, and no complete project proposals had been filed, the planning coordinator would issue the RFP. The setting of the deadline would be a sensitive issue. The process must put pressure on the transmission owner to develop and implement the projects expeditiously, but the timetable should allow sufficient time for the projects to be carefully designed and for environmental impact assessments to be prepared. If there were more than one transmission owner, they might have to work a number of things out between them (but in the U.S., they have some history and experience of doing just that). The idea is that the threat of an RFP process and the public embarrassment of missing a deadline should be sufficient to motivate the desired behavior.

The RFP process is a superior mechanism to impose discipline in the design and implementation of projects and adherence to costings and timetables. However, the costs involved and the difficulties of establishing transparent and objective evaluation criteria may outweigh the advantages. After all, this is only a "backstop" in the event that no merchant or regulated transmission projects have been proposed, and there are no generation and demand-side solutions. It should occur relatively rarely, especially if the amount of information and price signals that we have discussed is available to potential developers.

Subsidies distort signals

In theory, a competitive solicitation process could affect a number of investment decisions. For example, developers of distributed generation or merchant transmission lines might be tempted to wait until the solicitation process took place, in the hope they would be able to achieve a regulated subsidy or the regulator agreed to socialization of some or all the investment costs. Subsidies distort the effectiveness of price signals in the energy and transmission markets, and should not be permitted by regulators. In any event, we are talking about a situation where there has been a market failure. This means that we are very likely to be talking about projects with large economies of scale and situations where only transmission will do the job. Subsidizing generation will not produce the answer we are looking for.

Competition for ideas

The competitive process may, in some cases, consist of competition for innovative ideas, commercial deals, and financing. It will involve comparing both regulated cost recovery mechanisms and market-based prices. The proposed cost recovery mechanism could be based on rate-of-return regulation or performance-based regulation. Another freedom that will need to be factored into the evaluation process is the ability to allow bidders to structure their investments flexibly. Some will want to design, finance, build, own, and maintain the projects themselves, while others will be happy to lease them to another transmission owner or to the system operator. Following the practice adopted in Argentina, the selection of the winning bidder should be based on the lowest annual cost rather than the total cost.

Should transmission project development be preempted?

The fact that a market participant, developer, or incumbent transmission owner could proceed with an expansion project at any time is difficult for a planning coordination exercise to deal with. It means the information that the planning coordinator publishes and its assessment of the need for performance enhancement and transmission expansion may be overtaken by events at any moment. Moreover, the need for and the economics of projects that are being developed in the market or on a regulated basis could be affected by later projects. This is a problem for both market-based and regulated project development, especially in view of the long lead-in times involved.

Tempting though it may be to impose a rule to the effect that no one may propose or implement a project that might adversely affect another project that is in the course of development, it would almost certainly be untenable to do so. A later project might be more beneficial for the market or for consumers. In any event, a regulator would be put in difficult position with respect to a project that it had already approved for regulated cost-recovery. If the later project proved to be more beneficial to the consumer, either on a merchant or regulated basis, and the earlier project was no longer needed (in whole or in part), the regulator would almost certainly have to make the difficult decision to disallow the addition of the new facilities in the regulated asset base on the grounds that they would not be used or useful. This would be particularly difficult if some of the facilities had already been built and had already been included in the asset base. In all fairness, the regulator should allow recovery of the costs incurred even if a project is concealed in this way. As we have seen, FERC has given a clear indication of its thinking with respect to merchant transmission:

> *We also conclude that third parties should be allowed to build merchant facilities outside the context of the plan, subject to ISO review to ensure that such facilities meet technical requirements and do not reduce the overall transfer capability of the ISO's grid.*[26]
>
> *FERC's proposed standard market design embraces and encourages private, market-based expansion.*[27]

The counterargument to allowing all merchant projects to proceed is that if another project is proposed and implemented, it could immediately rule out other options that may have been identified as superior, from an integrated network and market benefit perspective, and the price signaling might not be able to avoid this. Even if current price signals show a need, transmission projects take a long time to develop and implement. The developer and the planning coordinator would base their decisions primarily on new connection or interconnection applications and plant closures, as well as forecast demand, congestion, prices, and market needs. Work needs to start in anticipation of potential congestion, adverse effects on reliability, and market power problems. Options need to be compared in advance. If some projects are not preempted, there will simply be a mess, and the consumer could suffer.

Then the question arises as to how the preemption would be achieved. The existing market participants and incumbent transmission owners could be required to not proceed with a project that was outside the planning coordinator's assessment under a prohibition contained in the market rules, which are typically given contractual force through the system operator. Obviously, this would not bind a third party developer that had not yet applied for membership or market participant status.

If the project were to operate independently of an existing system (*e.g.*, to connect captive generation to captive load), there would not be a problem. If the project needed to be interconnected with an existing system, the incumbent transmission owner could be required under the market rules to force the developer to be bound by the market rules as a condition of interconnection to its system. The most reliable means of preemption would be through legislation or regulation. However, a legislative or regulatory rule that interferes with commercial freedom in this way might be politically untenable, and might look like command-and-control central style planning by a back door route.

Advantages of a consensus-building approach

The idea of this open, consensus-based planning process is that it should reconcile all the various interests of the parties, or, at least, make them transparent. Markets thrive on transparency. If consensus is achieved, as we shall see in chapter 10, the solutions that are found are more likely to be implemented, and the merchant and regulated transmission projects will have the best chance of coexisting efficiently and economically.

Unless the prices are right, as we discussed in chapter 4, consensus will be difficult to build, because there will always be winners and losers from transmission investment, and there is no easy way to make sure that only the beneficiaries pay.

The outcome of a planning coordination process endorsed by an independent planning coordinator should create credibility for a project included in its assessment. It will smooth the path of the project with regulators determining need and with the agencies granting environmental consents and planning permits. If the process was well conducted and was inclusive of all stakeholder interests, it should help to identify environmental issues and impacts on local communities at a highly desirable, early stage. The process should also enhance the credibility of the project from the point of view of potential objectors (by choosing a solution that best addresses their concerns), help to minimize or mitigate the impact of needed expansion on the environment and local communities, and minimize the scope for dispute and delay.

The PJM process

The PJM regional transmission expansion planning process has achieved considerable progress in consensus building. Figure 8-5 indicates how the process is carried out, using a variety of tools to build a reliable system but it does not indicate the effort that the PJM ISO puts into ensuring that it has all the necessary information and that it consults as many PJM members and stakeholders as possible (including regulators).

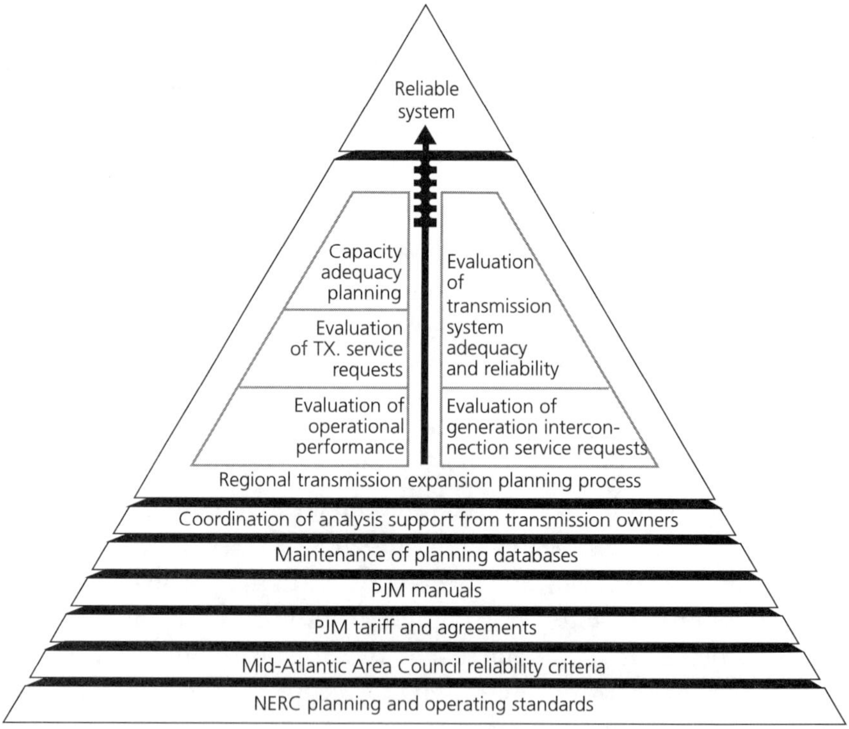

Fig. 8-5 PJM regional planning process (Source: PJM Interconnection LLC)

The process involves:

- ongoing adequacy reliability assessments
- impact studies relating to generator interconnection, firm transmission service requests, and advanced technologies
- meetings with the broadly-based stakeholder Transmission Expansion Advisory Committee
- taking projects proposed by merchant developers and incumbent transmission owners into account in a fully integrated process
- coordinated expansion plans across multiple transmission owner systems
- coordinated expansion plans based on all known needs, with the objective of identifying the most effective and efficient expansion plan for the region
- the evaluation of alternative solutions, including transmission, generation, and demand-side options, accommodating a wide range of decision drivers or criteria
- expansion plans that incorporate and reflect the widest possible range of stakeholder input into the process

PJM has not yet had to use a formal solicitation process in order to expand its transmission systems, because the transmission owners have signed an agreement to build the projects identified as a result of the process.[28] As yet, there has been no need for a solicitation process. The consensus-building exercise is not the subject of a structured procedure of the nature described more fully in chapter 10. Obviously, there is a limit to the detail of the stakeholder concerns that can be aired at large meetings. Even if the ISO invites written submissions, it may not receive as much input as it would like, particularly from consumers. This is a problem that the system operator of a large regional market is bound to have.

The PJM process is focused on identifying reliability criteria violations, but the transmission planning coordinators in PJM have been looking at operational needs and congestion with a view to including them in the process. It is clearly easier to develop specific criteria and tests for reliability and other operational issues. Congestion or "transmission for economics" is more challenging, because it is a moving target. PJM is very mindful of any potential negative impacts on the energy market, particularly because the reduction of congestion can create winners and losers. As a result, the exercise to broaden the scope of the process to cover congestion is being approached very carefully.

Builder of last resort

The final question that needs to be addressed in the planning coordination process is what is to happen if the market fails to respond to a clear need, and there is also no response to the planning coordinator's backstop solicitation process. Would it be possible to force an incumbent transmission owner or a market participant to expand the system on the basis that the regulator will virtually guarantee cost recovery on a regulated basis?

In Alberta, the transmission administrator (TA) can directly assign a project to a transmission facility owner (TFO). The TFO has the right to reject the assignment, to assign it to some other entity, or to issue an RFP to procure the facilities competitively. There is some uncertainty as to whether the TA is the builder of last resort under a direct assignment.

In several countries, incumbent transmission owners have a virtual obligation to build in order to serve any potential user that requests transmission service. This is the effect of NGC's Transmission License Condition C7D in Part II, as well as Section 211 and 212 of the U.S. Federal Power Act, 1920. However, the position of transmission owners in the U.S. has changed considerably under FERC Orders 888 and 2000.

No obligation to build in the U.S.

Prior to Order 888, the obligations of utilities to build transmission were based on an assumption that they were vertically integrated and would build transmission to meet their power supply responsibilities. They also had an obligation to build imposed at state level. The traditional planning process integrated generation and transmission expansion and, in some cases also demand-side, social and environmental considerations (under a paradigm called integrated resource planning or IRP). The objective was to achieve reliability at least cost. Order 888 accelerated the separation of transmission from generation. Where retail services were unbundled from transmission, FERC, for the most part, assumed jurisdiction over transmission. There was uncertainty whether there was an obligation to build in those states where this unbundling had taken place, but Order 888 provided that transmission owners had an obligation to build under the open access transmission tariff (OATT).

Under Order 2000,[29] the functional separation of transmission and generation was required, giving FERC jurisdiction over transmission. It removed any decision-making role with respect to transmission planning from transmission owners and prohibited them from preserving the right to build transmission in their service territories.

Under Order 2000, it appears that no one has responsibility for building transmission, and the state planning obligation has been eroded. The RTO has the obligation to plan for transmission expansion, but no obligation to build. The transmission owner is no longer the service provider under the OATT, and incumbent transmission owners are not even allowed to build without going through a competitive bidding process conducted by the RTO. Under FERC's proposed standard market design, in the event of market failure or a lack of response to an RTO or and ITP solicitation process "such that the independent transmission provider determines that additional facilities are needed, the affected transmission owner(s) would be required to expand or upgrade the transmission system."[30] It is not clear how the "affected transmission owners" might be identified.

The position if no one wants to build

The chances that the market and regulated transmission owners will not respond to a need identified by the planning coordinator are extremely slim. However, the solution does not lie in restoring central planning if this eventuality does arise. Even if the regulator were to enrich the regulatory approach to make the investment more attractive to investors, the uncertainties over the siting and permitting processes will remain, as will the difficulties with dealing with landowners and objectors. It is impossible to make a project risk-free. The key issue will be whether powers of expropriation or eminent domain to acquire the rights-of-way that are needed for the expansion can be obtained.

An incumbent transmission owner may be reluctant to take on a project as a provider of last resort. It may not want to run the risk of losing public support by developing an unpopular project that meets with widespread hostility, even though the project has clear social benefits. It is not clear that even if the regulator is sufficiently generous, there will always be a developer prepared to take on the project. The key ingredient will be good communications and presentation skills to the local communities to explain why there is no alternative but to build the needed transmission. It will be a message that no one will want to hear. If the project is really risky, should it be built?

Need for expropriation (eminent domain) powers

The choice of solutions to the need for a builder of last resort is extremely narrow. The obligation to build can only be placed on a transmission developer that has powers of expropriation or eminent domain, if they are needed, and has the necessary staff and the resources to design the appropriate project. Transmission project design is complex and involves bringing together knowledge of the existing system, the ability to carry out system studies, the evaluation of substation space and expansion potential, engineering expertise, knowledge of existing rights-of-way and good relationships with landowners and local communities. These are ingredients that will only be found easily in the incumbent transmission owners.

Right of first refusal

The issue has arisen of whether regulated transmission owners should be given a right of first refusal, particularly those that might find themselves in a position of builder of last resort. In the U.S., given the nature of the obligation of utilities to build transmission at federal and state levels, there have been attempts to interpret this obligation as a right of first refusal in certain circumstances. However, the policy of FERC seems to be clear. A right of first refusal was originally submitted as part of the nodal pricing system for New England, but FERC struck it down,and it was removed from the transmission expansion scheme.[31] FERC also struck down the provision in the GridSouth proposal that would have, in effect, created a right of first refusal for transmission owners.[32] A right of first refusal is hard to justify and does, in any event, beg the question of what happens if the right is not exercised.

Planning Tools

Inherent complexities of planning transmission

It is not a simple exercise to decide what expansion is needed, let alone what projects should be built to meet a particular need. Transmission reinforcement requirements are, and will likely remain, driven by the impact of new connections and interconnections as well as plant closures. In a deregulated industry, there are additional complications. The modeling that is involved is not easy. It involves inputting demand forecasts and anticipated generation figures that may well be uncertain, especially in a competitive market environment, calculating marginal costs and the costs of constraints, running power flow models under different scenarios, and attempting to measure the benefits in terms of reduced congestion, increased reliability, reduction in the costs of the exercise of market power, and reduced price volatility. Economists might describe the exercise mathematically as not particularly well behaved. The modeling may produce multiple solutions that are equally valid and might be explained or dismissed within the boundary of the model's error band.

Clearly, there is considerable uncertainty going forward, plans can quickly change, and projects may be cancelled at short notice. It is not a simple exercise to input the potential for reduction of the exercise of market power and increased competition into models for transmission expansion. However, work has been undertaken on several different approaches that appear to be very helpful.

Application of real options and game theory

Sam Lovick and his team have developed a technique for evaluating different types of proposals based on the application of real options[33] and game theory.

The difficulties of comparing generation and transmission proposals that we have discussed earlier are made more complicated by the fact that while generation and transmission projects may indeed be substitutes for each other, they may also complement each other. Therefore, the transmission investor is faced with the two-fold problem of investing in the face of uncertain future wholesale prices (leading to uncertain future transmission rentals), plus the problem that rents may be eroded by generation investment or augmented by it. Generation projects, as we have observed, are likely to be developed for private profit and, increasingly, on a merchant basis. Transmission projects, which as a practical matter remain largely regulated, are judged by whether they maximize social welfare and facilitate wholesale markets. Their evaluation approach quantifies and then prices these factors.

Real options theory arose from the observation that firms were avoiding some investment opportunities that appeared to give returns in excess of their cost of capital and therefore, seemed to be foregoing profits. The solution to this paradox starts with the recognition that firms have some choice over the timing of investment, and that investments once made are often irreversible in that investors cannot use the resultant assets for any other purpose. Over the course of time, investors will gain better information, improving their ability to make better investment choices. Investors, therefore, face the choice to invest now, or to wait for better information. If they wait, then they reduce the likelihood that the investment will turn out badly (because they have better information). Thus, there is some value in delay. It turns out that the value of delay explains the paradox, in that the expected return on the investment delayed is greater than the expected value of the investment made today, so the profit-maximizing decision for the firm is to delay.

Transmission and generation investments fit nicely into this framework. It is not difficult to see how the expected returns from such investment will improve by waiting for better information. For example, the dramatic decline in demand in California following the decline of the information technology sector would have profoundly affected investment decisions. In such cases, real option values are likely to be high—ignoring them will lead to inefficient investment. Demand uncertainty is not the only determinant of real option value. In the case of transmission investment, the uncertainty over generation investment is also very important. The problem is further complicated by the interdependency between the returns from generation and transmission. Sam Lovick and his team[34] approach this problem using game theory.

Using the prisoner's dilemma to evaluate the generation and transmission projects

In deregulated power markets, generation and transmission investments are made by different organizations and are not formally coordinated (as they were under previous "least cost planning" regimes). Since there are two or more independent players whose returns are interlinked, they can be considered, in economic terms, to be players in a game, and so amenable to game theoretic analysis.

The best-known application of game theory is the "prisoner's dilemma" in which two prisoners play a non-cooperative game. Each can either confess to a crime or plead innocent. The jail sentences they receive (or payoffs) depend upon what both prisoners do. This is summarized in Figure 8-6 that shows the "strategy" for each prisoner, and the payoff in years of imprisonment for the (Prisoner 1, Prisoner 2) pair. If neither confesses then they are both imprisoned for three years. If both confess, they are both jailed for five years. And if one confesses, but the other does not, the "confessor" makes a deal and is freed, but the other "non-confessor" is freed.

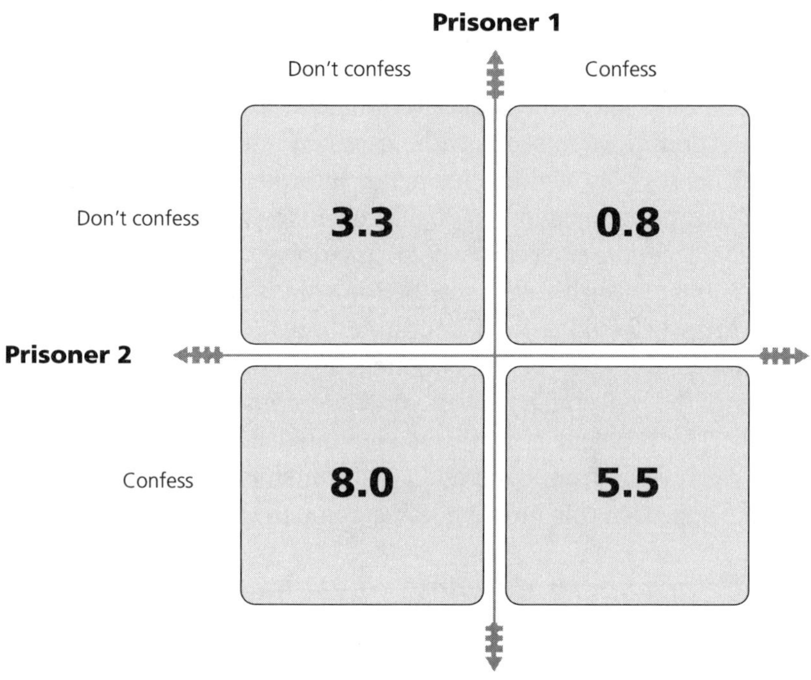

Fig. 8-6 The prisoner's dilemma (Source: Sam Lovick)

If the prisoners know the game and have just one shot, what do they do? The best joint outcome is for both to stay silent. That way, the total jail time is six years. But this will not be the result, because both prisoners know that if the other prisoner did not confess, they would do best to confess. It then turns out that the dominant strategy for both players is to confess, which means they both get five years (even though both would do better to not confess), and the joint payoff for all other combinations of strategies is better.

The same logic (if not precisely the same game construct) can be applied to the transmission-generation investment game. Each player—a transmission investor and a generation investor—can either invest now or delay the investment decision, and a payoff matrix can be constructed as shown in Figure 8-7.

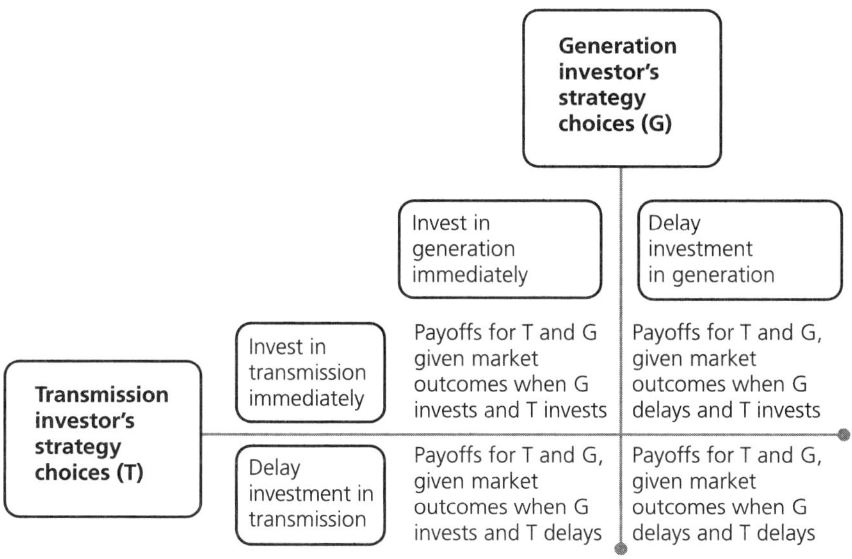

Fig. 8-7 Investor's strategies (Source: Sam Lovick)

With the game payoffs calculated, the transmission investor has to simply choose the investment strategy with the highest social payoff (highest social NPV, inclusive of any option value) for transmission investment.

A number of possible outcomes arise from this. For example, the transmission project should be delayed if a profitable generation project is available, and if the expected social gains from that generation are greater than the expected social gains from transmission. The interesting insights arise when the two projects complement each other, *i.e.*, where either project alone does not give rise to social welfare gains (or private profit, as appropriate), but where the two in combination do, or when the transmission project's social value is contingent on certain prior generation investment.

The outcome from the process is then an investment decision tree indicating when to invest and when to delay, which takes into account future uncertainty, the welfare dependencies between generation and transmission, and even the welfare dependencies between discreet transmission projects (*i.e.*, the benefits of one transmission investment may be dependent on a different transmission investment elsewhere either geographically or in time).

The advantage of this approach is that:

- In effect, it prices uncertainty
- It provides an investment decision rule as to when to invest or to delay making the investment decision
- It is useful for evaluating proposals for very different types of investment (the problem of comparing apples with pears), although it does not create particular projects or solutions
- It avoids the problem inherent in simple scenario analysis—how to best weight different scenarios in the final appraisal
- It avoids the need for competitive solicitation processes for generation or transmission projects except on backstop basis in the event of market failure
- It provides a means of establishing whether the investment is efficient, even if part of the investment is non-revenue-earning at some point
- Noneconomic costs and benefits, such as the impact on the environment and local communities, can be taken into account by modeling the uncertainty (in respect of project costs and timing) that these factors engender
- It can take into account early investment in generation, where the first mover advantage is greater than the real option value of the investment

- It can deal with situations where transmission investment may lead generation investment
- It should be capable of dealing with the situation where a transmission project is under way to allow for a new connection or interconnection that is subsequently cancelled
- It can build on the traditional procedures that evaluate when investment is needed to maintain or improve reliability

Cigré study

A study by the industry association, Cigré, proposes a planning model that formally takes into account issues affecting the environment and stakeholders in order to avoid a two-step process where the planning is carried out on a techno-economic basis and is then analyzed and changed again as a result of consultations with stakeholders.[35]

The study looks at five main stakeholder groups:

- those who manage the transmission networks (primarily, the system operator and the network owners)
- those who use networks to trade (demand customers and power producers)
- those who facilitate energy trading (market operator, retailers, traders)
- statutory authorities (electricity regulator, and other public agencies)
- a class that it identifies as environmentalists

The proposed methodology is based upon a range of views of the future and a number of perspectives on development priority. Recognizing that it is impossible to reconcile the interests of all stakeholders, a pseudo-optimization is created with the goal of avoiding wasteful effort, so that a bundle of projects that maintain a reasonable stakeholder balance and stakeholder acceptability is selected. Arguably, the transmission planners do this instinctively, but failure to maintain a position that is reasonably acceptable as far as stakeholders are concerned causes considerable cost and delay.

The modeling process involves the following steps:

- An internally consistent scenario is developed for each significant view of the future
- For each scenario, network analysis is carried out to determine the development options, their expected project lives, and the effect on other projects
- Projects are evaluated against planning standards and criteria, and defined as "must do, speculative, or hedging" actions
- Each scenario and each project is then placed into a matrix table and its effect on each major impact axis (network cost, environment, etc.) is quantified
- By consulting focus groups and the identified stakeholders, the importance of the impact axes is determined as a generic table that can be applied to all projects—the impact axes are used to determine the effect of each project on the stakeholders, and a judgment is made on the relative importance of stakeholders' interests in the decision-making process
- Using multiplication by factors representing the stakeholder importance, a table of projects weighted against stakeholder interests is prepared, and projects with the highest overall score in the weighted stakeholder table are identified on the basis that they are likely to offer the greatest overall benefits to the stakeholder group
- Since this analysis has been carried out only on each scenario, the best projects across all scenarios are then identified
- A final check against planning standards and criteria is made to ensure that the "must do" and "hedging" projects are included in the preferred projects

Comfort zone of the important stakeholders

A model that is capable of optimization of a bundle of solutions and the interests of a disparate stakeholder group looks quite an attractive tool. The study, however, does express the necessary degree of realism, recognizing that each stakeholder group will have a different view on what should be done. The difficulty is that within each stakeholder group there are bound to be different views as well. Although the inputs into the model may be the result of much consultation and discussion, a project will not be capable of implementation if it is not within the comfort zone of those stakeholders that are critical to its implementation, particularly if they include landowners and local communities whose interests are affected. The model is designed primarily to optimize the interests of power producers, demand customers, the system operator, the network owners, and environmentalists.

The study recognizes that "each stakeholder has a comfort zone and a no-go area." The boundary is seldom well defined, and it is not easy to identify when a stakeholder chaos boundary has been crossed.

> *It is therefore suggested that these human factors prevent true mathematical optimisation. The best that can be done is to represent each stakeholder as a quantitative axis in a multi-dimensional analysis and select a bundle of projects which lies on the safe side of the multi-dimensional chaos boundary surface.*

Advantages and disadvantages of a stakeholder optimization method

The practical experience of implementing transmission projects around the world indicates that the projects that proceed most smoothly are those where the stakeholder interests are taken into account in the preparation and planning stages. Indeed, the more time that is put into taking stakeholder concerns into account in the preparation stage, the less time the siting and permitting processes are likely to take in the next stage. The tool that the Cigré has developed will require software packages to be developed, as it appears to be the first of its kind. It has a number of advantages:

- It collects and classifies both "hard" and "soft" information
- The methodology requires an understanding of how the problems and solutions are perceived by the different stakeholders and for their concerns actually to be taken into account
- It provides a methodology to incorporate different judgments as to stakeholders' concerns in an analytical framework, as an alternative to the traditional economic valuation driven by the planning standards and criteria
- The methodology applies a structure to the research and studies required for the decision-making process
- It helps to create a consensus-building environment at the earliest possible stage that is consistent with the approach advocated in chapter 10

The methods described in the study do have some disadvantages:

- Multi-criteria methods may put pressure on stakeholders to reveal preferences or intentions they might prefer not to disclose
- Some of the methods are difficult for non-experts to understand
- If the evaluation is presented in an obscure manner, it might be perceived as a scientific way to disguise a decision already taken
- The outcome of the methodology will, essentially, only be a point of reference, around which there will, in practice, be a wide variety of divergent stakeholder views. Stakeholders are typically interested in their own interests and not in solutions that achieve an equitable balance of all interests
- If the most influential stakeholders do not accept the methodology, the input data, or the outcome, there is no guarantee that the identified projects will be easy to implement
- The methodology appears to give more weight to the interests of industry participants and not enough to the interests of local communities and environmentalists, which may mean that the identified projects may not be those that minimize the impact on local communities and the environment

Reference networks

A team headed by Professor Goran Strbac at the Manchester Centre for Electrical Energy, University of Manchester Institute of Science and Technology in the U.K., has developed the concept of a reference network as a tool for transmission expansion and regulation. A reference transmission network, in its simplest form, would be topologically identical to the existing transmission network, with the existing generation and load layouts and cost characteristics. It can take into account restrictions in building new facilities, such as height restrictions and existing rights-of-way or route corridor restrictions. It would operate at the same voltage levels as the real transmission system, but the individual transmission circuits or components would have optimal capacities. A reference network hence determines efficient investment in transmission, together with corresponding efficient congestion.

Assuming that transmission capacity can take any size, the reference network would have exactly the optimal amount of congestion with which the optimal amount of investment cost would be associated, such that the total investment and operating costs are minimized. Instead of optimizing the capacity of only a few new circuits proposed to be added under a more conventional planning process, the optimization of the capacity is carried out for all transmission circuits, both existing and new.

The calculation is carried out using an optimal power flow model to which investment costs are added. A time horizon of one year is utilized. This gives the optimal capacity of each circuit through the balancing of transmission and generation costs. The modeling applies security constraints and results in an optimal dispatch on the reference network with optimal transmission circuit capacities. Locational marginal prices can also be derived for the entire year. This gives the optimal capacity of each circuit through the balancing of generation and transmission costs. Congestion costs are calculated from the locational marginal prices.

Reference network identifies the beneficiaries

The reference network not only provides a very useful comparator, it can also help to define the beneficiaries of transmission investment, which would provide a very useful tool in many jurisdictions for deciding who should pay for new capacity additions. The advantage is that it creates a network against which the real transmission network can be *objectively* compared. The optimal (reference) investment costs and the optimal (reference) congestion costs can be quantified and then compared with those actually achieved on the real transmission system. The comparison can also indicate the "usefulness" of individual circuits, and, from this, the need for new investment can be identified.

Regulatory tool

The reference network has obvious application as a regulatory tool for quantifying the benchmark costs of transmission as well as capital expenditure budgets for new transmission facilities. It can also be used by regulators as a tool for setting cost and performance-related targets in the design of incentive schemes. In effect, the transmission owner would be made to compete with its own reference model. The concept of a reference network provides a consistent framework for network pricing and can be used on its own or in conjunction with locational marginal pricing.

Pricing based on the reference network allocates existing and new transmission investment efficiently among network users. It avoids cross-subsidies, both temporal and special. In simple terms, the mechanics involve determining the reference network capacity and investment costs and converting the circuit transmission prices into nodal transmission prices, where the charges are associated with nodal power injections, rather than comparing or monitoring the users' contributions to each line flow. By allocating circuit prices to nodes, prices become location-specific, and by applying circuit prices during the period when the optimal line flows are binding, prices become time-of-use specific.[36]

Modeling into the future

The main problem with the concept is that it involves modeling into the future. It is based on predictions of future demands as well as generation location and cost (bid) characteristics. Obviously, judgments have to be made as to the appropriate time horizon. It is not easy to obtain the necessary stakeholder input or to reach consensus on the forecasts and the inputs. However, the uncertainties associated with forecasts are an inherent part of any planning process.

Potential benefits of the reference network

The reference network concept has a number of other potential benefits apart from being a planning tool:

- It can be used by regulators to set performance incentives and to approve capital expenditure budgets and investment costs, as well as assisting in determining the need for the expansion
- Reference network-based pricing rewards useful investment and discourages overinvestment
- It can be used as a tool for cost-benefit analysis
- It can efficiently allocate investment costs and help to identify the beneficiaries of the expansion
- It can help to analyze the costs and benefits of and can accommodate alternatives to transmission[37]

The Role of the Regulator

A variety of roles

There needs to be clarity as to the role of the regulator in the planning coordination process, and there are various options. The regulator could be the:

- determiner of the need for the expansion
- ultimate decision-maker as to what precisely should be built
- ultimate decision-maker as to how the costs should be allocated
- entity that is empowered to review and approve the broad plan

- entity that is empowered to review and approve the detailed plan
- economic regulator for the purposes of approval of capital and operating expenditure allowances for transmission price regulation purposes
- resolver of disputes (or appointer of arbitrators)
- convener of consensus building and mediation processes
- observer or stakeholder (representing the consumer interest), until it comes to the time of the formal exercise of its powers with respect to the determination of need, the siting process, and cost allocation

Avoid micromanagement

Regulators around the world typically carry out more than one of these functions, and there is no particular reason that their functions should embrace them all. However, regulators should avoid micromanagement of the planning coordination process and involvement in the detail of the design of the projects. The principles of risk management should apply here as elsewhere—the risks of planning and implementation should be handled by those in the best place to manage them.

On this basis, for example, several regulators in the U.K., Norway, and New South Wales simply approve an overall capital expenditure plan, often by hiring consultants to question the need for certain items and to carry out a cost/benefit analysis. OFGEM makes an allowance for certain levels of expenditure when setting NGC's price control. NGC is then free to spend whatever it likes (so long as it meets its statutory and license obligations). Although there is one school of thought that the regulator would not be justified in incurring the cost of permanent planning staff for this exercise, there is another view that the regulator needs people on a permanent basis who know the assets, understand how the system operates, and can anticipate the issues that will arise. Consultants tend to come and go and may not provide the necessary continuity.

The regulator may also perform a valuable role in resolving disputes between stakeholders as to changes to the planning rules. In order to avoid increasing regulatory risk and deterring investment (discussed in chapter 6), the regulator's role should be backstop in nature and not frontline. Regulatory micromanagement will destroy the entrepreneurial innovation and efficiency that transmission expansion needs.

FERC has proposed that Regional State Advisory Committees should be established to enable state regulators to participate in RTO operations. One of the functions of these committees would be to work with RTOs to seek regional solutions in transmission planning and expansion, although it is not clear exactly how this would be accomplished in practice.[38]

Methods of Preparation—A Summary

Preparation follows the list of ingredients

At the end of the last few chapters, we have assembled a list of ingredients for our recipes for transmission expansion, including raising and setting agents. The next logical step involves laying out the methods of preparation in the recipes for both merchant and regulated projects. We have considered the issues relating to the design of various methods of preparation that will apply to both. Although they appear to be most relevant to regulated projects, there are a number of issues that are relevant to merchant projects, because of their need to interconnect with existing transmission systems and regulated projects. Many of the issues that relate to the preparation of merchant projects were discussed in chapter 5 in the context of the key ingredients enabling those projects to flourish.

Once we have laid out the methods of preparation in a logical order, the transmission developer will be in a position to design the project. This involves a variety of steps and skills, including the evaluation of system studies, substation space, rights-of-way use, and expansion potential, as well as detailed engineering expertise. Sizing the project is an art as well as a science, as it involves anticipating future use of the system and market developments. It is not easy to predict how market participants will value physical and financial rights with respect to their future transmission service and energy market risk management needs. Transmission owners and developers have an abundance of expertise in this area and do not need this book to tell them how to do it.

However, the design of projects does involve expertise in siting and permitting processes, and relationships with landowners and local communities. These are issues that are not fully within the control of transmission developers and are the subject of our discussions in chapters 9 and 10.

The first steps

Before carrying out the main step of the preparation—the planning coordination process—we should start by undertaking a few preliminary tasks.

Review. We need to review the existing planning and operating standards and the regulatory performance targets, as they are the starting point of most regulated transmission planning decisions.

Interrelationships. In carrying out the review, we must look closely at the trading arrangements and market rules, because planning, operating, and performance standards are highly interrelated with the substance of the market rules.

Who sets and enforces? We should decide who should set and enforce the standards. The expertise that resides within the industry should be used to assist in the design of the standards, but the ultimate responsibility for setting and enforcing the standards should reside with the regulator, because standards are a key ingredient for so many of the core functions of the regulator, particularly in tariff setting and protecting consumers as to quality of service. The system operator can be used as a frontline monitor, but the regulator will have to act as the ultimate enforcement agency and will also need to police the system operator. Fragmented jurisdictions in setting and enforcing standards should be avoided. A single agency will be more effective.

Changes. If we think that the standards need changing (perhaps because they can be relaxed in light of improved monitoring techniques or can no longer be justified because of their cost), we should make sure that they are replaced with standards that are capable of measurement and realistic in the context of transmission.

Use economic standards. Wherever possible, we should attempt to replace detailed technical standards with broadly-based economic standards (but with adequate performance targets to ensure that quality of service does not deteriorate), because technical standards for transmission involve low probability, high impact events that do not lend themselves easily to meaningful standards or targets.

The planning coordination process

Decide to coordinate. We should agree that central coordination of the planning process is desirable and obtain the agreement of stakeholders to participate in an effort to build consensus as to what investment is needed. This is a different model to the traditional command-and-control central planning process.

Decide who should coordinate. The decision of who should carry out the planning coordination lies between the system operator (RTO), a transmission company owning a significant part of the existing infrastructure, and an entirely independent planning coordinator. FERC has already determined that the RTO or ITP should have the role of planning coordinator in the U.S., and this makes sense because of multiple transmission ownership.

In other jurisdictions, where there is a transco and a single transmission system, the role should obviously be given to the transco. However, where there is multiple ownership of transmission, and generation and demand-side solutions are a real possibility in order to meet transmission expansion needs, the system operator would be a better option. This is because it will be viewed as the most impartial entity with respect to the choice of solutions to implement. A transmission owner could carry out the coordination (and would certainly be well-placed to do so), but the process would need to be subject to review by the regulator or the courts to ensure that no bias in favor of transmission solutions were present. An independent planning coordinator (such as the former transmission administrator in Alberta) would not be the ideal candidate, because it would not have continually updated information as to the performance of the existing assets that was as good as the information available to the transmission owners and the system operator. The transaction costs would probably be higher, and there would be doubts about its incentives to provide efficient transmission service.

Planning coordination. The planning coordination process would be carried out on an annual or biennial cycle looking over a five- to seven-year time horizon. It would involve a number of steps including establishing and maintaining an accurate planning database; collecting information and inputs from all stakeholders; assessing the degree of likely generation and demand-side responses and merchant transmission projects; the publication of information as to the performance of the transmission system, its adequacy and reliability and the current level of congestion costs; assessing market power problems; and obtaining wide stakeholder input as to needs, and local community and environmental sensitivities. In the usual way, the process would take into account applications for the connection or interconnection of committed generation and demand projects, transmission projects (regulated and merchant), planned interconnection projects, and long-term transmission service requests.

Consensus building. The planning coordinator would publish its assessment of the transmission performance and capacity enhancement needed, but would not design specific projects. It would discuss the assessment with transmission owners, market participants, and other stakeholders (and possibly also with the regulators). Their input would be taken into account in a process that endeavors to reach consensus and acceptance by substantial majority of the stakeholders that their views have been properly considered.

Project development. Once the assessment is substantially accepted, the planning coordinator will invite the incumbent regulated transmission owner (or owners) to design projects to meet the capacity and performance enhancements that have been identified, and to file applications with the relevant regulatory, siting, and permitting agencies by a specified date.

Solicitation process. If applications to the applicable regulatory, siting, and permitting agencies covering the needed enhancements have not been filed by the specified deadline, the planning coordinator will issue a request for proposals (open to regulated and merchant developers), for projects to assist in meeting the needed performance and capacity enhancements described in the assessment, inviting generation, transmission, and demand-side solutions. The bidding evaluation process would be conducted under transparent rules and objective evaluation criteria. The winning project or projects would be those that scored highest under the published evaluation criteria, for example, applying specified methodology to calculate the cost-benefit ratio, at least as a threshold.

Builder of last resort. In the highly unlikely event that there is no response to the solicitation process, the choice of a builder of last resort will depend upon which transmission developer would be able to obtain the necessary powers of expropriation or eminent domain, and whether it could be attracted by a regulated tariff that contained an enhanced rate of return and involved no usage risk. If the project is really risky, however, should it be built?

NOTES

1. These standards are inserted in NGC's transmission license, but are reinforced through the Grid Code that is binding on all system users

2. Sanford V. Berg and Jinook Jeong, "An Evaluation of Incentive Regulation for Electric Utilities," *Journal of Regulatory Economics* 3 (1991), pp. 45-55

3. OFGEM (the Authority) has approved Revision 6 of Issue 2 of the Grid Code for implementation from February 18, 2002. Under General Condition 4.3 of this latest revision, the Panel is to consist of:

   a) a Chairman and up to four members appointed by NGC;

   b) a person appointed by the Authority; and

   c) the following members:

   (i.) 3 persons representing those generators each having large power stations with a total registered capacity in excess of 5 GW;
   (ii.) a person representing those generators each having large power stations with a total registered capacity of 5 GW or less;
   (iii.) 3 persons representing the network operators;
   (iv.) a person representing the suppliers and retailers;
   (v.) a person representing the non-embedded customers;
   (vi.) a person representing the generators with small power stations and/or medium power stations (other than generators who also have large power stations);
   (vii.) a person representing the BSC (Balancing and Settlement Code) Panel (governance mechanism); and
   (viii.) a person representing the externally-interconnected system operators.

4. The year 2000 was a dry year in New Zealand and because of the consequent reduction in generating capacity, the market design in New Zealand was put under strain. There was a review as to whether the underlying design of the wholesale market was sound, because prices inevitably rose. Part of the problem for Transpower was a lack of a regulatory regime. The industry took steps to establish a single governance board. Transpower's view was that the board should be required to promote the public interest and be able to enforce the rules necessary to

create a more efficient market. Transpower New Zealand Limited, "Submission to the Post Winter Review of the New Zealand Electricity System" (October 2001)

5. E-mail to the author

6. Promoting Wholesale Competition Through Open Access Non-discriminatory Transmission Services by Public Utilities and Recovery of Stranded Costs by Public Utilities and Transmitting Utilities, Order No. 888, 61 FR 21,540 (May 10, 1996), FERC Stats. & Regs. ¶ 31,036 (1996). Open Access Same-Time Information System and Standards of Conduct, Final Rule, Order No. 889, 61 FR 21,737, FERC Stats. & Regs. ¶ 31,037, (1996)

7. Kiah Harris, "Impediments to System Expansion," presentation to the Sixth Electricity Conference, International Bar Association, Section on Energy and Natural Resources Law, Berkeley, California (February 26, 2002)

8. FERC Notice of Proposed Rulemaking, 18 CFR Part 35, Docket No. RM01-12-000 (July 31, 2002)

9. "Incentives For Transmission Enhancement," a position paper of the Electric System Reliability Task Force (July 9, 1998) attached as Appendix F to "Maintaining Reliability in a Competitive U.S. Electricity Industry," Final Report of the Task Force on Electric System Reliability, Secretary of Energy Advisory Board, U.S. Department of Energy (September 29, 1998)

10. José A. Rotger, "Reconciling Market-Based Transmission and Planning," presentation to Harvard Electricity Policy Group, San Diego (January 24, 2002)

11. FERC Order Providing Guidance on Continued Processing of RTO Filings, Docket No. RM01-12-000 (issued November 7, 2001)

12. FERC Notice of Proposed Rulemaking on standard market design, 18 CFR Part 35, Docket No. RM01-12-000 (July 31, 2002) ¶ 335-350

13. Australian Competition and Consumer Commission, "Regulatory Test for New Interconnectors and Network Augmentations" (December 15, 1999)

14. I am grateful to Robin Cohen for discussing the concept with me

15. As we discussed in chapter 4, depending on the design of the congestion management mechanism, the transmission owner may not be the entity that keeps the congestion rents. Under the New England market design, all FTRs are included in

mandatory auctions. Entities with auction revenue rights receive revenues from the FTR auction. Load primarily receives the FTR auction revenues, as do entities that have paid for new transmission upgrades that have increased the capacity of the transmission system. Entities that purchase FTRs through the auction or the secondary market collect the congestion rents. Under the PJM model, network and firm transmission customers are assigned the FTRs, and any additional FTRs are auctioned. The FTR holders collect the congestion rents. The transmission owners receive the revenues from the FTR auction

16. ISO New England, Inc. et al., Order Conditionally Accepting Congestion Management and Multiple Settlements Systems, 91 FERC at 62072 (2000)

17. There are plans to replace the transmission administrator with an ISO in Alberta

18. Department of Trade and Industry, Electricity Transmission License Standard Condition C7G, paragraph 1 (September 27, 2001)

19. Standardizing Generator Interconnection Agreements and Procedures: Advance Notice of Proposed Rulemaking, 66 FR 55140 (November 1, 2000), FERC Stats & Regs.¶ 35,540 (2001) at 35,801-02

20. ISO New England, Inc. et al., Order Conditionally Accepting Congestion Management and Multiple Settlements Systems, 91 FERC 61,311 at 62076 (2000)

21. FERC Notice of Proposed Rulemaking on standard market design, 18 CFR Part 35, Docket No. RM01-12-000 (July 31, 2002) ¶ 348

22. Eric Hirst and Brendan Kirby, "Key Transmission Planning Issues," *The Electricity Journal* (October 2001) p. 68

23. As Eric Hirst and Brendan Kirby point out: "The difference in lifetimes between the transmission project and its alternatives raises the issue of whether the alternatives should be assessed against the cost of *deferring* the transmission project for several years or against the full costs of *displacing* (eliminating the need for) the transmission project. If the transmission project will likely be needed in any case, although at a later date, the deferral approach makes sense." Eric Hirst and Brendan Kirby, "Key Transmission Planning Issues," *The Electricity Journal* (October 2001), p. 64

24. Jim Detmers, Armie Perez, and Steve Greenleaf, "Valley-Rainbow Transmission Project, Memorandum to California Independent System

Operator Board of Governors, Folsom, California (March 23, 2001), p. 3

25. Eric Hirst and Brendan Kirby, "Key Transmission Planning Issues," *The Electricity Journal* (October 2001), pp. 59-70

26. ISO New England, Inc. et al., Order Conditionally Accepting Congestion Management and Multiple Settlements Systems, 91 FERC 61,311 at 62076 (2000)

27. FERC's Notice of Proposed Rulemaking, 18 CFR Part 35, Docket No. RM01-12-000 (July 31, 2002) ¶ 337-350

28. See Section 1.7 of Schedule 6 to the PJM Operating Agreement and the Transmission Owners Agreement (available on the PJM website www.PJM.com)

29. FERC Order No. 2000, Docket No. RM99-2-000 (issued December 20, 1999)

30. FERC Notice of Proposed Rulemaking on standard market design, 18 CFR Part 35, Docket No. RM01-12-000 (July 31, 2002) ¶ 348

31. Carolina Power & Light Company, 95 FERC P61,282 (2001). In this Order, FERC directs the applicants to eliminate provisions that gave the transmission owners the right of first refusal to construct transmission facilities in their service areas. FERC found that these provisions unduly limited the decisional authority of GridSouth over transmission planning, presenting the possibility of discrimination by self-interested transmission owners favoring their own generation, and possibly precluding lower cost or superior transmission facilities or upgrades by third parties from being planned and constructed

32. ISO New England, Inc., 91 FERC P61,311 (2000). In this Order, FERC directs ISO New England to revise its proposal to eliminate any decisional role transmission owners may have in the current expansion plan and concluding that only the ISO should identify expansions and upgrades required for reliability reasons

33. Avinash K. Dixit and R.S. Pindyck, *Investment under Uncertainty* (Princeton, N.J: Princeton University Press, 1994)

34. Sam Lovick's team comprised Professor Robert Wilson of Stanford University, Dr. John Small, Julia Frayer, and others from London Economics' Cambridge, Massachusetts office

35. Cigré TF 38-05-10, "Optimal Network Structure in an Open Market Environment" (June 17, 2001), p. 16. See also short report under the same title and reference number (Nov. 27, 2001).

36. The transformation of the reference circuit to reference nodal prices can be carried out by using conventional sensitivity analysis. By definition, the sensitivities measure the impact of an incremental change in injection on a particular line flow. In this case, the process is reversed, where the circuit price (that corresponds to a line flow) is transformed into a nodal price (that corresponds to the power injection). Prices for demand and generation connected to the same node would have the same magnitudes, but different polarities. Although circuit prices are always positive (representing circuit annuitized investment costs), nodal reference investment prices can be positive or negative, depending on the impact the particular injection has on the flow in the binding circuit

37. Professor Goran Strbac, "Concept of Reference Network: A Tool for Transmission Regulation and Expansion," presentation to the Harvard Electricity Policy Group, San Diego, California (January 24, 2002). See also Goran Strbac and Ron N. Allan, "Performance Regulation of Distribution Systems Using Reference Networks," *Power Engineering Journal* (December 2001)

38. FERC Notice of Proposed Rulemaking on standard market design, 18 CFR Part 35, Docket No. RM01-12-000 (July 31, 2002) ¶ 554

CHAPTER 9
Timing: Obtaining the Permits and Rights-of-Way

Ingredients of this Chapter

Timing is critical

In any cooking process, controlling the timing is critical. In the context of transmission expansion, control of the timing seems to be quite impossible. It is possible to more or less stay in control but only until the process of obtaining the siting authority, governmental consents, planning and environmental permits, the land, and the rights-of-way begins. The whole exercise in most countries has a justified bad reputation that acts as a deterrent to investment. Why is it such as mess, and what can be done about it, short of embarking on the very difficult task of designing something entirely new? Even if there were some obvious solutions that could easily be implemented, there seems to be little political will to make radical changes anywhere in the world. We shall describe this whole process as permitting. This expression is shorthand for obtaining all types of approval, consent, permit, certificate, expropriation right, power of eminent domain, land, rights-of-way, easements, wayleaves, or servitudes relating to the need for the project or the siting, routing, planning, zoning, or environmental laws.

Organization of the chapter

This chapter will begin by describing why the permitting process is unpalatable to investors and why it is a mess in most countries. The two main reasons are the nature of the permitting process itself and the number and strength of the objections from landowners and local communities. We conclude that many of the difficulties are embedded in the culture of society in the relevant jurisdiction, and in most countries there is little political will to make radical changes. As a result, we shall have to look for solutions that are largely based within the existing legal, regulatory, and social frameworks. We then look at the inherent tensions between local interests and regional or national infrastructure needs and ask the question of whether national-level planning or sectoral planning by government is likely to assist. The survey of recent trends in several developed countries indicates that national-level planning by government is unlikely to solve the problems in the near term.

In order to find answers, we analyze the decisions that have to be made and characterize the decision-making processes. We then pose a simple question "who should decide?" We evaluate the various options in order to design the best recipe. The issues relating to obtaining land and rights-of-way pursuant to powers of expropriation and eminent domain are then discussed with special reference to private sector transmission expansion, projects in the national or regional interest, and the complexities created in a deregulated environment.

This is followed by a review of the situation in U.S. and an explanation of how fragmented decision-making at state level can be coordinated, by looking at the state of Massachusetts. After considering the FERC siting authority for natural gas pipelines, we discuss whether a similar solution is likely to be available for transmission lines, and, if not, what other solutions might be found for the U.S. The chapter concludes with some ideas as to how some smaller changes could be made to improve the efficiency of the permitting process and a case study of one of the longest permitting processes for an expansion project in a deregulated market, which ran for a decade.

The Permitting Process Is Unpalatable

There are three fundamental reasons that permitting is unpalatable to investors:

- the uncertainties about the outcome, cost, and timetable of the permitting process for a project
- the defects of the process itself
- the strong objections that transmission projects always seem to elicit

The uncertainties

Investors live with uncertainties and analyze them as risks. As explained in chapter 2, risks affect much of the decision to invest as well as the cost of capital. A firm will look for ways to manage these risks, either by its own efforts or by purchasing insurance or risk-management instruments. In the context of the permitting process, a developer finds it very hard to assess the risks and to obtain information on which to make risk-management decisions.

Why the process is a mess

There is no simple or elegant way to describe why permitting is a mess. Perhaps that is why governments, regulators, and legislators have not found a simple or elegant way to fix it. There are many reasons that contribute to the unpalatable nature of the current state of affairs around the world. The reasons include:

- the length and cost of the process
- the methods of public participation
- fragmented decision-making
- basic inefficiencies in the process (including over-judicialization)
- policy conflicts (and the lack of political will to make decisions)

Length and cost of the process

The time taken to obtain the necessary siting, planning, and environmental consents is inherently long. It is seldom shorter than two years and can take as long as 10 years for a major project.[1] This makes it very difficult for the planning and implementation processes for needed transmission facilities to keep up with the demands that are placed upon the transmission sector, if only in terms of meeting load growth and accommodating the retirement or construction of new generating plant. Implementing transmission projects to maintain or improve reliability and to mitigate generator market power quickly could be impossible if a new route corridor is involved.

Some of the reasons for the length of the permitting process are legitimate. It will take time to prepare environmental impact assessments, to listen to the stakeholders, to make sure that the right route has been chosen, and to find the best solutions in order to mitigate the impact on the environment and the local communities. However, the main reasons for the length of the permitting process arise from other defects. A common complaint emerges as to the length of the time that it simply takes to make decisions. Professor Malcolm Grant of Cambridge University has characterized this problem as the "procrastination culture." It goes without saying that time costs money, particularly in the case of development of infrastructure that always seems to be struggling to keep up with the changing demands placed upon it.

The Brighton-High Ridge line

The saga of the Brighton-High Ridge line is one of the longest in transmission expansion. The line closed two loops to form an open-sided, figure-of-eight 500 kV system in the Washington, D.C. area and was of strategic importance to the security of supply to the nation's capital. The initial filing with the Maryland Public Service Commission (MPSC) was made in 1976. There were three years of hearings, and, after a year of deliberations, the PSC adopted the hearing examiners' conclusions and allowed the project to proceed. There were a number of appeals, and the Maryland circuit court affirmed the decision in 1985.

In 1985, Howard County passed a zoning ordinance that made it impossible to build the line. This led to legal proceedings over whether the certificate of public convenience and necessity issued by the MPSC could be preempted by county-level zoning ordinances. In December 1987, the authority of the MPSC was upheld. However, in 1988, an appeal was lodged with the MPSC asking for the case to be reopened, because there was new evidence of the effects of electromagnetic fields (EMF) that had not been considered. After several hearings, the MPSC decided in 1990 that there were no health concerns. There was a further appeal, and in 1992, the final permission to construct was issued. The line was eventually energized in 1994, by which time the rest of the system to serve the Washington area had long since been built.

It is interesting that this case involved a long hard look at the issue of EMFs and that no potential health hazards were found. It is impressive that some very determined homeowners and property developers could provoke some seven years of hearings. The route was originally chosen, because the land in question was a swamp and attracted little opposition. In the 1980s, it was reclassed as "wetlands" and became a much more attractive place in which to live. The cost of transmission system congestion to consumers in and around the nation's capital as a result of the delay in building this line is incalculable, and the quantum would be daunting, to say the least.[2]

Cost can create an efficiency incentive

Estimating the cost of obtaining permits and costing the impact of delays are always difficult. A common estimate of the minimum cost of preparing only the environmental impact assessment for a major, but nevertheless straightforward project is U.S. $1 million dollars. Because the processes are usually so open-ended and uncertain, the development costs of a project are a serious deterrent, even though most developers know that incurring these costs is unavoidable and necessary. For this reason, most utilities try to utilize existing corridors to the maximum extent possible (which creates an incentive for efficiency in their use). However, there comes a point (as in Norway and Argentina) where all of these obvious steps have been taken, and a new line using a new route corridor is needed. The transmission developers must do all they can to keep skilled and experienced people available to deal with local communities, landowners, and objectors, and to provide as much continuity as they can so that they have the best chance of creating the sort of relationships that will allow their interests to be heard and taken into account. This will enable a project to proceed on the best possible basis for the local community and the environment.

Increased public participation

Public participation in the permitting and decision-making processes appears to be increasing, whatever the culture and openness of the country and its style of government and administration. Those who care deeply about the environment and local communities concerned about their health, safety, quality of life, and local amenities have joined the landowners, who are concerned with property values, in order to swell the ranks of objectors. They are increasingly well organized, better funded, able to share knowledge, and aware of their rights and the vulnerabilities of project developers, as well as being vocal and energetic. In a deregulated environment, the objectors can include other market participants that feel that their competitive or other interests may be adversely affected. The design of a permitting process that allows large numbers of objectors to participate is never going to be easy.

Methods of public participation

An efficient method of enabling local communities and other stakeholders to participate in the decision-making process is often lacking. The usual answer has been to judicialize the process, sometimes to such an extreme as to reinvent court proceedings. Public inquiries and regulatory hearings all take time and money and do not provide a forum that either objectors or developers find user-friendly.

John Kay published a graphic account of the inquiry into the Sizewell B nuclear power station public inquiry, describing the conduct of the Central Electricity Generating Board (CEGB), the protestors, and the role of the inquiry chairman, Sir Frank Layfield:

> *Layfield was left to arbitrate between the technically competent but flawed and one-sided presentations of the CEGB and the incoherent demands of the protesters. So the inquiry moved to its inevitable conclusion. It approved the CEGB's plans, with minor gestures to the opponents. If the object was to enable protest groups to imagine that they had had their say, the Sizewell inquiry may have served a purpose. As a means of assessing the costs and benefits of the project, it was a comprehensive waste of time and money ... If public inquiries into such issues are to be held, they should be investigative inquiries, in which the inquiry team has the power and resources to explore the issues for themselves. The conflicting proceedings of expensive barristers in 340 days of public hearings were as relevant to establishing the truth about Sizewell as the jousting of medieval knights.*[3]

There are also increased appeals and judicial review of decisions, the threat of which usually slows down the primary proceedings so that the decision-makers can protect themselves from litigation at a later stage.

This is the inevitable result of the higher levels of the available information, knowledge, awareness, organization, funding, and affluence of objectors, which we have already observed. It is perhaps also reflective of the fact that the legal and regulatory framework, as well as national and local plans, and policy guidelines, are often open to interpretation as a result of fragmentation of decision-making and policy conflicts.

Fragmented Decision-making

Fragmented decision-making can greatly add to the length of the permitting process and increase its cost. It is very common for different aspects of the same project to be the subject of decisions by the different agencies. For example, the energy minister or the federal or national regulator may undertake the primary approval process or the determination of need for the project. The municipalities or local authorities will usually issue the local planning permissions and zoning permits, while environmental, siting, or routing issues may be dealt with by a third agency or set of agencies, perhaps operating at state or regional level.

The Edison Electric Institute cites the example of the Chicago-Apple River transmission line project that began in 1996 as a proposal for a 39-mile, 230 kV line to link facilities in Minnesota and Wisconsin. Because the project crossed two states, involved an electric cooperative, and would have to cross a river and a national park, the reviewing agencies included the Public Service Commission of Wisconsin, the Minnesota Environmental Quality Board, the Rural Utilities Service, the U.S. Army Corps of Engineers, the National Park Service, and local governments. Project managers expect construction of a scaled-down 161 kV line to begin in 2003 or 2004.[4]

A local authority can certainly throw a monkey wrench into an approved transmission line project. For example, seven county committees denied permission to enter county land to conduct the preparatory work for the new Arrowhead-Weston 345 kV line that had already been approved by the Wisconsin Public Service Commission.[5] The developers were forced to reroute to bypass county opposition.

Because transmission line projects are linear in nature, multiple applications or filings with successive local authorities along the length of the route may be inevitable. In the Czech Republic, it is necessary to file documents with each local authority, together with details of the ownership of the land over which the line passes and whether the landowner agrees to its construction or not.

The fragmentation of jurisdiction can arise because of the way in which the legislation has developed. For example, certain environmental issues with water resources, federal land, coastlines, or fishery protection are often dealt with under separate legislation by different agencies.

Some jurisdictions have attempted to overcome the fragmentation of the decision-making by creating a "one-stop shopping" process (including several states in the U.S. and various civil law countries such as France and Argentina). Other jurisdictions have dealt with this problem by allowing the government to "call in" the decision to the relevant minister if it is a project of national or regional importance (such as the U.K. and New Zealand). However, even where a decision is "called in" in this way, the local authorities may still need to issue local planning permissions and zoning permits.

Basic inefficiencies in the process

Over-judicialization. There are many inefficiencies in the management and procedures of the permitting process. Some are the result of over-judicialization of the process, as we have observed, and the inefficiencies that are inherent in any litigious process. Some of the inefficiencies arise out of inherent uncertainties created by lack of published decision-making criteria, inconsistencies in rules and procedures, lack of discipline in making decisions within reasonable time scales, poor case management, and lack of information and transparency.

Lack of resources. A common problem is the lack of staff and resources of the siting authorities and other permitting agencies. The staffing budgets can often be small, and pay levels in the public sector can be so low that the permitting authorities often find that they lose their most able staff to the private sector. Wholesale privatization of the services of siting authorities and permitting agencies is a very sensitive issue where accountability and democracy are important, particularly at local level.

Judicial challenge. Another inefficiency is created by the risk of judicial review of the proceedings or appeals from the decisions of permitting authorities. If leave is granted to appeal or review a permitting decision judicially, it will have the effect of adding 12 months or more to the process, even if the challenge is unsuccessful. Objectors are increasingly using judicial review and appeals to cause delay.

The courts will try to deal with unmeritorious claims expeditiously and deter them through case law. The U.K. courts have, for example, narrowed the scope of challenges based on certain technical grounds such as non-compliance with part of a European Union (EU) directive or challenges based on breaches of the Human Rights Act. However, the courts do not seem yet to have applied the "so what" test. In other words, when granting leave to appeal or challenge the process under judicial review proceedings, the courts will assume that the grounds of challenge are well founded. They do not consider what the likely consequences would be and the extent to which anyone would be prejudiced if the decision were allowed to stand. As a result, it could be argued that the courts appear unable to balance the interests of the individual against the interests of society as a whole, and do not consider whether a much wider group of interests would suffer from the delay in the implementation of the project.

Policy conflicts

The costs and delays associated with permitting are exacerbated by government and regulatory policy conflicts, particularly relating to the lack of clear decisions as to the best way to deal with a variety of policy goals, such as:

- environmental and amenity protection
- economic development
- attracting new investment and investor protection
- consumer protection
- social welfare
- security of supply of energy, deregulation, and privatization of the electricity industry
- more generally, the protection of local community interests with respect to a project that is in the broader national or regional interest

Many governments do not set a complete policy in all of these areas, and, indeed, some of them do not set policy at all. If policy is set, it does not necessarily follow that it is transparent and even known within the different ministries or arms of government. There is also the potential for policy conflicts to arise between the government and the regulator, particularly where the roles and responsibilities of each are not well defined in the legal framework and where the accountability of the regulator is unclear.

If policies are set, it is inevitable that there will be tensions between them. It is not possible to achieve all policy objectives at once, and it is not easy for governments to set clear priorities. The two most obvious policy conflicts in relation to the development of any infrastructure relate to local versus national interests and energy needs versus environmental protection. These are difficult decisions for society. It is hardly surprising that they take time.

Sometimes the policy conflicts are entrenched in legislation. For example, the Massachusetts Siting Board's enabling statute states that its objectives are "to provide for a reliable energy supply for the Commonwealth with the minimum impact on the environment at the lowest possible cost."[6] These are all commendable objectives, but there are obvious tensions between them.

At local level, policy setting and coordination are extremely difficult to achieve, and there are bound to be conflicts between the policies of different local agencies and authorities as well as conflicts between the policies of local and national government:

> *Local plans are all too often out of date, and, coupled with the two tiered approach to planning policy built into the system [in the U.K.], conflicts and confusion are rife.*[7]

Strong objections are inevitable

It is perhaps because of the very intrusive nature of steel lattice transmission towers and their visual impact, rather than the health and safety issues created by the uncertainties surrounding EMFs, that gives rise to such strong objections to the building of new transmission assets. Transmission lines on single poles have less visual impact, but can be considerably more expensive than their steel lattice counterparts, if they are to have the same ability to withstand storms.

Transmission lines do not pollute the atmosphere in the way that certain types of power stations are able to, but they are resisted with almost greater hostility in many countries. Indeed, when announcements are made as to the retirement of generating stations, NGC has received immediate requests for the associated transmission lines and towers to be taken down. Much depends upon the culture of society in a particular jurisdiction as to the extent, nature, and strength of the objections and also how the permitting process deals with them.

Culture of society

The way in which decisions are made and the speed of decision making are very much rooted in the culture of a country. This culture is also reflected in the legal system, the regulatory framework, and the permitting process. This may explain why decision-making is more streamlined in civil law countries that have a background and culture of defining the relationship of the individual with the state and providing remedies to the individual in cases of abuse of administrative power. The state may well have a culture of providing for society. This may be the reason that when Electricité de France (EdF) built its nuclear plant, it took the trouble to take care of the local communities and created employment.

The cultural attitude to property rights also has a significant impact on the planning and decision-making processes. Many of these processes are the product of the historical importance of property rights that are perhaps waning in favor of concerns over the environment, amenity issues, and quality of life.

Where the culture is very open and there is considerable local democracy, the style of decision-making may be less directive and more consensual. It is not unreasonable to expect the process to take time in order to allow all objectors to have the maximum opportunity to be heard and to ensure that they feel that the impacts of the project on their local interests and the environment have been taken into account. Where national or local government or a particular minister is responsible for making the decision, this culture will engender considerable nervousness on the part of the relevant government minister (who is a politician dependent upon popular votes) in announcing a decision. Fear of media attention and backlash from those who are disadvantaged or feel unfairly treated will give rise to procrastination.

Even in a more closed society or where a government is more directive than consensual in its style, it is hard to justify streamlining a process in order to avoid delay in a manner that may impinge upon the ability of all objectors to have their interests properly taken into account. Whatever the culture, it is clear that there is little appetite for redesigning planning, siting and zoning laws, and procedures to speed up the process by limiting the opportunities for objectors to participate. The New Zealand National Development Act of 1979 included an attempt to streamline the permitting process for major projects of national interest. It was very unpopular, partly because it appeared to give the project an inherent advantage over the local interests and environmental protection. The incoming government rapidly repealed it.

Answers are not easy to find

The question then arises of whether it is possible, either by changing the legal and regulatory framework or by redesigning the procedures, to achieve a process that results in better and more efficient decision-making, without affecting the local community and environmental interests adversely. A further question is whether the uncertainties surrounding the outcome of the decision-making could be minimized, because they clearly deter investment.

The issues with respect to transmission expansion are inherently hard to deal with, particularly at a time when individuals and communities are becoming much more aware of their rights, as well as the need to protect the environment. They are demanding better answers to the question of how the national energy needs of society can be reconciled with local interests. The answers are not easy to find. As Judge David Sheppard of the New Zealand Environment Court said, "It is very difficult to know how loud is too loud and how safe is safe enough."[8]

Local interests vs. regional or national needs

One of the inherent difficulties of any linear project is that it is likely to pass through the jurisdiction of more than one decision-making authority. It may not benefit the consumers or voters in the areas of all of the authorities through which it passes and, of necessity, may well have adverse local consequences. This is why society needs to have a system for reconciling the impact of projects on local communities and the environment with national or regional interests.

In the U.S., this is a key issue for transmission and interconnector projects that pass through or across more than one state, particularly in view of the fact that the permitting process is carried out at state rather than federal level. Indeed, it is also an issue for generation projects. Officials in the states of Kentucky and Tennessee were reported to be making it more difficult to build power plants in their states. They have been deluged with requests from independent power producers to build plants in their states, because they sit at the crossroads of natural gas pipelines and transmission networks that could be used to send power to growing markets in the southeast. The potential capacity of the proposed plants would far exceed the needs of either state, even at times of peak demand.[9] From the point of view of society, the building of these plants could be a very efficient solution and one that minimizes the broader impact on the environment and local communities.

National or regional decision-making authorities may not adequately take into account local interests. However, it is possible for projects that are approved locally to be vetoed by another higher authority, which seems, on the face of it, to present a curious result. It is an obvious point that the more local the decision-making, the less weight is likely to be given to regional and national interests.

The Irish electricity utility, the Electricity Supply Board, made an application for a permit to build a line across the area of a local authority in County Donegal in order to transfer power between the neighboring regions. Because there was no benefit to the consumers in the area of the local authority, it imposed a condition requiring the line to be significantly undergrounded. No such condition was imposed by the neighboring local authority.

The Cross Sound Interconnector in New England suffered from a similar problem. The Connecticut Siting Council turned the original application down, in part, because it was unable to establish the need for the project in Connecticut, for which it was rightly criticized by many (including the Bush administration). When the project was ultimately approved, the Council countered some of these criticisms by explaining the benefits to Connecticut and the region. The Connecticut State Legislature Environment Committee later announced a one-year moratorium on the building of interconnectors between the State and New York using cables passing under Long Island Sound. The Environment Committee's bill would put a hold on any similar Cross Sound projects, while the State studies how such projects might affect the power supply, economy, and environment.[10] These cables would benefit consumers in the New York area. Reconciling these potentially-conflicting national, regional, and local interests appears to call for some sort of national- or regional-level planning coordination process.

National-Level Planning

Inclusion of transmission lines in national plans can be helpful

One simple solution for reconciling local interests with national needs is for the national government to publish a national- or regional-level development plan. Many transmission owners maintain that it is significantly easier for transmission to be expanded where there are well-established and published development plans at a local or national level. The transmission owners in Norway and the Czech Republic have made this point.

National development plans can be relatively rare in many developed countries, but local development plans are often the mainstay of local government and also of the relationship between local and central government. In theory, published development plans should create transparent policy objectives. When they are published by the local or national government that will typically be taking or strongly influencing the decision-making, the transparency should provide more certainty of outcome for all parties. However, a plan may not, of itself, stem the objections or speed up the process.

The logic for development plans to include transmission

There is logic for including needed transmission and distribution lines in national and regional development plans, because most development plans will need to take into account the changing needs for electricity service, that will place new or different demands on the transmission and distribution systems, which will require action to be taken. Comprehensive or integrated planning for the transmission needs of major growth areas such as new urban or industrial developments and city business districts would be particularly beneficial. It would be helpful to include the necessary transmission expansion in those plans, which would reduce the uncertainties surrounding the obtaining of permits, consents, and rights-of-way, because the issue of need (the public convenience and necessity) of a proposed new transmission facility would be argued, at least in part. It could also create a clear causal link between the benefits to and impact on the community and the environment, economic development at national, regional, and local level, and the infrastructure needed to support that development. There can be strong local and municipal support for new transmission lines when they prove to be necessary to allow a local development plan to proceed or local business to expand.

Recent trends in national-level planning

National-level planning has been associated with coercive regimes and communist governments. It fell into disrepute towards the end of the twentieth century, but there is some evidence of a modest increase in national-level planning in democratic, developed countries.[11] The EU policies have clearly served as a trigger for the introduction of some degree of national-level planning. Coupled with the EU Directive on the Completion of the Single Market in Electricity (the IME directive),[12] the EU policies should act as a catalyst for more regional planning[13] and coordination between regulatory authorities.

Some national-level planning has been in response to a high level of growth or increase demand for infrastructure services. In the past, it was usually adopted in response to a crisis (*e.g.*, World War II). The absence of a real crisis, according to Jerold Kayden, may be part of the reason that in the U.S., national-level planning has not been reintroduced.[14] Perhaps the size of the U.S. and the considerable authority that individual states possess over land use issues make national-level planning in U.S. simply too difficult to implement.

There seems to be little appetite to adopt national-level planning in order to minimize or override objections from stakeholders in infrastructure projects (although Sweden appears to be something of an exception).[15] The attention will be placed on the values and the objectives that are to be maintained and pursued and on what planning methodology will provide the best tools. As Rachelle Alterman observes:

> *National planning would then largely be seen as an instrumental rather than an ideological issue. Such planning will increasingly be viewed as a useful tool for democratic, advanced-economy countries that are seeking to improve their standing in the world of increasing economic and demographic competition, and are eager to answer voters' demands for better, global and local environmental management.*[16]

Sectoral planning

Sectoral planning may also be growing in strength, although it may be poorly connected with other associated planning. This may be a response to the need for increased national planning, which may be more easily accomplished on a sectoral rather than a comprehensive basis. There are new paradigms for coordination and integration of these plans with those emerging from other sectors. For example:

- There are governmental policies to weave together a few sectoral plans into a single program (*e.g.*, land use, transportation, and air quality)
- A statutory basis is given to the preparation of a comprehensive plan and a coordinated review of sectoral plans
- A comprehensive economic plan to coordinate sectoral plans might be adopted
- There may be more extensive sectoral planning that is subsequently integrated into a comprehensive national plan

- Comprehensive plans may be prepared by several ministries having responsibility for each of the relevant sectors from their own perspectives, which are then synthesized into a single plan

For the most part, national plans are not prepared or implemented in a command-and-control style. They are usually prepared in a collaborative manner, and the result is a flexible (as opposed to "blueprint"-style) advisory or policy guideline. However, incentives are often created in the form of making national funding available or increasing local legal powers, on condition that the guidelines are adhered to and that comprehensive local planning is properly carried out. There are exceptions to this trend reported from France, Sweden, the Netherlands, and the U.K., where major national projects are the subject of a quite detailed plan or directive, because there is a concern that local government or objectors might delay or block implementation.

Public participation

Sweden, France, and the Netherlands have fast-track procedures for projects in the national interest. The Netherlands restricts appeals and litigation to the initial "in principle stage." Once the decision to proceed is made, appeals are restricted in the detailed planning stage.

Resistance to national-level planning

While comprehensive national or regional planning that includes the electricity sector has some attractions, it is not clear that all countries view national-level planning as useful. For example, the U.S., the Republic of Ireland, and Sweden still encounter resistance to it. Apart from the command-and-control stigma that attaches to it, there will inevitably be concerns about remoteness from local issues and environmental impacts. National-level planning appears to be most useful in small countries and countries that find their land and natural resources under threat of exploitation.

Whether comprehensive planning will help to encourage needed transmission investment will depend on how long it takes, how well it is done, how well the public participation is handled, and what the scope is for judicial review or litigation. It must be in tune with changes in the culture and style of decision-making in the country concerned, as well as changes in the economy and society. This means that one size will not fit all. The most important consideration is what the goals are and whether national-level comprehensive planning will produce a better answer for society.

Regional planning is an obvious compromise

An obvious compromise lies in the use of a regional planning authority that would be able to overcome local parochialism, but would be closer to the landowners and local communities than national agencies. There have been proposals to move to a regional planning regime (based on local elected authorities) in the U.K. The U.S. established regional transmission groups (RTG) for regional planning, but they were not regarded as particularly successful. In the U.S., FERC proposes to mandate regional planning, making ITPs or RTOs responsible for coordinating the planning processes with input from Regional State Advisory Councils.[17] However, redesigning a process for transmission planning coordination on a regional basis would obviously be easier than designing a process for regional permitting or establishing a regional infrastructure development plan. It does not look as though national or regional planning will provide all of the answers as to how to make the permitting process more efficient and effective, attractive as the idea might be.

In order to solve the problem of how to reconcile local, regional, and national interests through a transmission permitting process, we need to look at the kinds of decisions that have to be made and how they might be characterized.

What Decisions Have to Be Made?

A variety of related decisions

There are a number of different, but related decisions that have to be made that may fall under the jurisdiction of several agencies in some countries. The typical decisions are whether to grant:

- Overall approval of the project on the basis of need for the expansion (which traditionally involves a broad cost-benefit analysis made from the perspective of the consumers as a whole)
- outline (in principle) permission for the project to proceed, with the detail to be worked out subsequently (as in the U.K.)
- planning, zoning, or local authority permits or waivers for a variety of activities (*e.g.*, access to roads, breaking open streets, widening gateways)
- environmental consents, on the basis of the anticipated environmental impacts and the proposed mitigation measures
- other consents, depending on the route, terrain, and nature of the project, from specialist agencies dealing with fisheries, forestry, wetlands, national or state parklands, agriculture, archaeological finds, historic monuments, etc.
- the obtaining of expropriation (compulsory purchase) powers or powers of eminent domain for the acquisition of land and rights-of-way

Characterization of the decision-making process

There are several different characterizations of the decision-making processes as they relate to permitting that are evident from the way in which they are carried out in different jurisdictions. The decision-making process to find the best way to reconcile national or regional needs with the impact on the environment and local communities to arrive at the optimal solution for society could be regarded as essentially:

- an adjudication (judicial or quasi-judicial)
- a dispute resolution
- a consultation
- the mitigation of adverse environmental and local impacts
- consensus building
- problem solving
- a combination of all or any of these

The process must necessarily involve elements of dispute resolution, consultation, mitigation of adverse environmental and local impacts, consensus building, and problem solving. On the face of it, the least likely of these options is a process that is an adjudication, because we do not need to determine who is right or wrong. We have cast the decision to be made as one of reconciling different interests. However, in many jurisdictions, the decision-making is characterized as adjudication, which explains why it is often conducted in a judicial or quasi-judicial fashion, even if it is not conducted by the courts.

It is, perhaps, cynical to regard the process as purely a consultation, but some objectors may feel that it is an accurate description of what happens in practice. However, it is a term that may help to describe the perception that once a government makes up its mind, as a political matter, that national or regional infrastructure is needed, it often gets built.

Objectives need to be identified

The design of the process should be carefully analyzed against what the objectives are, the tensions between them, and their relative priorities at the time. It also involves consideration of the inherent roles and responsibilities of national and local government and the various regulatory agencies in relation to infrastructure development.

Local and national democracy and accountability of the decision-makers to the people whose interests are to be protected are important issues. Those interests no longer simply extend to the traditional concerns over health, safety, the environment, and property values, but also to human rights, fairness, and proportionality of the government's actions. They very much relate to the relationship of the individual with the state.

These are issues that are at the heart of local and national democracy. This is the reason that many countries take the view that the ultimate decision should be made by national or local politicians, rather than unelected judges, bureaucrats, or arbitrators, because they are directly accountable to the electorate.

It is also necessary to consider the balance between the overall benefits to the community and the harm to the individual. These are not issues that can be conveniently put into a computer model to find the right design of a process. The objectives can only be broadly defined by reference to a number of policy areas. There are tensions and trade-offs between all objectives. There is no easy answer, which explains much of the reluctance of governments to make sweeping changes as to the way in which they handle these processes.

Complexities of a deregulated environment

As the need for transmission becomes more regional in scope and the market becomes more competitive, it is increasingly difficult to separate the decisions as to need and charging for the investment. It is very difficult for the regulator to ensure that price signals are in place to create incentives as to the efficient use of the transmission system without considering the need for new transmission facilities. Locational price signals that indicate when and where investment is needed are less likely to have an impact when the agency making the siting or need decision is not only different to the agency responsible for setting the prices, but where the agencies are also accountable to entirely different constituencies.[18]

If a transmission project is built on a market basis by a merchant developer, rather than a regulated transmission owner, consumers have no administratively-derived obligation to pay for the facility, and the prudence of the investment is not an issue, because the developer is taking on the entire risk.

For the purposes of the economic, cost-benefit analysis of the project and the determination of need, it is not always clear, in a deregulated environment, what evidence must be provided to decision-makers, particularly if the investment is being undertaken on a merchant basis. For example, does the merchant developer have to show that cheaper alternatives involving upgrading of existing facilities, distributed generation, and demand-side measures have been evaluated? If so, the determination of need would become very difficult, if not impossible, if the project were motivated by an opportunity to arbitrage price differences between markets.

The old paradigm of what constituted need, or the way in which the economic test involving the cost-benefit analysis is applied, was developed exclusively for regulated transmission. The whole approach needs to be rethought for a potentially competitive market in transmission. In response to the emergence of competition, the state of Ohio has essentially dropped the requirement that need be demonstrated for merchant investment. Few other states have made this reform. Some states, such as Massachusetts, have made this change for power plants, as generation has been subject to competition for longer than transmission. If merchant transmission trails behind merchant generation by several years, perhaps the legislative response may be doing the same thing.

The Supreme Court in Florida issued a decision that was interpreted to preclude anyone other than a Florida utility from applying to build an electric facility. The only exception is if the developer has committed 100% of the output to a jurisdictional Florida utility. This case appears to inhibit competition and the creation of regional markets.[19]

Separating the need determination from the environmental and local mitigation

There is an argument that the economic need determination should be separated from the assessment of the environmental and local issues. Moreover, the basis on which the determination of need, and therefore whether the project should be approved, may need to be revised. Originally, transmission was built to transport power from the generating stations of the transmission-owning utility to its native load or captive customers. In many state siting laws in the U.S. and in the regulatory and permitting reviews in many jurisdictions, the relevant "need" is that of the local, native load, or captive customers. Whether an industry has been deregulated or not, much more regional trading between utilities is now taking place, and transmission systems and interconnectors need to be expanded for this purpose.

The advantages would outweigh the disadvantages

The decision as to the need for the project could be made by the appropriate national or federal agency. This would avoid parochialism at the local level resulting in a project of value to a region being turned down. Once the economic decision as to the need had been made, the decision as to the actual route could be made at the local or state level. Deciding on measures to mitigate the impact on the environment and the local communities clearly requires local knowledge. "In principle" decisions to proceed would help to dispel some of the uncertainties of the permitting process for all parties.

This approach would also allow local and environmental considerations to play the most important role in determining the physical location of the project and the measures to mitigate its impact on the environment and local communities. In this way, both the national or regional interests and the local interests could be accommodated and reconciled.

The disadvantage of the approach is that it may be felt to preempt the decision-making authority of the local agencies and limit their ability to turn down a project on the grounds that the impact on the environment and the local communities was so severe that the project was not in the interests of society, despite the fact that it had passed the economic test as to need. In this case, however, it might be possible to establish some kind of test that would enable the local agencies to reject the project or refer it back to the national or federal agency for review.

Another disadvantage is that it creates, in effect, two processes that might involve duplication of effort and increased cost and delay. If this separation is made, it would be necessary to impose time limits for the making of the decisions in order to achieve the necessary degree of coordination and efficiency in decision-making.[20] However, as already indicated, the advantages of having a positive determination of the economic need for the project are likely to represent a significant benefit to the developer in the form of improved certainty of outcome and to the landowners, environmentalists, and local communities in the form of a separate process that really focuses on their concerns.

Who Should Decide?

The candidates

Decisions relating to siting authority, permits, and environmental consents are made by a variety of different persons and agencies around the world. Broadly, they can be grouped into:

- the legislature
- the national or federal government (or a minister)
- state, provincial, or regional government
- local authorities or agencies
- the federal and state regulators (of which there may be more than one, with different and potentially overlapping jurisdictions)
- a special tribunal or decision-making body established by law
- a special environment court

No simple answer

Although the question "who should decide?" is a simple one, it does not have a simple answer. The choice appears to be driven by two philosophical, but slightly conflicting objectives. One objective is the need to have an authoritative and credible decision-making body presided over by a judge or a regulator that is, above all, seen to be fair. The other often-quoted objective is the need for direct accountability to the individuals and communities whose interests will be affected by the decision, such as a government, minister, or local authority. Of the two objectives, fairness is much more important than accountability. In any event, accountability may not be so important if there is adequate opportunity for the stakeholders to have their own interests taken into account through a public participation process. Accountability through elected representatives is, arguably, only necessary if a member of the public has no real means to participate in a decision that affects his or her interests and, therefore, elects decision-makers to undertake the role on his or her behalf.

Fairness vs. accountability

According to Diedre Matthews of the Massachusetts Energy Facility Siting Board in her comments on this chapter:

> *It seems to me that the model is less important if the project has come through a "ground up" development process, in which as many local issues as possible are resolved before formal permission to construct is sought.*

She quotes the MetroWest case that we shall look at later in this chapter, which went through three years of public meetings before it was brought to the Siting Board.

If the "easy questions" are resolved in this way, and only the "hard questions" for decision-makers remain, it is probably better that the decision-makers not be elected.

In many countries, judges and regulators are not elected by the people, but are appointed through a political or bureaucratic process. Many constitutions ensure that the judiciary is independent and separate from government and the legislature, and although the judiciary has accountability, it is defined differently. In some countries, part of the judiciary is directly elected, and judges are always accountable through the many rules of law and procedure, their reasoned decision-making, and through media comment. The accountability of bureaucrats may be more diffuse. It depends upon their reporting lines, transparency of their decision-making, and how their decisions are reviewed and scrutinized.[21]

Avoiding conflicts of interest

It is an obvious precondition that the decision-makers must have no conflicts of interest. For example, the chair and panel members should not have affiliations with environmental, objector, or promoter groups. The independence of the U.K. Secretary of State in making planning decisions has been challenged in the courts under the European Convention on Human Rights on the basis that the Secretary of State is not sufficiently independent in assessing whether applications meet prevailing planning policies because he/she also sets or is capable of setting the policy. The House of Lords upheld the Secretary of State in the Alconbury case on human rights. It decided that in the light of European case law, there was sufficient ability to appeal his decision to the courts that met the requirements of the Convention. However, the philosophical issue still remains.[22]

Legislature

Although the legislature in democratic countries is elected and is therefore accountable, most countries have not given the lawmakers a formal decision-making role in the permitting process. Most legislative processes are slow and cumbersome. However, there is certain logic to the legislature making decisions in order to establish policy and resolve policy conflicts, even though it is typically the role of government. In the U.K., planning inquiries have often debated policy issues at length where there was clearly a policy vacuum, in order to determine the need for the project. The Sizewell "B" nuclear power station inquiry described earlier was a prime example of this.

In the Green Paper "Delivering a Fundamental Change," the U.K. government proposed that parliament should debate proposals for major projects and reach a decision in principle on whether or not the project should proceed before the planning inquiry process starts.[23] This was a response to the criticism that public inquiries have become embroiled in debating national policy. There was to be a three-stage process:

- the development and production of a national policy statement (including public consultation)
- parliamentary stage, that would consider the need for and location of the project
- a public inquiry into detailed aspects of the scheme and local issues

The parliamentary approval would not have granted planning permission, which would have been for the Secretary of State to consider following the public inquiry and the inspector's report, in the usual way.

In the face of considerable opposition from the Commons Transport Select Committee and concerns in the Procedure Committee of the U.K. House of Commons, the government decided to drop these proposals.[24]

The Green Paper proposals would have failed if they had merely moved discussion from one forum to another without making considerable savings of time in the overall process.[25]

The U.K. Hybrid Bill procedure (a parliamentary process) that has been used for major projects like the Channel Tunnel is not greatly liked by stakeholders or developers. It gives limited opportunity for stakeholders to be heard, and developers feel that it gives excessive incentives for lobbying and power to local authorities to exact high levels of compensation. However, it can be expeditious by comparison with the public inquiry process. The real problem is that a legislative proceeding that is efficient and not open to lobbying would always be difficult to design.

National or state government

The national government is usually empowered to make the necessary policy within which a project will be allowed to proceed. In civil law countries, the project may be the subject of a declaration of public interest that will make the path of its implementation easier. The effect of the declaration is similar to a government decision "in principle" that the project should proceed or an outline planning permission in the U.K. where the detail is worked up later. However, in many countries, government (at whatever level) actually makes all or most of the relevant decisions. The rationale is that ministers or politicians who make the ultimate decisions are fully accountable for them, because they are democratically elected.

Politicians at the national, state, provincial, or regional level should have accountability, in that they are elected in democratic countries. They will be supported by bureaucrats who tend to have little direct accountability. Through no fault of their own, neither the politicians nor the bureaucrats are likely have the requisite knowledge or the time to put into the decision-making that is necessarily involved in finding the best solutions to deal with the local and environmental impacts of a complex infrastructure project. There will be difficult trade-offs between national, regional, and local interests that will have to be made, and politics may cloud the judgment of issues that are technical and economic in nature. Even if the advice they receive is clear, politicians will tend to delay controversial decisions and avoid making bold policy statements, fearing political or media backlash, particularly if an election is pending or where they only command a small majority. Delay in decision-making will serve no purpose and will benefit neither the objectors nor the developers.

In some jurisdictions, such as the U.K. and New Zealand, the government can "call in " a decision from a local agency and have it made by the relevant minister. As Gary Sector points out:

> *Much delay is, however, caused by the involvement of the Secretary of State who makes decisions on about 300 planning cases each year. These involve cases of more than local importance, or where he considers a departure from the local development plan necessitates his consideration, or greenfield issues are at stake. For a developer to be subject to a call-in is no laughing matter. Aside from the expense of an inquiry, delay is the watchword.*[26]

Although there may be a need for some degree of accountability, if the means of public participation is very limited, the need for a knowledgeable decision-maker is no less compelling. The government could appoint experts to make reasoned recommendations, but this approach can be undermined if the decision-maker can overrule the recommendations.

Local authorities

The decision-makers in local authorities are also politicians and suffer from many of the same problems. However, they are close to the local issues, and most jurisdictions give them the primary decision-making role with respect to local land use. Local authorities in many jurisdictions have the authority to grant planning consents and zoning permits, sometimes with respect to the largest of projects, and even if the primary approval is given by national government. Massachusetts has a zoning exemption provision if a project is determined by state regulators to be in the public interest.[27]

The main difficulty with giving local authorities the sole or primary jurisdiction over national or regional infrastructure projects is that they will naturally tend to focus on the impacts of the project on the local community, without considering the wider regional or national interests.

Federal or state regulators

A national or state regulator[28] would have the requisite knowledge of the needs of consumers of electricity. Indeed, it is the role of the regulator to oversee the working of the electricity industry and to protect the consumer. If a transmission expansion project is regulated, the need for the project will require regulatory approval in some manner, if only for the purposes of deciding how its costs are to be recovered through transmission charges to be paid by users of the system (and ultimately, the consumer). The regulator is usually sufficiently neutral and independent to be able to stand back from the political issues surrounding a transmission expansion project to be able to make a decision that is fair.

Another way of looking at the question is to decide whether there is any justification for separating the decision-making process as to the approval of the need for the project and the process of deciding who is to pay how much for the facility. From the point of view of consumer protection, the two decisions are highly interrelated. This analysis converts itself rapidly into an argument that, as in the U.S. gas pipeline industry and in the transmission sector in Argentina, the regulator should decide the need because this will sharpen the incentive on the part of the developer to achieve the most cost-effective solution for the consumer. In other words, consumers are likely to hold the regulator responsible if it approves a project that results in a dramatic increase in charges.

A special decision-making body

Because the process involves making difficult decisions with respect to reconciling the national or regional need (that is largely economic) for a project with the impact on the environment and local communities, the creation of a special decision-making body with the requisite expertise appears to be the best solution.

The issues involved in siting and approving transmission lines can become complex and technical, mixing the disciplines of engineering, economics, law, and policy. Ideally, the process should allow for multidisciplinary decision making. The body could be presided over by a judge or regulator to give it the necessary authority and credibility. Its independence will ensure that it is seen to be fair.

This may sound like an easy solution, but the question of accountability of the special body will arise unless the view is taken that the means of public participation is so well designed and credible that accountability becomes less important, because the stakeholder voice will be heard directly. Then the only issue is whether the government, the regulatory agencies, and the public are prepared to give the special body the authority that it needs and to accept its decisions.[29]

An environment court-an example of a multidisciplinary decision-maker in New Zealand

We can look at a real and thoughtfully designed example of a multidisciplinary body rather than a hypothetical model.

Constitution of the Court. The Environment Court in New Zealand was established under the Resource Management Act 1991. It has 7 judges and 13 commissioners who are all members of the bench of the Court. The commissioners have technical or community qualifications and experience. The technical commissioners usually have engineering and land science qualifications, and the community commissioners have broad legitimacy to assess the public interest.

Role of the Court. The role of the Court is to decide if a project represents sustainable management of the environment. A broad public interest test is applied, as well as a test as to the sustainable well-being and management of the environment. The procedure is well disciplined, and the process is very thorough. The Court hears appeals from decisions made by local authorities on the merits of the case as well as on points of law. The Court is able to be more activist than the local authorities in the decisions that it makes, because of the public law concepts under which it operates. Non-legal representation is encouraged, and anyone can plead before the Court. Its decision suspends judicial review from the process. When an appeal is made from a local authority that has primary jurisdiction to approve a project, there is a full rehearing of all the issues. Although the existence of the Environment Court might appear to undermine the local authority processes, the view is taken that the local community should be given the initial chance to judge a project.

Call-in. A decision before the local authority can be called in for decision by the minister, either on his or her own initiative or on an appeal by an interested party. A national project might be called in to the minister in the interests of accountability. This procedure is rarely used. The decision of the minister can be appealed to the Environment Court. The minister is represented before the Court in the same way as any other interested party. The government is subject to the jurisdiction of the Environment Court in the same way as private parties. The Environment Court does not have complete jurisdiction over all environmental issues. Where decisions must be made, for example, under the fisheries or forestry legislation, the Court has no jurisdiction.

Role in legislation. The Environment Court has a role in framing subordinate legislation in each region and district to help manage the environment, particularly in relation to water treatment, emissions control, and planning permissions.

Jurisdiction of the Court. Where a public authority proposes a plan, there is a formal process once the plan has been exhibited and anyone can make submissions. The hearing will take place, and the public authority will make the decision. Appeals lie to the Environment Court on the contents of the plan, the rules, policy, and objectives. This requires some delicacy of treatment. The public authority is elected and is democratically given a policy-setting role, unlike the Environment Court.

The jurisdiction of the Environment Court is usually invoked on major infrastructure projects. The legislation directs that the Court has to consider whether the project is reasonably necessary for achieving the objectives of the project, whether adequate consideration has been given to alternative sites, routes, or methods, and whether it would be unreasonable to expect an alternative to be used. There are tensions between these goals, inevitably. Many of them essentially turn on the question of public interest.

No precedents. There is no precedent rule that is applied by the Environment Court. Each case is looked at on its merits. The view is taken that while precedent may help to achieve consistency of outcome, parties always see the issues in their cases differently. The fact that another group of people may or may not have objected or may or may not have reached agreement in a similar situation is irrelevant. The Environment Court is unwilling to concern itself with motives, which can be frustrating for proponents.

Case volume

About 70% of the cases have negotiated outcomes. About 15% are resolved by Court-annexed mediation, and about 15% go to a full hearing. A panel including a specialist Environment Court judge and one or more commissioners conducts the hearings. The commissioners receive some on-the-job training in mediation and other skills required. The Environment Court disposes of more than 1,100 cases a year (including cases settled following negotiation or mediation, as well as those decided following contested hearings). It is swamped with work. Attempts are being made to improve on the time that the process takes. The waiting time for a hearing did rise at one point to 18 months, and delivery of the decision would take 6 months after the end of the hearing. It would take 6 months to make an application and a year in the primary process.[30]

Private investors and minor upgrades. There has been a debate on whether the law should be adapted for private investors and whether an established network utility should be required to adhere to the same standards and obligations as other applicants. There is also an issue of whether consents should be required simply for the upgrading of existing facilities.

Inherent tensions in the objectives. There were extremely high expectations of the Resource Management Act 1991 that replaced more than 20 major statutes. Greater efficiency and improved public consultation were the objectives under an Act that was designed to achieve sustainable management of resources. Inevitably, there are inherent tensions in the objectives of the Act. It was assumed that local communities were the best judges of their own environmental problems and how to deal with them. However, this assumption increased the tensions. Lack of trained resources at local levels made efficiency and improved public consultation more difficult.

Criticisms of the Act. Implementation was criticized for being underfunded. There have been concerns in the business community over inadequate integration of social, economic, and environmental concerns; processes that were too costly, complex, and time-consuming; and regulations that were unfair and took away people's rights. Criticisms included local councils not working cooperatively, only limited guidance from central government, a lack of customer service focus by regulators, and misuse of consultative processes by those who wanted to delay development. A lack of national standards emerged as an issue.[31] The lack of resources and perceived delays in the Environment Court have had an impact on the overall perception of the success of the Resource Management Act, even though only 1% of the decisions of local authorities are appealed to the Court. It has embarked on a program to reform its case management systems to achieve greater efficiency.

Criticisms of the Environmental Court. The specific criticisms that are levied at the Environment Court include a perception that opponents can force developers into costly settlements by holding out the threat of Environment Court action (but arguably, this is inherent in any process where there is the possibility of an appeal to a higher authority). There are also concerns over inadequate resources of the Court, and the ability to waste time and money in traversing issues that are not central to the main issues. These are also common criticisms in most countries. This is a likely outcome whenever it is possible to challenge or appeal a decision, because decision-makers will want to "appeal-proof" the decision by investigating all possibly relevant areas or issues.

The business community has described the processes of the Environment Court as being designed for people without a well-founded claim to an interest in the issues at hand. On the other side of the coin, objectors have claimed that they were bullied with threats of costs likely to be awarded against them. Again, these are issues that are probably inherent in any process of this kind.

The value of the Court. The Court is praised for its use of mediation, and there is good evidence that it is useful and effective in resolving differences and in securing results that are good for the environment, communities, and developers. The Environment Court is well respected and does add value in finding efficient solutions to land use issues. It is an attractive model, and considerable thought went into its design.

Application to other countries. The concept of an environment court has been aired in other countries.[32] It suffers from the criticism that the creation of a specialist court is not necessary and undermines the jurisdiction of the existing court system. If one specialist court is created, there will be arguments to create others. There are economies of scale and scope in having a smaller number of more generalist courts. It is also a major undertaking to pass legislation to establish a new court and its jurisdiction and one that is likely to be slow and expensive.

In conclusion, the concept of an environment court provides a philosophically elegant model that, if it were well resourced, could provide a vehicle to find effective solutions.

So who should decide?

Role of the regulator. The economic determination of need is best made by the regulator in the case of a transmission expansion project made on a regulated basis, because it will have to analyze the need in order to decide the cost allocation issues. There is no basis for and, therefore, no need to make this determination in the case of a market-based project.

Special body. As far as the determination of the non-economic issues is concerned—the route, the environmental, and the local community issues—the ideal answer lies in a hybrid approach using a multidisciplinary entity comprising a credible, authoritative, and knowledgeable chairman, such as a regulator or a judge, with a panel of decision-makers drawn from other disciplines to assist with the necessary technical and other expertise that is needed. Government would have the ability to participate in the process in the same way as any other interested party.

Regional scope. The issues to be decided are strongly local in their impact, and this argues for a body that is closer to those issues than a national body to make the decisions. Ideally, in the case of a transmission line that crosses the boundaries of two or more local areas, the body would be regional in its composition, objectives, and jurisdiction.

Accountability. Accountability is difficult to manufacture.[33] The chairman and the panel could be answerable to a local or national politician or minister. They could be required to conduct themselves according to transparent rules, designed to enhance that accountability. Several jurisdictions require the government to endorse the decision of the chairman of an inquiry or public hearing. However, the whole process would be undermined if it were possible for the politician or minister to disagree with the finding of the chairman and his panel members and throw it out. In other words, the authority and credibility of the decision-making process would be adversely affected and result in political lobbying. The primary process would not be taken seriously. Provided that the stakeholders, landowners, and local communities have had an adequate opportunity to input into the process and that their input has been taken into account, the question of accountability of the decision-makers becomes less important. Allowing elected or accountable representatives to make decisions is much more important where the citizen is not involved in the decision-making process at all.

Judges should not be ruled out as decision-makers on the grounds of lack of accountability or of lack of specialist expertise. If a judge were to be given a presiding role, it would add a great deal of credibility to the decision-making process, because of the special analytical and judicial skills, fairness, and impartiality that he or she would be able to bring to bear.

Fairness. Credibility and fairness, however, are much more important than accountability in decision-making with respect to land use issues. If the decision-making body can achieve public respect and a reputation for fairness, then formal accountability matters less. In many countries, the judicial system is likely to achieve this respect, much more so than politicians. However, judges do not always command respect for their knowledge of the issues and being able to keep in touch with public sentiment. That is why a model consisting of a panel of multidisciplinary decision-makers, including a judge, is attractive.

Separate decision as to need from routing and local impact. However, it is unlikely that many countries will find the political will to create a new body designed along these lines. In this case, the answer would seem to be to separate the decision relating to the approval of need for the project from that of the issues surrounding the siting or routing and the mitigation of the impact on the environment and the local communities. The best answer would be to have the regulator approve the need for the project and the local siting, environmental, and planning agencies decide the siting or routing issues. In both cases, there would be a need to avoid the costs and delays of fragmented decision-making. Rules relating to timetables and inconsistent or onerous conditions would have to be developed and complied with strictly. If necessary, powers of expropriation or eminent domain to enable the developer to obtain land and rights-of-way could be granted by the regulator (under authority derived from legislation), once the need for the project had been determined.

Appeals

The decision as to who should decide must also take into account what the appeal process should be. Many jurisdictions limit appeals to the courts on grounds of an error in the application of the law only, or only permit judicial review of the decision-making process as to whether it has been fairly carried out. Appeals and judicial review do not always allow an appeal on the merits of the case. The theory is that if appeals can be made on the merits, the primary decision-making process may be undermined, because appeals will always be made. An appeal, even one without merit, can hold up a project for months, if not years, quite apart from increasing costs that may ultimately be borne by consumers. However, the possibility of an appeal does create an incentive to table as many issues as possible at an early stage and to resolve them.

Powers of Expropriation and Eminent Domain

Legal base

The mechanisms to give transmission developers powers to acquire the necessary rights-of-way and land to build new transmission facilities vary quite considerably around the world. However, there is always a legal base (usually in legislation) for these powers that may be conferred by the siting authority, a government minister (Secretary of State), or the regulator. When we refer to rights-of-way, we also mean easements, wayleaves, and servitudes.

Restructuring usually involves a change in legislation

When the industry is restructured, the laws giving powers of expropriation, compulsory purchase, or eminent domain usually have to be revised to reflect the new structure of the industry. On deregulation, many countries (*e.g.*, the U.K. and Argentina) have taken the opportunity to update their laws to allow private sector companies and market participants to exercise the powers (subject to certain stated safeguards and procedures), if the regulator determines that they should be permitted to do so. This can be done either by making a specific provision in the license or concession of the transmission project company or by the regulator being empowered to grant these powers on a case-by-case basis, for example, when it approves the need for a specific project.

It is helpful to rationalize the processes for obtaining both the consents and the rights-of-way. For example, the FERC siting authority over interstate natural gas pipelines enables FERC both to approve a project and to give the developer the necessary powers of eminent domain to obtain the rights-of-way.[34] The same approach has been adopted in Argentina where the regulator, Ente National Regulator de la Electricidad (ENRE), approves a project and confers the necessary expropriation rights. In the U.K. and other countries that use licenses as a regulatory tool, the nature and scope of the powers and how they are to be exercised are set out in the law, but the powers themselves are conferred on transmission and distribution utilities through their licenses. These reforms are also necessary to encourage market-based expansion by market participants, even if the transmission charges are not regulated, otherwise they will be at a disadvantage relative to landowners that they have to deal with and the regulated transmission-owning utilities. Although, in theory, an entrepreneurial investor could attempt to negotiate with landowners to acquire all of the necessary rights-of-way and land, the uncertainties as to the cost and delays could deter investment.

These powers may not always be used in practice, because transmission owners and developers are very skilled at handling negotiations with landowners that result in an agreed solution. A negotiated outcome is always desirable to create ongoing relationships with affected landowners. The powers may also not be necessary for certain types of controllable power flow technologies or where new facilities are colocated on rights-of-way that have already been acquired (provided the rights allow for the additional use). However, developers find it valuable in the more difficult cases to negotiate in the shadow of these compulsory acquisition powers.

Private sector expropriation rights

The main intellectual difficulty that arises relates to the concept of giving powers of eminent domain or expropriation to the private sector or to market participants who may wish to develop transmission projects. Many countries hesitate to allow anyone other than an arm of government or state-owned utilities to exercise these powers unless private entities are fulfilling what is clearly a public purpose and, perhaps, also unless they make their facilities available on a common carrier basis. The powers were based on the concept that the public sector should be able to exercise expropriation powers. They were extended to utilities because they were providing a public service and were controlled directly through government ownership or indirectly through regulation.

The public sector is not perceived to have the same profit motive as the private sector, particularly where the forces of competition are at play. Governments may have an irrational fear about granting expropriation rights to private sector market participants investing in infrastructure for their own profit-motivated purposes. For example, the Isle of Man Cable Company that is developing the interconnection between England and Wales and the Isle of Man did not obtain expropriation powers, because they were regarded as inconsistent with the merchant nature of the project. However, there is no reason that the powers should not be granted to private sector market participants if proper legislative and regulatory safeguards concerning their exercise and the payment of fair compensation to landowners are put in place. It is important to create a level playing field with the incumbent utilities that have these powers.

As Ashley Brown points out, if these powers are only available to local utilities (as in Florida[35]), the incumbent transmission owners will have an advantage in that they will be able to purchase the necessary real estate rights and land at lower prices than their competitors, who lack these powers. As he explains:

> *In effect, the state is bestowing a very significant market advantage to incumbents at the very time that public policy is geared towards promoting and sustaining competition. Rules that are a relic of the old paradigm in electric markets merit reconsideration.*[36]

This analysis may provide another argument for combining the processes to deal with the approval of the need for the project and the granting of the eminent domain or expropriation powers.[37] The agency or regulator that approves the project will certainly satisfy itself that the developer is able to carry out the project responsibly and can be trusted to exercise the powers appropriately.

Projects in the national interest

Where a specific project or a certain category of projects is designated as being in the public or national interest, there may be a special regime for the expropriation of the necessary land and rights by the developers. For example, the regime for the U.S. railroad development was capable of overriding sub-federal and state rights. The regime for establishing cell towers in the U.S. has been written into federal law.[38]

Compensation

Agreeing compensation for the necessary rights-of-way takes time. The objective is to achieve an outcome that is accepted by the landowner as fair. In Malaysia, a local land office deals with compensation for all types of infrastructure development. It notifies each of the affected landowners and arranges for the utility to pay compensation at a rate set for each type of impact (*e.g.*, tree felling, destruction of crops, etc.). The rate is the same for all types of infrastructure development. In other countries, there is a methodology or a process for establishing what compensation should be paid on a case-by-case basis. There may be a predetermined appeal process as well. From the point of view of the developer, it is much easier if access to the land can be given as soon as the eminent domain or expropriation powers are exercised (as in England and Wales), and the amount of the compensation is decided later. Clearly, this should take place expeditiously, but it is not always the case.

Terminable rights-of-way

The outcome of the acquisition of rights-of-way is not always as certain as it might be. The uncertainties, inevitably, center on the quantum of the compensation that should be paid, but there is also uncertainty as to the type of rights that may be granted. In some jurisdictions, terminable wayleaves or rights-of-way (fixed or indefinite terms) may the best that can be negotiated or granted under the exercise of the powers. This obviously begs the question of what happens when the rights terminate, because it is likely that the renewal process will have to start again *de novo*. Inevitably, this creates potential for litigation.

Some transmission owners have inherited a substantial number of terminable rights, because they were cheaper to obtain than permanent rights. They relied upon the public utility status of the transmission owner and the inertia of landowners not to press the issue with them on renewal or termination. Many transmission owners have a substantial number of unregistered, terminable rights that, when fully disclosed, could make the raising of finance difficult.

The best way to resolve these issues effectively is by creating transparent rules, guidelines, and timetables and enforcing them through legislation, concessions, contracts, or regulation.

Handling relationships with landowners

Several transmission owners have teams that are dedicated to establishing good working relationships with landowners who are or are likely to be affected by the existing transmission assets and any potential expansion. They take the view that it is not sufficient simply to negotiate the acquisition of a right-of-way, easement, wayleave, or servitude. It is also necessary to maintain that relationship into the future for the purposes of not only construction but also ongoing maintenance once the transmission facilities have been build. However, many utilities say that good relationships with landowners may become more difficult to achieve as the industry is deregulated. For example, shortly after restructuring in England had been completed, NGC was served with a notice by the local authority in the Docklands area of East London terminating a wayleave. After a difficult public hearing, NGC was granted an easement for only 20 years.

The Situation in the U.S.

Jurisdiction unclear

In the U.S., the process for obtaining the necessary permits and rights-of-way for transmission lines is described as "siting." The process is carried out at the state level, notwithstanding that FERC has jurisdiction over transmission tariffs and interstate trade. In fact, as we saw in the supplement to chapter 7, the boundaries between the jurisdictions of FERC and the State regulators with respect to transmission tariffs have not been particularly clear.

The original design of the siting processes tended to be driven by three informal objectives:

- to establish the "need" for the project
- to establish an efficient, decision-making process that would consolidate the reviews by all state government agencies and minimize the need for local approvals
- to create a coherent and transparent process with the opportunity for formalized input by the public

The processes have made sense to both the utilities and the potential objectors. Utilities were happy with a mechanism that established that proposed investments were needed and that streamlined the proceedings before a variety of different agencies into one mechanism. The potential objectors required a mechanism that established need, particularly if it resulted in diverting the utility's capital into environmentally-friendly investments such as renewable resources. Like the utilities, they were content with a process that provided a transparent, definitive forum in which they could pursue their interests, without having to replicate their efforts in several different processes.

State processes

New transmission facilities require permits and consents at state level. By statute, most states forbid construction of new utility facilities unless the owner first receives a certificate of public convenience and necessity from the relevant agency. Although the difficulties in obtaining these certificates are much greater when a new right-of-way is required, some states require a certificate to be obtained even when new right-of-way is not involved. The law on this point currently varies from state to state.

Only 28 states have enacted siting laws, and there is a great deal of diversity among them. The diversity covers how "need" is defined, the criteria or standards that are applied, the agencies that are involved, and how the states interact with each other. The expropriation or eminent domain powers do not all flow from the siting process. In a majority of states, these powers derive from the legal status of being a utility rather from the fact that the developer holds a permit to build.

Approximately half of the states require formal permits (certificates) through proceedings of the Public Utilities (Service) Commission (PUC) or Facility Siting Council. These proceedings can examine many aspects of a project, but they do not require findings on energy policy. Approximately 20 states specifically examine alternatives to transmission line construction. The U.S. shows a remarkable degree of "one stop shopping" for siting permits. All but six of the states that have siting laws have a single agency with the authority to issue siting permits. There is much criticism of the costs, delays, and uncertainties involved in the U.S. processes. A good process is to be found in the state of Massachusetts, which shows how a potentially very fragmented decision-making process can, with some effort, be successfully coordinated.

The Massachusetts process

The state of Massachusetts has an Energy Facility Siting Board and a number of state environmental agencies. The Massachusetts

Environmental Protection Act (MEPA)[39] pulls the process together, and the State Government has created a special unit (the MEPA unit) to carry out the coordination.

Environmental impact report. The proponents of the project are required to submit an environmental impact report (EIR) to the MEPA unit, which must assess and approve the report before any of the environmental agencies can issue permits. It has powers to request further information and typically will ask a lot of questions. This is the unit's only role, and it serves the purpose of avoiding the need to make multiple filings of an EIR with different agencies.

Role of the Board. The Energy Facility Siting Board determines the need for the line, evaluates the alternatives, and, if it is satisfied, will issue a project approval. If a developer has an insuperable difficulty in obtaining a permit or approval from a state or local permitting agency or if an approval is appealed, it can apply to the Board for a certificate of environmental impact and public interest. In all of its work, the Board will look at the environmental issues and push the proponents to develop the project in a manner that has the least environmental impact, at least cost to the consumer. The Board has heard relatively few cases each year on transmission and distribution lines, because there is an exemption for small projects using existing rights-of-way for no more than 10 miles and at voltages under 115 kV. This is designed to encourage new lines and upgrades using existing rights-of-way. The cases that it hears are mostly distribution support projects that serve local consumers.

Coordination. There is coordination with the state Public Utilities Commission (PUC) that regulates transmission charges to consumers (which are a bundled with charges for energy and distribution). Because of this coordination, the PUC is unlikely to second-guess whether a line is necessary or not when an application is made to put the assets into the rate (asset) base of the utility or to exercise eminent domain powers. The chairman of the Siting Board is the chairman of the PUC and

PUC commissioners have three of the nine seats on the Siting Board. The other six members of the Board are the secretary of Environmental Affairs, the State Economic Development director, the Commissioner of Energy Resources, and three members of the public appointed by the governor of the state. The Board is multidisciplinary.

The Board achieves coordination in various ways. It can issue a permit on behalf of another agency if there is delay in issuing that permit or if a permit that has already been issued is burdened with onerous conditions.

Applications in parallel. Applications for all of the necessary permits and approvals can be made in parallel. Applications will need to be made to local agencies for zoning relief, street opening permits in towns, and to the Conservation Commission if wetlands are involved. The Department of Environmental Protection of the state government issues permits primarily dealing with air and water pollution. There are other departments or agencies that deal with fisheries, wildlife, agriculture, and parks, where specialized consents or permits may be necessary.

If the project is large, the developer could apply for Siting Board approval of the route and then carry out the environmental process with the MEPA unit. The certainty of knowing that the need for the project had been approved would clearly be useful to a developer about to undertake major costs in developing an EIR.

An application has to be made to the PUC for permission to exercise powers of eminent domain to obtain rights-of-way and land. The filing can be made with the PUC and the Siting Board in a single docket (a single case), to ensure that the applications are coordinated. The eminent domain powers would then follow automatically from the issue of the permits. Typically, this process is used for zoning relief.[40]

Time taken. The time that the process takes, despite the number of agencies involved, looks short by comparison to NGC's experience with the North Yorkshire line, which we shall look at later. In the case of a reasonably non-controversial project, an adjudication by the Siting Board might take between 8-10 months, and a similar time period (or perhaps slightly less) would need to be allowed for obtaining the other environmental permits in parallel. It would take between 3-6 months to obtain the permits and zoning relief from the local agencies, although it is always necessary to carry out local consultations with stakeholders in advance, before coming to the Board.

In the MetroWest case, National Grid US developed a transmission project designed to improve reliability in the Framingham/Marlborough metropolitan area in central Massachusetts. National Grid US looked at a number of options (demand-side management, distribution improvements, generation, transmission improvements) and decided that transmission improvements were necessary. Three or four years were spent negotiating with various communities, and National Grid US decided to renovate an existing line and put in about five miles of new 115 kV line on existing poles. As it turns out, the Siting Board had no jurisdiction, because the new line was too short. A second circuit on the existing poles had been approved back in 1976, so there was no state-level proceeding, but the local permitting process (to put a new circuit on existing poles) lasted a very long time.

Need within the state

In another case involving Massachusetts, the Supreme Judicial Court interpreted the state law to say that the Siting Board is required to make (and justify) a finding of Commonwealth need (as distinct from a utility need) for a project before approving it.[41] The Massachusetts legislature later amended the statute in response to the decision. The Electricity Restructuring Act in 1997,[42] enabled the Siting Board to find "need" for transmission lines and other energy facilities within the state if there were an assurance of improved reliability, cost reduction, or environmental improvements. The Board has issued a decision approving a gas pipeline solely on the basis of economic and environmental benefits, without any improvement in reliability. The statute has also been amended so that power plant proponents are not required to demonstrate need at all.

Interstate transmission. There is flexibility regarding interstate transmission. The Board would be unlikely to take a narrow view of the question of need if the transmission line were proposed across part of Massachusetts that did not serve Massachusetts consumers directly. If there were clear benefits and a shortage of transmission capacity could be demonstrated, the project would probably be approved. Despite the Electricity Restructuring Act in 1997, the statute in Massachusetts still contains some remnants of the central-planning paradigm involving an assumption that transmission facilities will only be built by vertically integrated monopolies on a regulated basis.[43]

In the U.S., few states have statutes that specifically require the relevant agency to consider not only the needs of the state but also the needs of its neighbors. Some of the statutes, as we have seen, will require a project to be turned down if the benefits do not occur within the state, regardless of the benefits elsewhere in the region. In practice, however, many states probably would look at regional needs, but they are not necessarily required to do so. The difficulty with this approach for transmission expansion is that the major projects and interconnectors are likely to be regional in their benefits and, in the case of interconnectors, purposely so.

FERC's siting authority with respect to natural gas pipelines

There has been significant expansion in gas pipeline capacity in the U.S. that has been largely made possible by:

- FERC being the siting authority
- informal regulatory incentives

FERC does try to push decision-making to the state and local level by making extensive prefiling consultations with state and local agencies a prerequisite for a quick review process. FERC has made it clear that it expects developers to go through state environmental permitting.

FERC could be the siting authority for transmission

Several owners and developers of transmission facilities have argued that FERC should be given the same authority over the siting of transmission lines. Indeed, draft legislation was prepared for this purpose.[44] Both regulated and merchant transmission companies have advocated this approach.[45]

Given the success of this approach with natural gas pipelines and the need for transmission expansion in the U.S., the idea appears to be a good one. However, there seems to be little political will to interfere with states' rights that are embedded deeply in the American Constitution. The idea also calls into question the nature of the authority that would be given to FERC. The questions that are debated are whether it would be the primary or only siting authority for all transmission lines, whether its jurisdiction would be limited to interstate transmission lines, and whether it should only have some kind of backstop authority if the necessary certificates and permits are being delayed at state level. The difficulty with any backstop approach is that it may undermine the credibility of the primary process. One possible solution might have been for FERC to have the primary authority, but to delegate it to state agencies in suitable cases.

If FERC were to be successful in restructuring the operations and trading arrangements of the entire American power sector into large independent transmission providers or RTOs, the transmission planning process would be conducted on a regional basis (discussed in chapter 8). Out of necessity, this would put pressure on the states to coordinate with each other in issuing certificates and permits for interstate transmission lines. This might be a catalyst for reform.

Models for a regional siting process

As discussed earlier, a regional approach to permitting seems an obvious way to approach the task of reconciling local and broader national interests.

Peter Fox-Penner has suggested this model:

> *For new high-voltage lines, I propose that the state and local siting processes that approve such transmission facilities should be combined into a single regional approval process, an electric planning and siting commission (EPSC). This regional body will be controlled by the states – with one federal representative – and will allow for one-stop shopping for regional facility approvals, without requiring traditional federal pre-emption or eminent domain.*[46]

He points out that the idea of regional transmission siting boards is not new and that the establishment of multi-state bodies to preempt state and local law may require something akin to federal eminent domain as a means of enforcing the decisions of the EPSC.

There are several other options:

- As we have already discussed, FERC could become the siting authority in the same way that it has the authority in regard to natural gas pipelines. Apart from the political controversy and opposition from many of the states and property rights groups, a further disadvantage of the approach is the remoteness of FERC and the process itself from local interests, even though FERC requires gas pipeline developers to go through state environmental permitting
- A joint board of federal and state regulators could be established under the Federal Power Act. However, the decisions as to the composition and voting on the joint board could be controversial. The board would give the states an active role in the decision-making rather than reducing them to mere parties. It would

achieve a balance between local sensitivities and the broader regional and market interests

- Where multi-state facilities are involved, the state agencies could enter into a special arrangement or compact that would define the means for making decisions. In this way, a regional siting agency could be formed. This approach has not been particularly successful in the past in the U.S., and a number of details would have to be dealt with, such as whether the arrangement were to be binding or voluntary, permanent or ad hoc, who convenes the agency, who is able to enforce decisions, and to whom appeals are directed
- A model state siting code could be established in the U.S., although there is no guarantee that all states would agree to adopt a model code. It would provide uniformity, but all decisions would still have to be made on a state-by-state basis[47]
- Finally, we could follow the line of reasoning in the discussion earlier—who should decide? We concluded that it is desirable to separate the decision as to need and the decision with respect to the siting and routing issues that impact the local communities and the environment. This would give FERC the role of determining need, from which powers of eminent domain would flow and would give the state agencies the role of dealing with the routing and the impacts on the local communities and the environment

The implementation of this last option has the most chance of succeeding, given the current climate in the U.S. It would be a much more effective solution if rules relating to timetables and requirements to avoid inconsistent or unduly onerous conditions could be developed and strictly complied with. Non-compliance could be dealt with by rules that deem approvals to have been given without conditions if timetables are not met. Appeals could be made to FERC over inconsistent or unduly onerous conditions.

Improving the Efficiency of the Permitting Process

Small steps

The question then arises of whether it would be possible to take some smaller steps to improve efficiency of the permitting process without completely redesigning it.

Better coordination

Fragmented decision-making is inefficient and frustrating for applicants and objectors alike. This can arise in many cases. For example, the legislation may prescribe that for a class of applications affecting, for example, certain coastal waters, the decision is to be made by a specialist agency or the government itself. A minister or Secretary of State may be able to call in a particular application from a local authority that is considered to have national significance. The local authority's judgment of related applications can be unbalanced by the removal of other aspects of the project from its jurisdiction and its general credibility as a decision-maker may be undermined. Better communications, shared information, and joint committees between agencies would help and should not be too difficult to put in place, if the various decision-makers are willing to make a modest amount of effort.

Lack of coordination between different ministries or arms of national or local government is also a recipe for delay, expense, and frustration. The Ministry of Energy, for example, may be empowered to make the decision on the need for a new transmission line, but may even be unaware of the policies of the Ministry of the Environment that may affect the project. Clearly, some simple, interministerial coordination mechanism or, at its most basic, an interdepartmental committee for national infrastructure projects could do much to help the process along.

The Basslink joint advisory panel

An example of an attempt to coordinate the process for a particular project comes from Australia. A special arrangement was put in place to deal with Basslink project to build an interconnector between Tasmania and Victoria. The joint advisory panel (JAP) consisting of representatives of the governments of Tasmania, Victoria, and also the federal government was established. NGC (the developer) put forward an integrated impact assessment statement (IIAS) covering the tower design, landscaping, and marine impacts. The JAP checked it for completeness, and it was then presented to the public. Drafts of the IIAS could be submitted for informal feedback before finalization. The process also involved public consultations and public hearings chaired by a judge. Well-defined time scales (in terms of weeks) were established, and the whole process could have taken less than 12 months if the expected level of objections had not been exceeded. The JAP would then make recommendations to the three governments, and the proposal would either be approved as it stood or with conditions. The planning ministers in each jurisdiction were required to accept the recommendations, but there was no time scale for this. The project has major facilitating status as a project of national importance. There were also aboriginal land rights issues that had to be solved, and there was a tariff of payments to be made to local communities.

Local government plans

At local government level, the use of local development plans that recognize the need for transmission as well as other infrastructure could be a useful tool both to identify problems and potential solutions and to achieve a greater degree of coordination. In many jurisdictions, local authorities are required to prepare, publish, and comply with development plans. In some countries, the quality of local development plans varies. Indeed, in the U.K., some local authorities still have to complete their local plans over 10 years after the requirement of the

Town and Country Planning Act in 1990 came into force. In some jurisdictions, transmission and distribution companies do provide useful inputs into local plans. Another approach would be to require the local authorities to publish a statement of core policies setting out the local authority's overall planning strategy, detailed action plans, and an authoritative local planning map.[48]

The decision-makers could help developers and objectors alike by publishing more guidelines as to how to run the processes efficiently, what they expect of developers and objectors,[49] and the range of conditions they might attach in common cases. In the U.S., FERC and in New Zealand, the Environment Court and Ministry of Energy all publish guides on better stakeholder involvement and mediation. We shall look at the FERC approach (which relates to gas pipelines) in chapter 10.

Other small steps that could be taken relate to achieving some consistency in the rules relating to building under lines. In some jurisdictions, building under transmission lines is strictly forbidden. However, the transmission owner may be responsible for ensuring that anything that is built under a line is removed (and this may even involve the transmission owner paying compensation for the removal of the building, as in Malaysia), regardless of who built the offending structure under its lines.

The Massachusetts model provides an example of one method of achieving coordination, but there are obviously others. What is important is that no further fragmentation of decision-making should take place as a result of hurriedly drafted and poorly thought through legislation dealing with specialist topics involving land use and the environment.

Better case management

Project developers and objectors may agree that the decision-making processes are inefficient and could benefit from better case management and procedural rules. For example, it is common for environmental impact assessments to be attacked on procedural as well as technical grounds, which adds to the length of the process. The procedural rules ought to be clear and simple, so that procedural points do not arise. The focus should be on the technical content of the assessment. It would also improve efficiency if the issues that were open to public participation were strictly relevant to the case.

Most observers are astonished at the time and money that are involved in permitting processes. There must be ways to minimize the costs and to expedite these necessary decision-making processes. Lawyers are usually blamed for resisting reforms in litigation and judicialized procedures, because they have vested interests at play. However, there is an example of reforms in England and Wales in the context of civil litigation that might spark some ideas as to what can be done.

Woolf reforms

In 1994, Lord Woolf,[50] a very senior and respected judge in England and Wales was appointed to conduct a review of civil litigation with a view, among other things, to:

- reduce the cost of litigation
- minimize the complexity of the rules
- reduce the potential for delay and uncertainty
- improve fairness (especially where the parties have unequal financial resources)

The outcome of his review was published in July 1996 under the title "Access to Justice."[51] Far-reaching rule changes were published in 1998[52] and implemented on January 1, 1999.

The key ingredients of the reforms were new procedural rules that were:

- designed to encourage and incentivize the parties to avoid litigation and to make offers to settle. Failure to accept an offer could result in a financial penalty for the party that refused it, if the offer was matched or beaten at trial
- less adversarial and more cooperative in their approach to dispute resolution
- less complex

- stricter in the enforcement of procedural orders
- mandatory as to promoting mediation and other forms of alternative dispute resolution
- designed to impose and enforce shorter timetables
- innovative in efficient pretrial case management (including mandatory case management conferences and agreed protocols of best practice under judicial control, clearer statements of claim, shorter witness statements, and streamlined disclosure of documents)
- able to impose on expert witnesses a primary duty to the court (*i.e.*, impartiality) and to make them less partisan (a single witness, rather than multiple expert witnesses jointly instructed by the parties was encouraged to minimize the cost and delay caused by lengthy pleadings and cross-examinations)—the court has the discretion to impose limits on the number of expert witnesses per party, to prohibit expert witnesses, or to take the evidence in writing only
- designed to reduce cost and the ability to create delay
- overall, more interventionist on the part of the judiciary that is now required to exercise its powers to avoid litigation and hearings wherever possible, and to reduce the potential for cost and delay

It is still early to draw many conclusions as to how successful these reforms have been in achieving Lord Woolf's overall objectives. However, there are many good ideas in the new rules that could be applied to the proceedings and processes for the grant of approvals, permits, consents, and certificates for transmission expansion projects. The reforms have not been applied to these processes in the U.K., and the criticisms of the public inquiry system are similar to the criticisms of the civil litigation system before the reforms. The New Zealand Environment Court is also reforming its case management systems. Those countries that do not have a culture that favors litigation and judicialized, adversarial decision-making or dispute resolution processes may find the ideas interesting. One word of warning, however—it does take time to think through and implement reforms of this nature, even if they are quite modest. Nevertheless, a modest step will be better than living with the status quo.

Informal discussions

There is considerable scope for informal discussions with a view to reaching agreement on a wide variety of issues outside the formal process that could help to expedite the process. At the London Heathrow Terminal 5 inquiry, joint data groups were used to agree data outside the inquiry itself. Consultants were then able to put forward their interpretations of the agreed data. The deputy inspector was only called in by a joint data group if there was a disagreement that could not be resolved. He had no power to arbitrate, but was seen to be independent and his views were persuasive. Similarly, there were *ad hoc* joint working groups to discuss specific issues (*e.g.*, noise) outside the main inquiry, and the deputy inspector performed the same role. The conclusions of the groups saved considerable inquiry time.

Timetables for decisions

Another problem may be the time that it takes for decisions to be made following the end of the formal proceedings. Several countries have time limits imposed by law for the making of decisions once the consultation and hearing process has taken place. In the Czech Republic, the construction authority has 30 days to make a decision or 60 days if the application is complex. It can stop the process to gather information or refer an issue to the court for a decision. By contrast, in the U.K., there is no time limit as far as the Secretary of State is concerned.

Reducing the uncertainty

For a developer or a promoter, certainty is probably more important than speed or cost. As already discussed, a process that involves making a reasonably early decision as to the need for the project or a decision in principle that a project should proceed, should result in the promoter or developer being happier to spend time and money on mitigating the adverse impacts on the environment and the local communities, taking into account the views of objectors, and on finding the best solutions. Compensation packages, either on an individual or a community basis, should be easier to negotiate and agree. It does not necessarily mean that the developer will get away with paying less. It means that the transaction costs will be lower for both parties.

Precedent and published decision-making criteria

The lack of published policy obviously creates uncertainty. Published development plans (if they exist) may be a useful tool in this respect. The use of decided cases or precedent on a basis that is binding can also help to create certainty of outcome and minimize delays. However, most lawyers seem to agree that this is a more difficult issue in the area of the environment and local impact, because the circumstances surrounding each project and each local community are unique, and, therefore, from the point of view of the protection of the environment and local interests, it would be inappropriate, or even unfair, to follow precedent. In any event, technological advances can be rapid, and the means of mitigating adverse impacts are always improving. It would be unwise to encourage a developer to use old technology (if new technology were more cost-effective) because of the application of precedent.

Rationalizing the rules and conditions to achieve consistency

Permitting suffers not only from fragmented jurisdictions, but also fragmented and disparate rules and procedures. The building of the fourth line from Comahue in Argentina was beset by a problem with a single landowner who managed to take a case to the local court, which granted an injunction against the developer of the line on the grounds of trespass and "visual impact." The regulator had issued a certificate approving the project and had granted the necessary powers to expropriate the land and rights-of-way in accordance with Argentine administrative law. As a result of the injunction, the developer had to choose a different route to avoid the landowner's property and to take down towers that had already been put in place. There was no machinery for the developer to recover this cost or obtain relief from the obligation to build to a specified timetable. This lack of coordination of remedy under administrative and civil law should be ironed out.

Other small steps that could be taken relate to transparency, not only of the process, but with respect to the conditions that might be attached to the permits or consents. For example, it cannot be rational for one local authority to give consent for an overhead line and for the neighbouring jurisdiction to require the line to be entirely undergrounded in its territory, if there are no compelling reasons.

The Case of the North Yorkshire Transmission Line

Finally, we should look at a case study to learn some lessons as to what can happen in practice, and what solutions would be helpful. The story of the North Yorkshire line in the U.K. is by no means a chapter of accidents (indeed, it was well handled by NGC), but it is a story with a long duration. The whole permitting process took more than 10 years.

Section 37 of the U.K. Electricity Act of 1989 was designed to create a single process for obtaining both the governmental consent for building new electricity infrastructure and for obtaining necessary planning permissions under the Town and Country Planning Acts. While Section 37 deals with the major issues, NGC in fact required 30 or 40 local planning approvals (*e.g.*, for access to roads, gate widening, etc.). If the Section 37 approval does not give consent for undergrounding, NGC would have to go to the local authorities to obtain access under roads and also for constructing sealing end compounds.

The need for the North Yorkshire line came about as a result of the signing of a connection agreement in November 1990 between NGC and Teesside Power Limited for the connection of the new large combined cycle gas turbine (CCGT) plant. The connection was at a point on NGC's system where it would add to congestion in the North of England and the difficulties of exporting power to the load centers in the southeast. NGC's system designers came up with a plan in the late spring of 1990 and applied for Section 37 consent for the line and substations later on in the same year. Unfortunately, there was no alternative but to construct part of the line across an area of natural beauty. Environmental impact assessments were filed at the same time. In 1992, a public inquiry was convened that took four months.

NGC had proposed alternative routes for part of the line. In 1995, the Secretary of State said that he was minded to grant approval, subject to two sections over which he was uncertain. NGC agreed to underground one of them and looked for an alternative route for the other section. At this point, NGC began to obtain the rights-of-way by negotiating with landowners.

A second public inquiry dealt with other issues not dealt with at the first hearing, such as the sealing end compounds (for the undergrounding), and this took two-three months. An organization called "Revolt" applied for judicial review of the planning standards (maintaining they were too high and that the line was not needed). The issue never went to hearing, although it did cause delay. The Secretary of State was, perhaps, not very comfortable about making the decision, but eventually did so in 1999. The decision was expressed to be reviewable.[53] However, the decision omitted part of the line, and NGC had to go back to obtain consent for the rest of it. A condition was imposed that construction work could only be carried out in dry weather. NGC finally started construction at the end of 2000, but in early 2001 an outbreak of foot and mouth disease in Yorkshire meant that all work had to be suspended.

For the underground cable sections, NGC used compulsory purchase (eminent domain) powers (which it had already been given on a permanent basis under its license) to obtain the necessary easements. The wayleaves that it was able to obtain are reviewable and determinable after 10 years. NGC has wayleaves for about 80% of its network that, before restructuring, probably would have cost no more than U.S. $1.5 million in total. Now NGC has to pay that amount for five or six towers, because NGC is seen as a profitable company with a deep pocket, and landowners pay more attention to these issues than they used to.

The main tactic of the objectors was to cause delay in the hope that the line would become unnecessary. They had good access to information and plenty of able people, many of whom were retired. The objectors subsequently organized themselves to assist the landowners and local communities in Australia in objecting to the Basslink project in Australia with which NGC is involved.

The Recipe for a Good Process

Recipe on a blank sheet of paper

We have looked at a number of different issues that explain why the permitting process is unsatisfactory. If we could start with a blank sheet of paper, the recipe for a good process to reconcile national or regional interests with the impact on the environment and local communities would look at something like this:

- Transmission would be taken into account in regional and local development plans
- The regulator would at an early stage decide the issues relating to the need for a regulated project and whether, in principle, it should proceed (because it will need to do so anyway for the purposes of cost recovery through regulated charges to users and consumers)
- There would be no process for determination of need with respect to merchant transmission projects
- A special purpose, multidisciplinary body established on a regional basis, presided over by a judge or a regulator, would provide all other permits on a "one-stop shopping" basis
- The process would be non-adversarial, problem-solving, and consensus-building in character, and a real effort would be made to take into account the impact on the local communities and the environment
- Fairness and credibility would be regarded as more important than the accountability of the members of the multidisciplinary body
- Politicians would not make or second-guess decisions, and national, state, and local government would participate in the process in the same way as any other party

- The process would be handled efficiently through good coordination and case management, through rules that were rational and consistent, with timetables for the procedural steps to be carried out and for decision-making requiring strict compliance
- The multidisciplinary body would encourage early engagement of local stakeholders, publish guidelines for participation in the process, employ trained mediators, encourage settlements of disputes, and maximize the transparency of the process in order to create more certainty as to outcome, without applying a rigid system of precedent

Recipe for the real world

We do not live in an ideal world and do not have a blank sheet of paper. We cannot assume that there is real political will to recognize that things are broken and need fixing. We can, however, try to achieve improvements through the small steps that we have discussed.

The conclusions from this chapter and the lessons that have been learned from case of the North Yorkshire line are consistent:

- Dealing with multiple agencies increases cost, creates uncertainty, and has the potential for delay
- There should be really effective coordination among multiple agencies if they must, of necessity, all retain some jurisdiction over various aspects of the project
- Public hearings or public inquiries are not the best forum for solving problems. Structured, informal consultations with stakeholders, assisted negotiation, and consensus-building (explored in more detail in chapter 10) work much better
- It takes time to prepare for and convene public hearings or public inquiries, and multiple hearings should be avoided
- Effective rules with respect to timetables for decision-making would be really useful
- Politicians should not be asked to make decisions, but if they have to do so, they should only act on impartial, informed advice

- Governments (national, state, provincial, local) should participate in the process in the same way as other stakeholders
- The decision should cover all the issues that need to be dealt with
- The steps in the process should be predicable, transparent, logical, fair, and efficient. This will cut down on delays caused by appeals and litigation
- The conditions on decisions that are imposed should be consistent, fair to all parties, and reasonable
- The system for expropriation of rights-of-way (eminent domain) should be coordinated with the permitting process (if the powers are not otherwise conferred) in order to enable private sector, non-public utility, and merchant developers to obtain the necessary powers as well as regulated utilities
- It is useful for transmission owners and developers to build strong links with local authorities so that the local authorities understand what they are doing and why they are doing it, *e.g.*, they can also make useful inputs into local authority development plans
- It is important to put together a team of people who are effective communicators with landowners and objectors, who have the credibility and ability to create good relations with them throughout the process (continuity being important)
- A real effort must be made to ensure that the concerns of landowners and other objectors are taken into account

The next chapter looks at how the concerns of objectors can be taken into account most effectively through consensus-building to find the best solution to reconciling the national, regional, and local interests of society. It also explains how consensus-building can be used to provide a good and implementable answer without undertaking the non-trivial task of redesigning the permitting process, even minimally.

NOTES

1. It took NGC 10 years and two public hearings to obtain planning permission for the North Yorkshire Line. By contrast, it only took 4 months for the Armenian power authority to build an interconnector into Georgia in the later months of 2000. It calls into question what consents it needed in Armenia, but, in any event, the Georgian authorities refused to grant the necessary license, and it was not used to transfer energy between the two countries. Amiran Mgeladze and Tengiz Chkhivadze, "You Never Know, Whether they Are Hugging You Or Strangling You: From The Great To The Ridiculous...There Is Just One Transmission Line" (2001). The period in Norway is in the order of 7 years

2. Maryland PSC, Case 7004. The case is still open for evidence of EMF health hazards to be filed. I am grateful to Alan Beringer for providing this description

3. John Kay, " Meeting of Closed Minds," *Financial Times* (November 28, 2001)

4. Edison Electric Institute, "Making Competition Work" (January 2001)

5. Marathon County Circuit Court of Appeals, Wisconsin, Docket No. OICV0009000. In Minnesota, the issuance of a site permit supersedes and preempts all zoning, building, or land use rules, regulations, or ordinances promulgated by regional, county, local, or special purpose government. Minnesota Statutes, Section 116C.61, Subd. 1

6. Massachusetts G.L. c. 164, section 69H

7. Gary Sector, "Planning Ahead," *New Law Journal* (January 11, 2002)

8. This was said in conversation with the author

9. "New Power Plants Under Fire," *The Wall Street Journal* (January 9, 2002)

10. Connecticut Siting Council, Docket No. 208

11. Rachelle Alterman, ed., *National-Level Planning in Democratic Countries: An International Comparison of City and Regional Policy-Making* (Liverpool: Liverpool University Press, 2001)

12. Commission of the European Communities, Directive 96/92/EC Concerning Common Rules for the Internal Market in Electricity, *Official Journal* L 027, 30/01/1997 pp. 0020-0029

13. Some of the planning is carried out in order to take advantage of regional development grants

14. Rachelle Alterman, ed., *National- Level Planning in Democratic Countries: An International Comparison of City and Regional Policy-Making* (Liverpool: Liverpool University Press, 2001), chapter 2

15. Sweden appears to be an exception

16. Rachelle Alterman, ed., *National-Level Planning in Democratic Countries: An International Comparison of City and Regional Policy-Making* (Liverpool: Liverpool University Press, 2001) p. 37

17. FERC Notice of Proposed Rulemaking, 18 CFR Part 35, Docket No. RM01-12-000, ¶ 335-350 and 551-555

18. Ashley Brown, "Electric Facilities Siting: Options For The Future," paper presented to the National Governor's Association (November 2001)

19. Tampa Electric Co. et al. vs. Joe Garcia et al. 767 So. 2d 428. The case also calls into question whether siting agencies should allow competitors the standing to protest applications made by others

20. There is some precedent for this in the U.S. Telecommunications Act of 1996, with regard to the siting of wireless facilities. Section 704 (a) of the U.S. Telecommunications Act of 1996, titled "National Wireless Telecommunications Siting Policy," imposes a number of obligations on state agencies with respect to the siting of wireless facilities. In particular, it provides that any state or local government or instrumentality thereof shall not regulate the facilities if they comply with the Telecommunications Commission's regulations concerning the environmental effects of radio frequency emissions. Any person adversely affected by any final action or failure to act by a state or local government or any instrumentality thereof that is inconsistent with the subparagraph may, within 30 days after such action

or failure to act, commence an action in any court of competent jurisdiction that shall hear and decide the action on expedited basis

21. I have Judge David Sheppard to thank for these observations

22. *R (on the application of Holding and Barnes plc) v Secretary of State for the Environment, Transport and the Regions and R (on the application of Alconbury Developments Ltd) v Secretary of State for the Environment, Transport and the Regions and R (on the application of Legal and General Assurance Society Ltd) v Secretary of State for the Environment, Transport and the Regions [2001] All (D) ER 116 (May)

23. Department of Transport, Local Government and the Regions, Green Paper, "Planning: Delivering a Fundamental Change," HMSO (December 12, 2001). A subsidiary paper was issued from the same source, "New Parliamentary Procedures for Processing Major Infrastructure Projects," HMSO (December 17, 2001)

24. House of Commons Committee on Transport, Local Government, and the Regions, Thirteenth Report of Session 2001-02, published on July 3, 2002. The Deputy Prime Minister announced in his statement on Sustainable Communities, Housing, and Planning on July 18, 2002 that "I will speed up the planning of major infrastructure projects by setting out the government's objectives in clear policy statements and changing inquiry processes to make them more efficient. I have accepted the Select Committee's arguments that Parliamentary procedures for major infrastructure projects are not the best way forward."

25. Christopher Williams, "Planning for Major Projects Beyond the 20th Century," *Estates Gazette* (January 5, 2002)

26. Gary Sector, "Planning Ahead," *New Law Journal* (January 11, 2002)

27. Massachusetts General Laws, G.L.c. 40A, section 3

28. The question as to which of the federal or state regulators should make the decision depends upon their respective jurisdictions, particularly over transmission tariffs, interregional trade, and the oversight of the electricity markets at regional level

29. It is unlikely that the U.K. government would be prepared to do this

30. According to Judge David Sheppard of the New Zealand Environment Court, an overly judicialized process is not necessary. In his view, it is desirable to separate the (policy) questions of who should build and whether the cost is acceptable from the (justiciable) questions of whether the project is needed and whether the site is suitable

31. Denise Church and Claire Johnstone, "Sustainable New Zealand: Lessons Learned From The Development and Implementation of The Resource Management Act 1991," background paper for Resource Renewable Institute/California EPA Workshop, Sacramento, California (April 5-6, 2001)

32. Malcolm Grant, "Environmental Court Project, Final Report," to the Department of the Environment, Transport and the Regions (February 2000) ISBN 85112.360

33. I would like to thank Dan Bavly for his input into the issue of accountability of bureaucrats and civil servants

34. Natural Gas Act, 15U.S.C.S § 717 (2002)

35. The Supreme Court of Mississippi affirmed the order dismissing the condemnation petitions of Mississippi Power & Light, which was issued by the Special Court of Eminent Domain of Walthall County. The Supreme Court concluded in its finding that there was no public necessity for the taking of the land, because Mississippi customers would not benefit from the construction of a transmission line for the primary purpose of transmitting power from Franklin County, Mississippi to the eastern part of New Orleans, Louisiana, which falls under the service territory of Louisiana Power & Light. Mississippi Power and Light Company v. Louis A. Conerly, et al., 460 So. 2d 107

36. Ashley Brown, "Electric Facility Siting: Options For The Future," paper presented to the National Governors' Association (November 2001). It is unrealistic to expect market participants to register as utilities in order to be able to take advantage of the powers of eminent domain that are afforded to utilities in a state, because the disadvantages of registration usually outweigh the advantages

37. In a small minority of states in the U.S., project-specific eminent domain can be obtained as a result of the approval of project by the siting agency

38. Section 704(a) of the U.S. Telecommunications Act of 1996, titled "National Wireless Telecommunications Siting Policy"

39. G.L.c. 30, Section 62 et. seq.

40. Massachusetts General Laws, G.L.c 40A, section 3

41. Point of Pines Beach Association Inc. v. Energy Facilities Siting Board, 419 Mass. 281 (1995)

42. Chapter 164 of the Acts of 1997

43. I am indebted to Diedre Matthews for explaining this to me

44. The Murkowski Bill, 106th Congress, S.2098

45. José Rotger, TransÉnergie US, "A Framework for Federal Certification of Transmission Projects," presentation to Harvard Electricity Policy Group (June 2001). Mark K. Lewis, "Federal Siting Authority for Interstate Electric Transmission Lines: Transmission Capacity Cannot Grow If New Lines Cannot Be Built," *The Electricity Journal* (October 2001), pp. 36-39. Raymond Coxe, "New Paradigms for Siting Transmission in Competitive Electric Markets," presentation to Harvard Electricity Policy Group (April 10, 1997)

46. Peter Fox-Penner, "Easing Gridlock on the Grid: Electricity Planning and Siting Compacts," *The Electricity Journal* (November 2001), pp. 11-30

47. Ashley Brown, "Electric Facilities Siting: Options For The Future", paper presented to the National Governors' Association (November 2001)

48. Department of Transport, Local Government and the Regions, Green Paper "Planning: Delivering a Fundamental Change," HMSO (December 12, 2001). A subsidiary paper was issued from the same source, "New Parliamentary Procedures for Processing Major Infrastructure Projects," HMSO (December 17, 2001)

49. FERC has published the helpful guide called "Ideas for Better Stakeholder Involvement In the Interstate Natural Gas Pipeline Planning Pre-Filing Process," addressed to the industry, agencies, citizens and FERC staff, prepared by the FERC staff OEP Gas Outreach Team (August 2001). This kind of guide could form the basis of a more detailed description of the process and the range of conditions that might be attached

50. Lord Woolf is not related to the author

51. The Right Honorable Lord Woolf, Master of the Rolls, "Access to Justice," Final Report to the Lord Chancellor on the Civil Justice System in England and Wales, HMSO (1996), ISBN 0113800991

52. The Civil Procedure Rules (SI 1998 No. 3132)

53. Secretary of State for Trade and Industry, decision re Lackenby-Picton-Shipton overhead line (July 2000)

CHAPTER 10
Presentation (Listening to Stakeholders)

Ingredients of this Chapter

Good presentation makes all the difference

One of the most important skills of a contemporary chef is presentation. Even the most humble dish can look much more appealing if it is well presented. The same is true of transmission expansion projects. Where the stakeholders, landowners, and local communities are well looked after, the processes for dealing with permits, environmental consents, and the acquisition of rights-of-way will not only be smoother and swifter, the answer for society is likely to be better. The key ingredients of good presentation of any major infrastructure project are to engage the stakeholders, landowners, and local communities early, to empower them to participate constructively, rather than to alienate them in some adversarial dispute resolution process, and to take their concerns into account.

We begin by looking at the need to engage the potential objectors, the industry experience engaging with stakeholders, and the design of a stakeholder engagement model for project development. After discussing various forms of assistance that might be given to objectors, we will describe mediation as the ultimate engagement model in some detail. We shall discuss its advantages and disadvantages, and conclude that it has considerable attractions and may enable us to side step the difficult issues we discussed in chapter 9.

Engaging the Stakeholders

Visual impact of transmission towers

Nobody likes looking at transmission towers. Despite all of the fears surrounding the health risks associated with electromagnetic fields where a clear causal link to transmission has yet to be proved, the visual impact of a transmission tower, even if it is an elegant concrete pole, is what bothers people most. Transmission substations and switchyards are not attractive to look at either, and transformers can be noisy. It is hardly surprising that although they do not pollute the atmosphere, nobody wants transmission facilities in their back yard, because they affect what society calls amenity and quality of life and not just property values.

Objections are on the increase

Originally, the objections to major projects were based on adverse impacts on property rights explicitly protected by law. There are also other reasons. Cutting public and social services, the distrust of large institutions created in the 1960s, individualistic attitudes of the 1970s and 1980s, and the failure of institutions to solve or contain certain social and environmental problems have all led to increased objection to major infrastructure development in general and visually-intrusive transmission line projects, in particular. Some have argued that objector groups represent a kind of the "new environmentalism."

> *These groups have not only forced the cleanup of existing hazards and prevented new ones but also, when taken together, began to influence consumer (and thus business) habits, to create pressures for more preventative approaches to pollution, to expand opportunities for citizen participation, and finally, to extend environmentalism to sectors of the society traditionally unused to thinking ecologically.*[1]

The well-known acronym, NIMBY, stands for "not in my back yard" and NIMBYism is a description of the movement to object to anything built locally. Stephen Bird argues that:

> *In the end, NIMBYism is about the failure of democratic society to make decisions about facilities that benefit the greater good yet also impact and impose risk upon local communities.*[2]

He also argues that one of the most common reasons for land use conflicts are elements of environmental risk. Environmental impact statements and the way in which they are drafted have changed the way that projects and actions are conducted. These documents act as fodder for delaying tactics. They are highly technical and may do little to address the actual issues that concern citizens. Psychological "dread factors" can easily cloud distinctions between true and perceived risk.[3] President Clinton's executive order of 1994 directed all federal agencies to consider environmental justice when reviewing siting issues.[4] However, the concept of social injustice has emerged, characterized by undesirable land uses being sited in areas or neighborhoods of low incomes, high minority presence, and minimal political power. Wealthy communities have tended to be more successful than poorer communities in fighting major projects.

The lack of fairness and injustice in the siting and permitting process are increasing concerns. Benjamin Davy is of the view that:

> *[T]he adversarial and result-oriented framework for facility siting should be replaced with a system that encourages not only to search for a fair balance of interests, but also to accomplish a fair balance between different justices, different rationalities.*[5]

Transmission owners' experience in creating good relations with landowners and objectors

For the most part, transmission owners around the world have an impressive attitude and record in dealing with affected landowners and communities fairly and sensitively. Statnett, for example, consults the communities and landowners at the earliest possible stage and takes into account the concerns of known objectors and landowners in their own internal planning processes. The company prides itself on a team that has land use qualifications and long experience of dealing with stakeholders. Continuity is important, and Statnett's team has built up commendable relationships with landowners and local communities.[6] In India, Powergrid is flexible and makes diversions in order to avoid local opposition. It is very good at dealing with communities who tend to resist power stations, but become less troubled by transmission. Statnett, Powergrid and NGC all publish documents relating environmental procedures on their websites.

Transgrid's successful process

Transgrid in New South Wales, in effect, let the communities decide for themselves which route the Queensland-New South Wales interconnector (QNI) should take. The approach involved the implementation of a process that began by defining only the end points of the proposed interconnector. The local communities and landowners were then invited to give their views as to the most suitable route by drawing on maps in offices set up for the purpose at intervals along the potential corridor, so that none of the local communities had to go very far to give their views. The consultation process, which began in 1994, took more than two years. It was handled by staff chosen for their skills in community and landowner issues. They were kept in place throughout the consultation process in order to maintain continuity and credibility. Farmers were also engaged by the utility to help in the consultation process (which created a second income for them). There were many one-on-one consultations as well as public meetings. People would come to the local office and draw on maps to indicate the route they felt the interconnector should take. This exercise resulted in nine different routes being proposed. Eventually, a route was chosen that probably affected the least number of people. It was felt by many to have been a worthwhile exercise, and the industry was seen to have listened to the landowners and local communities. This process meant that the industry was not on the defensive when it put the proposed interconnector out to consultation and was constructively seeking assistance to find the best answer.

Industry experience is consistent with the literature

This experience is consistent with the literature. Herbert Inhaber argues that whenever a proposal has a sense of being a "faite d'accompli," it becomes virtually impossible for a community to feel as though it is possible to have any input or involvement in the process. Any potential for consensus building is virtually destroyed. He suggests four steps to create solutions:

- Create a volunteer community, not a victim community
- Use simplified explanations of the effects (not lengthy, highly technical analyses)
- Use compensation-setting devices
- Allow community control of certain aspects[7]

Designing a stakeholder engagement model for project development

It should be possible to develop a stakeholder engagement model that creates the best chance of designing a project that was sound enough that a decision, at least in principle, could be made that the project should proceed. The process for a major project involving a new route might begin with a coordinated planning process being carried out, along the lines discussed in chapter 8. This process should have involved as many of the major stakeholders as possible, whose input should have been taken into account. However, it is unlikely that all of the potentially affected landowners and local communities will have taken part in the process. Whether the transmission expansion project is the direct result of a coordinated planning process or it is a pure market-based project, the method of presentation of the project to the landowners and local communities will be critical.

Following the Transgrid QNI model, the proposal to build a line between two points (but without specifying the route) could be sent out for consultation, with a clear, non-technical explanation of the need for the line. Local communities and landowners would then be invited to give their views as to the most suitable route within a specified time period, and a mediation process (which we will discuss in more detail later) could be used to arrive at the right route. If no agreement is reached within a specified but reasonable time period, then the proponent would develop a route and make a proposal to the appropriate permitting or siting authority within a fixed period. The determination of need could be made by the appropriate agency (if separate) at the same time.

As we have observed, in many countries there is general agreement that it would be counter-productive to attempt to streamline a permitting or consent process. The result would be seen to adversely affect the rights of stakeholders, landowners, and local communities to a full hearing and to careful consideration of their interests and the environmental impacts. The question, then, is what is the best method of engaging the stakeholders, landowners, and local communities, giving them a full hearing, engaging their interests and taking the environmental impacts into account. The two most obvious methods are stakeholder assistance and mediation.

Assisting Stakeholders

Funding

Some project developers would like to be able to assist objectors by funding professional advisers for them. Some of the technical and economic issues, particularly in a deregulated market, become extremely complex. It serves no purpose if stakeholders are forced to spend their time making objections on the basis of understandable ignorance and a fear of the unknown, especially as they can scarcely be expected to understand all the complexities of the detail. It is always quicker and easier to negotiate a deal with a well-informed party than one that lacks knowledge. Howerver, many jurisdictions prohibit funding of this kind on the basis that it might encourage developers to attempt to "buy" stakeholder acquiescence.

Presenting information

It would be helpful if utilities could inform the local communities and landowners clearly and convincingly about the realities of the health and safety issues, the impact on property values, and the way in which the adverse effects on the environment will be compensated or mitigated at an early stage. The objectors would then be empowered to play a more effective role in the process to protect their interests, and a better result could be achieved for everyone. Presentations and written materials obviously need to be user-friendly, non-confrontational, and neutral.

NGC publishes annual statements of its environmental policy in the U.K. and U.S., with details of its objectives, targets, and performance. It sponsors studies, educational programs, recycling, and pollution prevention. There is a wealth of information that is readily available on the effect of its operations on the environment and on wildlife.

Guidelines published by siting and permitting agencies

It could be useful in some situations for the decision-making agency to issue a user guide as to what is involved and how stakeholders, landowners, and local communities should conduct themselves. An example of this is the guide prepared by FERC staff (described in the document as the Office of Energy Products (OEP) Gas Outreach Team), called "Ideas for Better Stakeholder Involvement in the Interstate Natural Gas Pipeline Planning Pre-Filing Process", which is addressed to "Industry, Agencies, Citizens and FERC Staff."[8] The guide encourages the industry to start early and involve all of the stakeholders at the outset, to demonstrate their commitment, to maintain open communications with agencies and stakeholders, to make route information easy and understandable, and to explain mitigation compensation benefits. The citizens are encouraged to get involved early and stay informed, to do their homework, and to know the players and the process to become partners. The role of FERC is also explained, and a commitment to providing information and training is given.

Computer simulation models

A more sophisticated tool to empower landowners and local communities would be an easily utilized computer simulation model to allow them to theoretically test environmental, social, economic, and cultural decisions or plans. If an appropriate model could be built, it would give communities the ability to make theoretical decisions that would help to focus attention on the likely outcomes and trade-offs that could be made in order to create benchmarks or solutions that could safeguard their interests and the environment. Denise Church and Claire Johnson give examples of the use of simulation models in New Zealand. They suggest that the technology should be kept simple and made accessible. There should be a good facilitator to ensure that the community has someone to help it use the model.[9]

Publicly-funded stakeholder advocates and legal aid

In some countries, special offices that assist local communities and landowners (such as the Environment Defenders Offices in Australia and the Offices of Rate Payer Advocates in the U.S.) have been established. These institutions provide legal and other trained staff to defend or promote the interests of consumers or stakeholders in various regulatory proceedings and decision-making processes. New Zealand has introduced government-funded legal aid for environmental groups taking cases to the Environment Court. In the U.K., the Central Electricity Generating Board (CEGB) was not permitted to fund the Friends of the Earth at the Sizewell "B" nuclear power station public inquiry, even though it had offered to do so.

Compensation

There are usually well-established systems for compensating landowners for the adverse effects on their property rights. The question of compensation for local communities also arises. Typically, it is provided in cash to local authorities or municipalities or in kind (*e.g.*, affordable housing, improved roads, new school buildings).

Where the siting or planning authority resides at local level, at least in the initial stages, the local authority or decision-making body might be tempted to use delaying tactics in the way in which it conducts the proceedings in order to extract the highest value of compensation.

Some jurisdictions frown on the making of payments to local communities. Others positively encourage it, such as in Norway, where there is even a tariff or "going rate." There is a view within the industry that the tariff could even be increased a little in order to achieve more acceptability of transmission lines to local communities.

There is an intellectual difficulty with the concept of compensation, because it implies that the adverse impacts on the environment, landowners, and local communities can somehow be bought and sold.[10] However, it is increasingly seen as a good solution. On the other hand, landowners and local communities need some sense of proportion as to the costs involved and the quantum of compensation, otherwise the transaction or development costs will increase exponentially.[11]

As Inhaber notes:

> *If we ask people how much removing the smoke from a stack is worth, they might say thousands or even millions, when they know that they will not see a bill for that amount in their mail. If they were told that their taxes would go up by the amount that they specify, then a given community will decide on an amount much closer to the true social cost.*[12]

Reverse auction mechanism

Inhaber suggests a "reverse Dutch auction." Regulators offer increasing sums of monetary compensation to the identified sites until one community determines that there is enough compensation to offset the negative effects of the project. Alternative sites might have to pay a levy to the community that agrees to take on the project. Taxes or levies could also be considered for communities that stand to benefit from the project.[13]

Stakeholder control

In the reverse Dutch auction, the power to select the compensation amount creates control over the process. The emphasis on control is important. If a community feels it has no control, the most powerful forms of opposition emerge. Charles Piller observes: The most hopeful aspect of NIMBYism – a determination to be part of the process: in effect, to end alienation – can help to build a healthier and more robust democracy.[14]

U.K. and U.S. compensation systems

The compensation system in the U.K. (described as planning obligations) is given statutory force under section 106 of the Town and

Country Planning Act, 1990. Any person with an interest in land may—by agreement or unilaterally—enter into certain obligations with the planning authority in consideration of the grant of the planning permission. The arrangements are notoriously slow to negotiate, even for the simplest developments, far from transparent (with the perception that a developer is able to buy planning permissions behind closed doors), and completely lacking in predictability of outcome and cost. The U.K. government has been looking at three options:

- to enforce a more rigorous "necessity test"
- to give flexibility for local authorities to negotiate planning obligations within the law
- to introduce "impact fees" following the U.S. model

The third option is the approach for which the government expressed preference.[15]

Development impact fees have had broad application across the U.S. for some time. Compensation is paid on the site-specific basis using "a rational nexus" methodology. This is set consistently with the impact of the development on a perceived need. The "rational nexus" has been the subject of considerable case law, with the result that impact fees are now only used for carefully defined purposes, because it is difficult to cost intangible adverse impacts, such as the loss of landscape value, using a tariff. In the U.S., transparency is always important, and the methodology for the calculation of impact fees must be both clear and predictable.[16]

However, all compensation systems seem to be problematic and could introduce additional costs that could ultimately be borne by the consumer. Those in favor of compensation argue that market forces could demonstrate to society the true social cost of building a new project. However, Inhaber notes "the auction system will not change the relative position of rich versus poor all that much."[17] It is also difficult to design the scheme that would cope adequately with a regional project that also provides benefit outside the region.

Case of the Cross Sound Cable

Paying compensation is not a solution to all problems, nor is it easy to please all the stakeholders. In the case of the Cross Sound Cable, the shellfish bed leaseholders made a big issue of the potential impacts on shellfish. Faced with a difficult situation, the Connecticut Siting Council pressed TransÉnergie to settle with the fishermen in order to avoid having to decide the issue. The settlement backfired in the media and public forums. Politicians made hay out of the fact that the fisherman had been paid off. The payments were made to look like bribes rather than payments for traversing the shellfish bed rights-of-way, interruptions in farming activities, and relocating the shellfish. The settlements would have been beneficial to the industry, because TransÉnergie would have funded extensive bed restoration and created the first oyster hatchery in Connecticut.

The first application was rejected by the Connecticut Siting Council, and the settlements with the State Department of Agriculture (Aquaculture Bureau) and with an environmental organization were voided. The original settlement with the shellfishermen is still in force, but the payment is now substantially reduced as the new, approved route in the shipping channel passes through only a few hundred feet of beds, not the three miles originally proposed.[18]

These forms of assistance may be helpful, but none of them is likely to achieve the ultimate objective. The goal is to find the best solution to the problem discussed in chapter 9—reconciling the need for a project in the national or regional interest and its impact on the environment and local communities. For that purpose, we must turn to mediation and consensus-building.

Consensus-Building and Mediation

Mediation in land use issues is gathering momentum

By far the most promising technique that can be used to both minimize delay and improve the outcome for the developer and the objectors is the use of consensus-building and mediation. In many countries, mediation in land use issues is almost unknown or is very much in its infancy. However, the pioneering work of Professor Lawrence Susskind of the Massachusetts Institute of Technology and the Consensus Building Institute in Cambridge Massachusetts has demonstrated, with relevant case studies and practical experience, that consensus-building techniques have great potential for resolving issues in land use disputes, and, equally important, if not more so, they provide a tool for arriving at better solutions that would allow transmission expansion projects to proceed.[19] In essence, because the facilitators and mediators ask the objectors to identify and agree on what all the issues are, there is a much higher level of understanding and empowerment, which makes the whole process much more effective. The solutions they design together become much more appropriate and acceptable to the local community and landowners.

The Environment Court in New Zealand is empowered to encourage mediation and regularly does so. The non-legal members of the Court receive on-the-job training in mediation skills. As explained in chapter 9, the use of mediation by the Court has been admired, and there is evidence that it is useful and effective in resolving differences and securing results that are good for the developers, the communities, and the environment.

Public vs. private disputes

Perhaps consensus-building and mediation in land use issues is in its infancy, because environmental disputes with respect to national or regional projects were thought of as public rather than private. The parties to a private dispute "own" it and have the right to settle it. Public disputes over transmission projects affect the public interest and possibly also the rights of others, who may not be party to the dispute. No one has the right to settle a dispute on behalf of someone else without permission. The matter was further complicated by the fact that a single agency might not have complete jurisdiction over the entire extent of the project. The traditional view was that the concerns related to whether individual rights had been adversely affected, in which case adjudication was needed (and not mediation) to resolve the issue.

However, consensus-building and mediation are gathering momentum in many fields, including land use issues and utility regulation.[20] Given the difficulties of designing an efficient and fair permitting process that were discussed in chapter 9, they do provide an appealing approach to funding solutions, because they require no laws to be passed and no rules to be changed. The process can be organized on an *ad hoc* basis if the key stakeholders are prepared to participate. It is also not usually subject to open meeting regulations or judicial review.

What is meant by consensus-building?

Consensus-building is a process for seeking unanimous agreement, involving a good faith effort to meet the interests of all stakeholders. The process concludes either when full consensus is reached or when most of the stakeholders agree that they can accept the solution after every effort has been made to meet the interests of all parties. It makes sense for the stakeholder groups to seek unanimity, but to settle for overwhelming agreement that goes as far as possible toward meeting the interests of all stakeholders.

Mediation is, in a sense, a means to build consensus, but the terms are often used interchangeably. There is a difference between mediation and facilitation. A *mediator* is empowered to seek a basis on which consensus can be achieved. A *facilitator* simply manages or facilitates discussion of the issues and asks questions to enhance understanding, without seeking solutions.

A fundamental principle of consensus-building that must be established at the outset is that stakeholders have the right to expect that no one will ask them to give up their interests for the sake of reaching consensus. However, they also have the responsibility that accompanies this right, to actively propose solutions that will meet the interests of all other stakeholders as well as their own.[21]

Three planning models

Lawrence Susskind, Mieke van der Wansem, and Armand Ciccarelli contrast three planning models:

- technocratic
- advocacy
- mediation[22]

The technocratic model, where the planning and siting process is apolitical and technically-oriented toward an efficient allocation of resources at a specific point in time, is criticized for the assumptions that it makes as to the authority of those in charge to develop and implement both policy and plans. The model is viewed as favoring the wealthy and those with political clout.

The advocacy model was developed in response to the failures of the technocratic model (largely in relation to urban renewal in the late 1950s and early 1960s in the U.S.) and is based on representation of particular interest groups. The main aim is to redistribute resources to ensure greater equity and improved quality of life for those least able to fend for themselves. However, it necessarily involves a highly judicialized process that can be lengthy and expensive. The outcome will be uncertain and perceived to serve the interests of only those who are well organized and represented.

By contrast, the mediation model involves balancing the interests of all stakeholders to achieve an informed consensus as to a solution that is both fair and efficient. A neutral mediator who pursues mutually acceptable agreements enhances the probability that an implementable plan will result.

The Centre for Effective Dispute Resolution in the U.K. argues, " A major attraction to the public sector is mediation's ability to take into account the needs of diverse interest groups of stakeholders."[23]

It is clear from the success that many transmission owners have had in proactively handling stakeholder concerns that the use of professional mediators would have assisted greatly on a number of occasions.

Facilitating a dialogue as a first step

The simplest form of assisted negotiation involves facilitating or managing the process to keep the dialogue going and to create an environment conducive to joint problem solving. The parties might describe this as facilitation rather than mediation at this stage, in that they will also ask questions to aid improved understanding. Facilitators may be able to encourage productive discussion. From this role, it is a short step to a greater substantive involvement as a mediator, who, without taking control, assists the stakeholders to move from "a zero sum mind-set" to greater willingness to participate constructively in order to find a solution capable of implementation. Obviously, complete neutrality is the key to success.

The main steps

The main steps involved in the mediation process are as follows:

- convening stakeholders
- clarifying responsibilities
- deliberating
- deciding
- implementing agreements

The steps are nonlinear. As we have observed, in large projects there is an advantage in starting a consensus-building process at an early stage, before the development plans are cast in concrete and before the environmental impact assessment is finalized, so that the developer has the best chance of meeting the stakeholder concerns.

Step 1—Convening and conflict assessment

First, an agency, public official, regulator, or even the court typically brings together the stakeholders. In the context of transmission planning, this may have to be done by the regulator, the RTO or planning coordinator or the developer. The more stakeholders that are brought together at an early stage, the greater the likelihood that an acceptable and implementable agreement can be reached. In this way, the public interest will also be better served.

The issue of identifying the stakeholders and how they are to be represented is extremely important at this stage. The mediator or facilitator must find out who else is affected, who is key to implementation of the resulting agreements, and who can undermine reaching agreement and implementation. The question of how many representatives each group will need to reflect the interests must be dealt with, as well as who else needs to be kept informed. It is helpful if it can be agreed at an early stage, as to how representatives are selected, the use of alternates, and how to accept newcomers. Stakeholders may prefer to self-select their representatives, or they can nominate representatives and let the convener or agencies select the appropriate number. At the first meeting, the mediator or facilitator will ask the participants if all of the viewpoints are adequately represented.

Following confidential interviews, a neutral party (not necessarily a mediator) then prepares a conflict assessment (on a no-names basis), summarizing the concerns of all stakeholders in their own terms. This is a very powerful tool for three reasons. First, it will lay out, in a very transparent manner, what the underlying issues really are and the problems that must be solved. Second, it imposes a discipline on the parties to justify the issues that they raise. They can scarcely ask for concerns that cannot be substantiated to be taken into account. Third, it can be used to identify all of the parties that should be brought to the table. Mediators and stakeholders have agreed that this tool can dramatically increase the chances of settlement.[24]

The stakeholders consider the assessment report and recommendations and decide if mediation should proceed. The willingness of the stakeholders to participate, the nature of the problem solving to be pursued, the legal and procedural constraints, and the relationships or coalitions between the stakeholders will determine whether mediation is appropriate.

Legitimizing stakeholder representation. This first step is extremely important, as it will involve legitimizing stakeholders and deciding whether *ad hoc* representatives should be recognized. The key ingredients will be those stakeholders that are crucial to the implementation of the outcome of the consensus-building process. If they cannot be involved, for whatever reason, then mediation is unlikely to be able to work. Skilled mediators or facilitators may be able to persuade even the most skeptical of stakeholders, and also those with the greatest interest in adopting delaying tactics, to come to the table. They will argue that surely the stakeholder will want to know what the others are saying and whether a better outcome for that stakeholder might be achieved by participating. Another key ingredient is the identification of the necessary funding.

Conflict assessment is an extremely effective tool. The importance and usefulness of the conflict assessment cannot be underestimated. A mediator or facilitator carries out confidential interviews with the stakeholders, listens to their concerns, and finds out who else should be interviewed. This exercise will identify hopes, expectations, and constraints. Once the stakeholders have been interviewed in this way (and advertisements can be placed in newspapers seeking interviews with other potential stakeholders), the conflict assessment report is issued on an entirely no-names basis. Often it will present a matrix of categories of interest, identifying the categories of stakeholder and the issues that are in conflict. It will be sent out to all parties for review and comment, so that every care is taken to ensure that it adequately expresses all of the issues, even though no names are attributed.

The conflict assessment is both a process and a product. It is a flexible tool and presents an opportunity to explain consensus-building and what is involved. The conflict assessment report that is sent out would include an agenda and timetable as well as a work plan and the budget implied by the work plan. It asks the question to stakeholders: "Would you come to the first meeting, if that is how we decide to proceed?" The assessor will make a recommendation to the convener whether consensus-building can go ahead or not, or at least, whether a first organizational meeting should be held. The assessor would not necessarily be the ultimate mediator and, indeed, would offer to not seek appointment if the stakeholders wished.

Deciding to proceed with mediation

Essentially, Step 1 initiates a discussion among stakeholders about whether to have a consensus-building dialogue. If there is substantial agreement or acceptance of the conflict assessment report (as a result of the review and commenting process), it can be very helpful. The report should identify the stakeholders' willingness to participate, highlight areas of common and opposing interests, identify potential opportunities for mutual gain, and identify obstacles to reaching agreement, as well as estimating the potential for a successful outcome. The report can even make recommendations as to the objectives and design of the mediation process; covering how decisions will be made, the responsibilities of the parties, and how they will communicate with each other.

The Consensus Building Institute Inc. has developed a checklist on the subject of how to determine if an assessment might lead to a consensus-building process. The list consists of:

- options—what are the stakeholders' options "away" from the table?
- problem definition—is the problem well-defined and are the issues clear?
- representation—are all representatives identifiable?
- commitments—are stakeholders committed to the process?
- deadlines—are there limiting deadlines?
- resources—are there sufficient resources (time and money)?[25]

If the recommendation is positive and the decision is made to proceed with consensus-building, usually the original convener hires a credible and non-partisan mediator. If the convener is not trusted (*e.g.*, because it is the developer), it may be difficult for the stakeholders to trust the process and believe in the neutrality of the mediator. For this reason, the best convener may be the regulator or the permitting or siting authority.

Step 2—Clarifying responsibility

The second step deals with the administrative aspects of the process. The participants must sign onto an agreed agenda and set of ground rules, defining roles and responsibilities and decision-making procedures. It will identify the role of the mediator and the stakeholders themselves, in that it will set the ground rules of their participation. This step also involves setting the agenda and dealing with the all-important communication issues.

Step 3—Deliberating

The third step helps to provide the information necessary to enable the parties to find solutions. The mediator would not offer technical advice that might lead to counter-arguments and the prolonging of disputes, but rather would ask focused, informed questions that are designed to create opportunities for the parties to explore ideas. The idea is to encourage them to articulate ideas that may lead to solutions. They are asked to focus on interests, not positions, to avoid inflated demands. A working draft will be used to stimulate discussion on a full range of proposals on the basis that there will be no agreement until all issues have been addressed. This avoids strategic withholding of agreement until other issues are resolved, which occurs when they are dealt with one at a time. All of the options will be faithfully recorded, and agreements and disagreement will be summarized. The mediator will continually search for a package that everyone can live with and will conduct bilateral discussions to seek out new solutions.

One of the ground rules might be that no negotiations are to be conducted outside of the process. No decisions will be made until the final meeting. The document reflects what has been agreed. It forms a useful basis for the parties to obtain corporate approvals and for review by constituents. It also facilitates the later documentation of the outcome in binding agreements and pleadings. This step is characterized as one of deliberating. The focus will be on transparency and exploring innovative solutions, rather than seeking commitment.

Step 4—Deciding

The next step is to make the necessary decisions, often by reference to a list of previously agreed, objective criteria or indicators, by which agreement must be gauged or evaluated. The decisions need not be unanimous, but should receive overwhelming support (provided that clear efforts have been made to address dissentient concerns). The focus here is to try to maximize win-win solutions for all stakeholders, so that everyone gains from the process. It is important to keep an agreed record of all of the decisions that are reached. As explained earlier, unanimity and a package of measures that everyone can live with should be the objective, even if it ultimately only commands the support of a substantial majority of the parties. It must be accepted that there has been a good faith effort to meet the concerns of all of the stakeholders.

Step 5—Implementing agreements

Finally, the decisions must be ratified by the stakeholder constituencies and implemented. The stakeholders must be held to their commitments. Sometimes, the result is only advisory[26] and must be adopted by ministers or public officials. Roles and responsibilities for implementation need to be agreed, and an implementation monitoring procedure should be put in place. The agreements can make special provision for adaptation to changing circumstances, as implementation may take a significant amount of time.[27]

Consensus-building involves a different approach to negotiation

Consensus-building is unlikely to succeed if an old-fashioned approach to negotiation is adopted. The traditional approach is characterized by Susskind as asking for more than you want, demanding concessions before you reveal what you are prepared to concede, and making sure that "they don't get what they want." He describes the consensus-building approach to negotiation as a "mutual gains" approach, which he characterizes as a "low-cost way of meeting other people's interests in return for their reciprocity."[28] The approach involves estimating what the absence of agreement means to each of the parties. Anything better than this will have value. It is not a question of trading concessions for concessions.

He describes four steps:

- prepare
- create value
- distribute value
- follow through

Prepare

Preparation involves estimating your "best alternative to negotiated agreement" (BATNA) and those of the other stakeholders. If possible, the BATNA should be continually reviewed and improved. Preparation also involves clarifying your negotiating mandate, understanding your interests and those of the other stakeholders, and, importantly, preparation in order to suggest mutually-beneficial options to other stakeholders. Probably 60% of the outcome is a function of the effectiveness of the preparation.

Create value

Creating value involves exploring the interests of all stakeholders and whether things can be done to "expand the pie" before deciding how to cut it. This is the time to create and explore innovative solutions on a "what if" basis, for which purpose it is necessary for stakeholders to critically reevaluate their own interests and those of all the other parties. This could well require cooperative moves and coalition-building. If it is perceived as a sign of weakness to suggest possible solutions, the mediator or other neutral participants could be used as a means of communication.

Distribute value

At this stage, the time has arrived to discuss and decide the principles on which the "pie could be divided", and the mediator or neutral participants could be used to suggest possible distributions and solutions. It is important for the stakeholders to behave in ways that build trust at this stage, because the drafting of the potential agreements will begin. It is important also to keep an eye on the implementation stage. If a stakeholder is alienated because of a perception that someone else may have played a trick to persuade that stakeholder to take less than what was fair, the ability to reach and implement an agreement will be jeopardized.

Follow through

Once agreements are reached, monitoring arrangements become important, because the signing and implementation of the agreements will take time. It is obviously helpful if it can be made easy for the stakeholders to live up to their commitments. Work on improving relationships at this stage is valuable. It is also useful to agree to use neutral parties to resolve disagreements.

What is a successful negotiation?

A successful negotiation can be characterized in three ways:

- a satisfying outcome for the stakeholders whose interests are all well met—a long-term commitment that works for all parties
- an efficient result, both in terms of time that has not been wasted and possible solutions that have not been left on the table
- an ending that is amicable, where relationships are enhanced and future dealings between the stakeholders will be easier

In short, land use disputes can only be satisfactorily resolved by informed agreements resulting in outcomes that all of the stakeholders want over the long term.

The roles of public bodies in mediation

The roles of the public bodies in land use disputes are extremely important. The public bodies will include the regulator, the siting authority, any government agency or department that is involved, and the ultimate decision-maker. The roles that these public bodies can perform are many:

Convener—often a public body will take the role of convener. It is important to have an impartial and credible initiator of the whole mediation process (such as the Environment Court in New Zealand). Although the project developer may have the keenest interest in resolving issues with all the stakeholders, there will be a natural suspicion of its motives, and it will tend to lack credibility. The convener essentially seizes the opportunity for meaningful dialogue and should be able to set an optimistic tone for the process. It can provide logistical support and help to frame issues, as well as underpinning the initiation of the process with legitimacy and authority.

Partner—the public body can be a helpful partner in the mediation process, demonstrating a willingness to work together, to teach, and to learn. It can contribute scientific and technical information and other resources.

Stakeholder—as a stakeholder, the public body can articulate the broader public interests. It can point out the constraints (legal, money, time), and work to ensure consistency with the legal and regulatory framework. It can help to frame solutions and decisions that can be defended within the agency.

The advantages and disadvantages of consensus building

There are both advantages and disadvantages to consensus-building.

Advantages. The advantages are many and diverse. Consensus-building:

- avoids the costs, delays, and hostility of litigation
- achieves better understanding, communication, and discussion of the issues
- creates opportunities for "win-win solutions" and joint gains
- builds trust and improves long-term relationships
- dispels cynicism and lack of involvement in public, national, or regional decision-making
- achieves more efficient use of resources and better compliance with the agreement that is reached
- resolves underlying issues
- develops a common or shared base of knowledge
- reduces criticism that public officials or governments are out of touch with the local communities in making decisions in the national or regional interests
- informs and empowers disadvantaged interest groups
- achieves greater acceptability of and satisfaction with the decisions that are made

Disadvantages. Consensus-building is not always a recipe for success. It may:

- not be faster or less expensive than any other process
- require intensive use of human and other resources
- be unable to alter stakeholder competitiveness or create incentives to negotiate[29]
- only achieve "lowest common denominator" agreements that are probably far from ideal, in that they will contain only those terms that are noncontroversial and that all parties can easily agree, which may represent little progress in resolving the major issues
- lack a code of ethics for mediators and fail to resolve what constitutes good practice and accountability

- result in litigation, if agreements are not accepted or are inappropriate
- allow powerful stakeholders to apply political pressure

Creating a framework for mediation in transmission projects

Given the apparent success of assisted negotiation or mediation in a number of land use disputes, it would be worthwhile to consider whether it would be helpful to design a mediation scheme for transmission expansion projects and to institutionalize it in the legal or regulatory framework. The intention would not be to make mediation mandatory, but to set out ground rules for mediation to be used, either voluntarily by the parties or to be facilitated by a regulator, competent authority (which might be a court), or public official in appropriate cases. This would legitimize the use of mediation and help to overcome some of the disadvantages to its use, identified earlier. For example, it could create a climate that might encourage consensus-building and provide the ground rules and a code of ethics. The idea would be to supplement, rather than replace, existing laws and regulation in order to allow mediation when it is appropriate, rather than to mandate it.

Legislation for mediation

There are examples of legislation or regulations on general alternative dispute resolution (ADR)[30] procedures, and there are also more specialized examples with respect to land use mediation. In Canada, the Office of the Provincial Facilitator (OPF) was created to mediate land use disputes in the Province of Ontario in 1992. Proceedings before the Alberta Environmental Appeals Board are mediated. In the U.S., there are examples of institutionalized land use mediation in Oregon and Florida. Connecticut has enacted a law called "An Act Concerning Mediation of Appeals of Decisions of Planning and Zoning Commissions."[31] In essence, it allows appeals of decisions of planning and zoning commissions to be stayed by agreement of the parties when mediation is conducted. The Act allows any aggrieved party to participate and sets a six-month time limit for the mediation to be concluded. It also sets out certain ground rules as to starting, conducting, and concluding a mediation, which must be conducted by an impartial third party, using generally-accepted mediation principles. Everyone must agree to the mediation, but a party can withdraw from it at any time. The mediator can ask anyone needed to resolve the dispute to participate in it. A number of procedural rules are incorporated.

In New South Wales, in the context of rights-of-way, the Land Acquisition (Just Terms Compensation) Act requires negotiation of compensation in good faith. If the negotiations fail, the applicant can go to the Land and Environment Court, from which there is no appeal.[32] Alternatively, a system of mediation can be invoked, and the court has a list of mediators. The parties can either agree on a mediator, or the minister can make a direction. The process can be quite cost-effective and works more than 50% of the time. A mediator can call for experts. Litigation is very expensive. Mediation is attractive to landowners, because they think there is a chance they would do better under this system.

Key mediation design issues

The design issues that are critical to the successful implementation of mediation (and the achievement of stakeholder agreements that are capable of implementation) are:

- Who should mediate?
- What role should the decision-makers play?
- How should complex technical issues be handled?
- Should confidentiality be protected, or should transparency be the goal?
- In what cases should mediation be mandatory?
- What role should tribunals or the courts play in procedural and substantive review of negotiated settlements?

- What role should tribunals and the courts play in enforcing and ensuring compliance with agreements reached during mediation?
- Should mediated or negotiated agreements have precedent value?
- What statutory basis, if any, is needed to conduct mediation within the context of administrative hearings?[33]

To this list could be added the issue of the code of conduct, accountabilities, and standards mentioned earlier and a host of other detailed requirements. For example, the Alberta Environmental Appeals Board rules require parties to come to prehearing meetings (which are designed to facilitate resolution) "fully prepared for a useful discussion of all the issues involved...and authorized to negotiate and make decisions with respect thereto."[34]

Who should mediate?

Perhaps the most important issue of those listed is the first—who should carry out the mediation? The obvious requirement is that the mediator should be entirely neutral[35] and command confidence. This will rule out most, if not all the stakeholders, including the project developer, which in the past may have attempted, in good faith, to make progress by facilitating discussions designed to result in a negotiated agreement. As Susskind and Weinstein explain:

> *The perception of the mediator's neutrality is critical – it allows a bond of trust to develop between the mediator and the parties involved. This bond of trust enables the mediator to receive confidential messages from the stakeholders; these in turn, provide clues about the direction that bargaining must take in order to achieve a settlement.*[36]

Land use planners can and have mediated land use disputes successfully, even where they were directly involved in the case. However, the process may have been one of facilitating consensus-building processes, which encouraged collaborative decision-making as a normal way of handling land use conflicts. However, in more difficult cases, outside professional planners will be more effective. There is no reason that an ISO or RTO should not put its planning staff through formal mediation training. The fact of the independence of the ISO or RTO should ensure its neutrality.

It would also be difficult for the decision-makers themselves to act as mediators, particularly if they are required to ratify the agreement that is ultimately reached. The stakeholders would be tempted to influence the mediator as to the outcome of the ratification process, rather than to participate constructively in finding a solution. Statutorily-appointed decision-makers such as ministers, judges, and tribunal members cannot remain neutral as mediators and at the same time carry out the role of ensuring that policy, laws, statutory tests, and regulations are fully met. If a tribunal acts in a quasi-judicial role (*e.g.*, an appeal board), this may be less of a difficulty.

However, there are examples of governments holding an effective and trained in-house mediation capability distinct from other departments. There are trained mediators in Massachusetts, Florida, and Oregon. Parties are attracted to using their services, because they are inexpensive and readily available. Often, the choice is not between in-house and outside mediators, but between using the in-house mediators and not agreeing to mediation at all.

Where in-house public officials are used, the government cannot use the mediator to voice its own interests, nor can it put the mediator in the invidious position where he or she must oversee a superior. With appropriate safeguards, this arrangement can work well, particularly because the mediator will have greater familiarity with the material being discussed. However neutral a mediator really is, the stakeholders will ultimately judge the neutrality more by behavior than by anything else. The more difficult, higher profile, or contentious cases are better dealt with by outside professionals.

How should complex technical issues be handled?

The technical issues in transmission can be complex, multidisciplinary, and the subject of a broad diversity of views. Dealing with technical issues by the appointment of expert witnesses who then battle the issues out under examination and cross-examination conducted by lawyers is clearly not a model that will work in the context of mediation. Indeed, Lord Woolf indicated in the report, "Access to Justice," (discussed in chapter 9[37]) that this model is not even attractive in the context of general civil proceedings. Lord Woolf suggested that a single expert should be appointed to deal with technical issues in a particular case, who would owe his or her allegiance to the court rather than to any of the parties to the proceedings. The same approach is appropriate in the case of mediation.

Facilitation or mediation skills can be usefully brought to bear to assist the stakeholders to agree to a shortlist of three technical experts from which a final choice can be made. The stakeholders can contribute to the compilation of a longer list of suitable experts. Every participant in the process is given the right of veto over a number of names, but is required to participate in a process that arrives at the shortlist and the final decision as to which expert is to be appointed.

It goes without saying that the expert must be well qualified and neutral. An expert that is highly biased or opinionated would not be credible. It is helpful if the expert, once appointed in this way, were to undertake the role of not only rendering an unbiased opinion on technical issues but also the role of educator and interpreter of the complex language of the project, in order to help the stakeholders understand the issues and verify them.

Other ingredients in the design of a mediation scheme for transmission expansion

The other ingredients in the design of the mediation scheme for transmission are relatively straightforward:

- While transparency should be the goal, a successful outcome may depend upon commercially sensitive information being disclosed to the mediator that he or she can agree to keep confidential or divulge on a no-names basis
- It is unwise to make mediation mandatory, because the nature of consensus-building is based on the concept of a voluntary attempt to reach unanimous agreement (but there is no reason that it should not be formally encouraged)
- The courts or tribunals should not play a role in reviewing the mediation process or the outcome, but if agreements can be reached and implemented, they could be enforced in the courts in the usual way
- As indicated in chapter 9, a system of precedent is not helpful in finding the best solutions that will allow a transmission expansion project to go forward—every case will be different, and every stakeholder and local community will have its own views as to the issues that will need to be taken into account
- The legislative models described earlier provide a useful basis on which to conduct mediation, if only to establish the ground rules within the context of administrative hearings and to create a climate that encourages the parties to attempt to build consensus. The beauty of mediation is that there is nothing to stop it taking place informally, without a statutory framework, if the parties so wish

Transmission projects are a suitable case for mediation

It is clear from the experiences of transmission owners in their dealings with landowners and other stakeholders that informal mediation has been applied successfully. This indicates that more formal mediation carried out by professional mediators should work even better, especially if it is used early enough, when constructive dialogue can take place and trust can be built. Clearly, mediation is highly skilled and is more likely to be successful if trained mediators are used.

Consensus-building to improve utility regulation

Jonathan Raab reminds us that consensus-building can be used in regulatory proceedings, for both adjudications and rule-makings. He quotes a number of impressive case studies where consensus-building has been used successfully to achieve good solutions for all the parties, including the regulators. He looks at how the success in what he describes as "supplemental" consensus-building efforts should be measured against traditional processes and asks the question of whether it is sufficient to look at whether a consensus has been reached and whether the participants have saved time and money, or whether it is necessary to look more deeply at both the process and the results.

He concludes that consensus-building can improve the legitimacy of adjudicatory proceedings, because they give stakeholders an opportunity to design their own remedies and plans that create better outcomes for them. It can improve the practicality of the solutions that the stakeholders find to satisfy their own interests. The case studies indicated that the parties were able, creatively, to resolve some extremely difficult issues. Practical plans probably would not have emerged from traditional, regulatory litigation processes. The parties may not save time and money because consensus-building is relatively resource-intensive, but the long-term net benefits over traditional adjudication appear to be both positive and significant.

There are similar findings with respect to regulatory rule-making processes using consensus-building techniques. Legitimacy was the key issue in these proceedings, because, for the most part, new rules were seldom welcomed, but the stakeholders preferred the consensus-building process and were prepared to live with the outcome. They understood the difficult trade-offs that needed to be made. The process also assisted in reaching agreement on structural issues and implementation. Again, Raab found that the long-term benefit was that the policies and rules were improved.

Raab describes four advantages to the use of consensus-building in electric utility regulation:

- Instead of technical information being used as a weapon, the parties seek out the best technical information and share it
- The proposed solution reflects the parties' own concerns and experience to a greater extent
- The parties are not constrained by precedent or legal restrictions (that might have constrained the regulator) to find practical and efficient solutions that better meet their needs
- The level of technical detail that can be achieved in arriving at the solutions is extremely difficult or even impossible to achieve in a contested regulatory proceeding

Raab finishes his very readable book with eight principles for consensus building in electric utility regulation, which are as follows:

- Initiate consensus-building as early as possible
- Include all stakeholders
- Secure direct involvement of the regulator whenever possible
- Provide adequate resources
- Do not exclude contentious or sensitive issues from consensus building efforts
- Consider assisted negotiation
- Structure consensus-building processes to supplement traditional adjudicatory and rule-making procedures
- Modify traditional procedures to accommodate consensus-building opportunities better

The official rules of the game do not matter

Mediation does not depend upon a particular legal or regulatory framework to be in place to enable it to take place. Because it is entirely *ad hoc,* if there are legal or regulatory constraints, mediation can transcend the boundaries. It avoids the difficult question of what changes should be made to the existing legal and regulatory framework. It is superior to a public hearing or inquiry where the outcome is unknown, because mediation will establish at the outset what everyone is striving for.

The outcome is likely to be a better proposal, but it will only be a proposal that must then be adopted and implemented. However, because consensus has been built, implementation is more likely to follow. A public hearing or inquiry will not give a better answer.

Solving cross-jurisdictional problems

There is no reason that consensus-building should not be used explicitly to reconcile national or regional policy with the impact on local interests and the environment. It should not be criticized on the grounds that it will give undue weight to local and environmental issues. Representatives from government departments and regional agencies can be invited to the table and to participate in the consensus building process. It would also be possible, and may be very helpful, for the national, regional, and local agencies to act as joint convenors of the consensus-building process.

Consensus-building is worth a try in transmission projects

In whatever way the transmission permitting process is characterized, whether it is informed problem-solving, dispute resolution, mitigation of adverse impacts, or consensus-building, the chances are that mediation will arrive at a solution that is the best that can be achieved and, equally important, is implementable. Mediation can cope with a large number of stakeholders,[38] several tiers of issues, and complex details. It can create a rational process out of a highly-charged political environment. It ensures that not only stakeholders, landowners, local communities, and developers all have an equal right to be heard, but it puts in place a mechanism that demonstrates to everybody, almost on a daily basis, that their input is actively canvassed and taken into account. In a sense, the stakeholders would be making a prior commitment to input into an effort to build consensus. They would empower the decision-makers, who have acknowledged that they do not have all the answers and need the input of stakeholders, to make a decision. The result is joint input that leads not simply to consensus, but to informed consensus. If it is well handled, a mediation process will be regarded as:

- fair
- efficient
- wise (in retrospect)
- stable (in the sense that it can withstand a change of leadership or government)

As a result, the outcome is more likely to be implementable. It wins on all counts.

Recipe for success

The "Facility Siting Credo" provides a useful summary of the recipe for success in siting and permitting a transmission facility:

- Institute a broad-based participatory process
- Achieve agreement that the *status quo* is unacceptable
- Seek consensus
- Work to develop trust
- Choose the solution that best addresses the problem
- Guarantee that stringent safety standards will be met
- Fully address all of the negative aspects of the facility
- Make the host community better off
- Use contingent agreements that spell out what will be done in case of accidents, interruptions in service, changes in standards, or the emergence of new scientific information about risks or impacts
- Seek acceptable sites through a volunteer process
- Consider a competitive siting process
- Work for geographic fairness
- Set realistic timetables
- Keep multiple options open at all times[39]

NOTES

1. Christopher J. Bosso, "After the Movement: Environment Activism in the 1990s," Environmental Policy in "The 1990s: Toward A New Agenda," Norman J. Vig and Michael E Kraft, eds., 2nd ed. (Washington, DC: *Congressional Quarterly*, 1994), p. 44

2. Stephen Bird, "A Sysiphus Complex At Logan Airport: Runway 14/32 in the Context of Nimby's and Lulu's," Graduate Credit, Urban Politics, Govt. S-1461, Harvard University, John F. Kennedy School of Government (August 6, 2001)

3. Herbert Inhaber, *Slaying the Nimby Dragon* (New Brunswick, NJ: Transaction Publishers, 1998), p. 18-19

4. Executive Order 12898,59 FR 7629 (February 16, 1994)

5. Benjamin Davy, "Fairness as Compassion: Towards a Less Unfair Facility Siting Policy," in *Risk* (Concord, NH: Risk Assessment and Policy Association, Franklin Pierce Law Center, 1995) (accessed January 11, 2002), http.www.fplc, edu/risk/vol 71 spring/dairy.htm

6. The industry association, Cigré has developed a formal methodology to take into account the interests of environmentalists at the techno-economic stage of the planning that is described in more detail in chapter 8

7. Herbert Inhaber, *Slaying the Nimby Dragon* (New Brunswick, NJ: Transaction Publishers, 1998), p. 20

8. FERC Staff OEP Gas Outreach Team, "Ideas for Better Stakeholder Involvement In the Interstate Natural Gas Pipeline Planning Pre-Filing Process: Industry Agencies, Citizens and FERC Staff" (August 2001)

9. Denise Church and Claire Johnston, "Sustainable New Zealand: Lessons Learned From The Development and Implementation of the Resource Management Act 1991," background paper for Resource Renewal Institute/California EPA Workshop, Sacramento, California (April 5-6, 2001)

10. Charles Piller, *The Fail-Safe Society: Community Defiance and the End of American Technological Optimism* (New York: Basic Books, 1991), chapters 6-7

11. John O'Looney, *Economic Development and Environmental Control: Balancing Business and Community in an Age of NIMBYs and LULUs* (Westport, CT: Quorum Books, 1995), p. 208

12. Herbert Inhaber, *Slaying the Nimby Dragon* (New Brunswick, NJ: Transaction Publishers, 1998), p. 91

13. Herbert Inhaber, *Slaying the Nimby Dragon* (New Brunswick, NJ: Transaction Publishers, 1998), p. 41-70

14. Charles Piller, *The Fail-Safe Society: Community Defiance and the End of American Technological Optimism* (New York: Basic Books, 1991), p. 205

15. Department for Transport, Local Government and the Regions, "Reforming Planning Obligations: A Consultation Paper-Delivering a Fundamental Change", HMSO, (December 2001)

16. For a description of the problems with an impact fee system, see J. Barry Cullingworth, *Planning in the USA: Policies, Issues and Processes* (London: Routledge, 2002)

17. Herbert Inhaber, *Slaying the Nimby Dragon* (New Brunswick, NJ: Transaction Publishers, 1998), p. 64

18. I am grateful to José Rotger for the explanation

19. Lawrence Susskind, Paul F. Levy, and Jennifer Thomas-Larmer, *Negotiating Environmental Agreements* (Washington, D.C. and Covelo, California: Island Press, 2000). Lawrence Susskind, Ole Amudsen, and Masahiro Matsuura, *Using Assisted Negotiation to Settle Land Use Disputes: A Guide Book for Public Officials* (Cambridge, Massachusetts: Lincoln Institute of Land Policy, 1999). Lawrence Susskind, Mieke van der Wansem, and Armand Ciccarelli, *Mediating Land Use Disputes, Pros and Cons,* (Cambridge, Massachusetts: Lincoln Institute of Land Policy, 2000)

20. For a description of consensus-building in utility regulation, see Jonathan Raab, *Using Consensus Building To Improve Utility Regulation* (Washington, D.C. and Berkeley, California: American Council for an Energy-Efficient Economy, 1994)

21. These definitions and principles were devised by the Consensus Building Institute, Inc., Cambridge, Massachusetts

22. Lawrence Susskind, Mieke van der Wansem, and Armand Ciccarelli, *Mediating Land Use Disputes: Pros and Cons* (Cambridge, Massachusetts: The Lincoln Institute of Land Policy, 2000)

23. The Centre for Effective Dispute Resolution, *Resolutions,* no. 28 (Summer 2001)

24. Matthew Taylor, Patrick Field, Lawrence Susskind, and William Tilleman, "Using Mediation in Canadian Environmental Tribunals; Opportunities and Best Practicies," *Dalhousie Law Journal*, vol. 22, no. 2 (Fall 1999), pp. 51-124

25. The Consensus Building Institute, Inc., presentation at "Mediating Land Use Issues Workshop," at the Lincoln Institute of Land Policy, Lincoln House, Cambridge, Massachusetts (December 6-7, 2001)

26. For example, the Bureau d'audiences publiques sur l'environnement (BAPE), which is the Quebec environmental hearings board, operates as an advisory board to the Provincial government. Its role is to inform and confer with the public on environmental issues and projects

27. These steps and the mutual gains approach to negotiation are described in more detail in Lawrence Susskind, Sarah McKearnan, and Jennifer Thomas-Larmer, *The Consensus Building Handbook: A Comprehensive Guide to Reaching Agreement* (Thousand Oaks, California: Sage Publications, Inc., 1999), and in course materials provided at the "Mediating Land Use Disputes Workshop," at the Lincoln Institute of Land Policy, Lincoln House, Cambridge, Massachusetts (December 6-7, 2001)

28. Lawrence Susskind, presentation at "Mediating Land Use Disputes Workshop," at the Lincoln Institute of Land Policy, Lincoln House, Cambridge, Massachusetts (December 6-7, 2001)

29. D. J. Amy, *The Politics of Environmental Mediation* (New York: Columbia University Press, 1987)

30. ADR means alternative dispute resolution. It does not involve formal litigation or arbitration, but may take the form of mediation, informal arbitration or expert determination.

31. Public Act No. 01-47 for Substitute Senate Bill No. 1037

32. Land Acquisition (Just Terms Compensation) Act 1991 (NSW). The objective of the Act is to guarantee that when land affected by a proposal for acquisition by an authority of the state (such as Transgrid) is eventually acquired, the amount of compensation will not be less than the market value of the land (unaffected by the proposal) at the date of the acquisition. In other words, the purpose is to ensure compensation of just terms for the owners of the land when their land is acquired by an authority of the state. The Act also provides clauses that are designed to encourage the acquisition of land by negotiation and agreement instead of compulsory acquisition process. These provisions are important and must be followed before the minister will agree to compulsory acquisition

33. Matthew Taylor, Patrick Field, Lawrence Susskind, and William Tilleman, "Using Mediation in Environmental Tribunals: Opportunities and Best Practices," *Dalhousie Law Journal,* vol. 22, no. 2 (Fall 1999), pp. 51-124

34. Matthew Taylor, Patrick Field, Lawrence Susskind, and William Tilleman, "Using Mediation in Canadian Environmental Tribunals: Opportunities and Best Practices," *Dalhousie Law Journal,* vol. 22, no. 2 (Fall 1999), p. 74

35. Lawrence Susskind and A. Weinstein, "Towards a Theory of Environmental Mediation," 9 *Boston College of Environmental Affairs Law Review* (1980-81), pp. 311-347

36. Lawrence Susskind and A. Weinstein, "Towards a Theory of Environmental Mediation," 9 *Boston College of Environmental Affairs Law Review* (1980-81), p. 347

37 The Right Honorable Lord Woolf, Master of the Rolls, "Access to Justice," Final Report to the Lord Chancellor on the Civil Justice System in England and Wales, HMSO (1996), ISBN 0113800991

38. Advocates of mediation admit that if the number of stakeholders is very large, litigation may be cheaper

39. The Facility Siting Credo was co-authored by Howard Kunreuther, Lawrence Susskind, and Thomas D. Aarts. It was designed and produced by Mr. Aarts under the auspices of the Wharton School's Risk and Decision Processes Center. It is available from the Publications Services, University of Pennsylvania, USA

CHAPTER 11

It's All in the Taste and the Presentation

Ingredients of this Chapter

It is necessary to draw together our ingredients, raising and setting agents, and methods of preparation, and apply what we have learned about the control of timing and of presentation in order to create recipes for successful transmission expansion.

We shall begin by summarizing the conclusions from the earlier chapters:

- when investment decisions will be made
- maximizing the utilization of the existing assets without significant investment
- ISOs, transcos, ITCs, and the allocation functions between them
- key ingredients from market design and transmission pricing principles
- conditions that will facilitate merchant expansion
- regulatory incentives
- important elements of planning and performance standards
- planning coordination process

- lessons learned with respect to the effective handling of the permitting and siting processes
- best methods of addressing the concerns of stakeholders

We shall then lay out recipes for different types of transmission expansion projects:

- small projects involving controllable power flows
- large projects involving controllable power flows (and radial AC lines)
- small alternating current (AC) network projects
- large AC network projects

We shall need to design the recipes in the context of the parties that will actually use them:

- merchant developers
- regulated non-incumbent transmission owners
- regulated incumbent transmission owners

This will enable us to take into account a variety of situations in different jurisdictions where the incumbent transmission owners may or may not be able to raise the finance to build new transmission capacity.

Finally, we shall attempt to anticipate what the recipes will produce. We will draw the inevitable gastronomic conclusion that it is all in the taste and the presentation.

Summary of Conclusions

When will investment decisions be made?

A decision to expand a transmission system will be taken when there is a clear signal that the marginal benefits (the reduction in the costs of congestion and losses) are greater than the marginal cost of building new capacity.

However, there are a number of other conditions that may well apply:

- Investment will occur if the cost of capital for the investment is less than the return on the investment (over the life of the asset)
- The value of additional power to be transported should be greater than the cost of the line
- Investment in transmission must be more cost-effective than transporting the fuel to a generating station

The goal of transmission expansion is not to eliminate congestion costs, because it may be cheaper to tolerate a certain amount of congestion than to eliminate it completely. For the purposes of our recipes, our goal is to overcome the obstacles to transmission investment rather than to create an infinite transmission system. However, congestion costs do provide signals as to the need for new transmission facilities, and, as we saw in chapter 1, congestion costs can explode quite suddenly, at the expense of electricity consumers.

Getting more out of the system to minimize investment

At an early stage, we should look at the contents of the store cupboard before going out to buy new transmission lines. It may be that there is spare capacity in the existing assets that could be utilized more effectively. There may be potential for reduced congestion and increased reliability that can be realized by new maintenance practices and technologies, or a modest investment in the upgrading of existing facilities to increase capacity and flexibility.

Incentives can be put in place to encourage, for example, outages to be taken at times when the value of transmission is at its lowest and to ensure that maintenance and upgrades are focused on the key circuits that have the greatest value to market participants. The National Grid Company plc (NGC) transmission services (TSS) incentive arrangements have provided a very effective model. The cost of rewarding the transmission owner to use the existing assets more efficiently will be very small by comparison to the benefits to consumers in terms of the savings in congestion costs, improved reliability, reduced market power, and minimization of the need for the construction of new facilities.

Transcos, ISOs, and ITCs

There has been good operational experience with transcos, independent system operators (ISO), and independent transmission companies (ITC). The transco model works particularly well where there is only one transmission system and no urgent need to attract non-incumbent investment in transmission. It is extremely effective in capturing economies of scope between system operations, on the one hand, and transmission ownership and asset management on the other. In other words, it is the best model for maximizing the capacity and flexibility of the existing assets that we have just discussed. It may be difficult to regulate a potentially large monopoly. However, OFGEM has provided a model for regulating NGC that works well and is transportable.

If there are multiple, potentially competing transmission owners, a model involving ITCs and an ISO will be the answer. The transco will suffer from lack of credibility, because of the potential bias of the system operator to favor its own transmission business, both in operational time scales and in the planning coordination process. The lack of bias in the ITC/ISO model overcomes concerns that a transco might not manage real-time operations, congestion, and imbalances, and the definition and administration of transmission rights, on an entirely non-discriminatory basis. These elements are key ingredients of a workable, competitive market that will facilitate transmission investment.

Allocation of responsibilities

The different industry structures and the detail of the trading arrangements or market rules have created considerable diversity as to the allocation of functions between the various industry participants. From the perspective of transmission expansion, in allocating these functions, it is important to remember the general principles as to when it will be economic to invest. As we have discussed, the decision to expand the transmission system will be taken when there is a clear signal that the marginal benefits (reduction in the costs of congestion and losses) are greater than the marginal costs of building new capacity. If those who are able to expand the system do not bear any of these costs, they will receive no signals and have no incentive to invest.

The key ingredients from market design and transmission pricing principles

The ingredients from market design and transmission pricing are critical from the point of view of efficient use of the grid, minimizing the need for expansion, and creating the necessary price signals for expansion. Our list of the best ingredients includes:

- locational marginal pricing in order to manage congestion and to integrate the energy market, with the pricing of scarce transmission (covering both congestion and losses)
- combined locational marginal costing, real-time pricing, and security-constrained economic dispatch
- nodal rather than zonal pricing, with the use of trading hubs, if simplification makes sense
- a system of financial transmission rights (FTR), of which the ideal flavor is the point-to-point FTR, that is sufficiently long term to meet the needs of investors (unless the new facility is a controllable, radial line, in which case physical rights may be sufficient, although they will need to be subject to use-or-lose-it rules and closer regulatory scrutiny)
- a fair and efficient mechanism for dealing with the allocation of FTRs to the existing transmission capacity and new capacity additions that will facilitate new entry in both generation and transmission
- a transparent auction system for allocating and trading FTRs

- FTRs that are both future obligations and options and cover losses (when the systems are fully developed to compute them)
- an efficient transmission-pricing scheme consisting of three parts—a transmission connection or interconnection charge, a transmission usage charge, and a transmission access charge
- an efficient pricing methodology for the recovery of sunk and fixed costs that allocates them to transmission users in a manner that does not distort new investment decisions and prevents fixed costs from being treated as a variable

The ingredients that will facilitate merchant expansion

Merchant investors will need certain ingredients to be in place so that we can place more reliance on the market to meet the needs of society for new transmission capacity. They include:

- the ability to assess who the potential users or beneficiaries will be and to be able to enter into contracts with them (to compel those who benefit to pay)
- the ability to realize or deal with economies of scale
- market-based prices
- locational marginal pricing
- a mechanism established in advance for property rights (physical or financial) to be allocated to the investor, representing the additional transmission capacity that the expansion has created
- a coordinated planning process that creates transparency, generates useful information for potential investors, facilitates market-based transmission, and takes it into account at the earliest possible stages
- the absence of regulatory hurdles and micromanagement
- a level playing field with incumbent, regulated transmission owners and network service providers
- the ability to enter into agreements on satisfactory terms with the system operator for the relevant control area in order to cede operational control of and integrate the project with the existing system

(for reliability and operational purposes) and with the relevant transmission owners with respect to the terms for interconnection and consequential reinforcement of the existing system

Regulatory incentives to invest in transmission

The list of best regulatory ingredients to make the most of existing capacity and to encourage the building of new transmission facilities is the following:

- The price cap (RPI–X) formula with a negative X factor or a cost pass-through element (where necessary) is the best of the existing approaches to attract investment. Yardsticks should be used in setting the X factor, if relevant comparators can be found
- A review period of 5-10 years will create more investor certainty and reduce regulatory risk. There should be consistency and clarity of accounting treatment from one period to the next
- An incentive scheme should be put in place to make sure that the existing transmission facilities are utilized to their maximum and most flexible potential. In the case of a transco, NGC's TSS arrangements would provide a useful model
- Incentives should be set on a comprehensive basis to ensure that average performance levels in all areas are maintained or improved. Certain minimum performance standards should be imposed in key areas to maintain minimum service levels
- The design of schemes with high-powered incentive effects should provide greater benefits to the consumer. They must include real potential upside for the utility. If the downside is significant or risky, the scheme will be seen as unfair, and it will deter investment. Carrots always work better than sticks
- The transmission owner must be given the regulatory flexibility to create prices that encourage efficient decisions as to plant location and the use of the transmission system
- There must be clear rules relating to revaluation of the regulated asset base. Revaluation downward should take place only if

it can be proved that the investment was never justified. It should not be possible for stakeholders to challenge an asset value without an exceptionally strong case

- Even if the rate of return is implicit in the price cap formula, regulators should pay attention to it. They should also recognize that transmission business risk has increased. They should allow accelerated depreciation and more front-end loading to the payback. Transmission investors will be deterred by a combination of their continued low beta (the measure of the risk of their business described in chapter 2), low rates of return, and straight-line depreciation over the long life of the asset in question. The cost of capital is still the key ingredient, even under incentive- or performance-based regulation, and a mechanism should be incorporated to adjust it in the light of prevailing market conditions
- The menu-based regulatory approach has many attractions, because it should deal well with the asymmetry of information that the regulator suffers. The utility should have the best information as to the extent to which it can improve its performance and achieve cost savings or efficiency gain. However, the approach is risky for the regulator, if it gets the menu wrong. If it is done well, a menu approach will create a regulatory incentive scheme with an incentive for the utility to choose the highest-powered incentive that it can realistically expect to deliver. It will create a mutual gain for the utility and the consumer
- Given the size of the investments and the long lives of the assets, a significant degree of regulatory commitment is necessary. It should be seasoned with a mechanism that achieves the necessary degree of regulatory flexibility, but on a basis that is predictable. If the regulator wishes to change its priorities and goals, it should give significant advance notice to avoid increasing regulatory risk and deterring investment

Because of the short-term effect of regulatory incentives schemes, relative to the length of the lives of the assets over which investors will expect a return, investment decisions must be planned and approved by regulators in an active process. There seems to be no alternative to regulatory rules requiring transmission owners and developers to present investment plans for regulated expansion to the regulator for approval and then to implement them.

The greatest contribution that a regulator can make is to not abdicate responsibility for the long-term viability of the system to transmission owners and system operators by placing them under aggressive incentives to achieve short-term cost savings and efficiency gain, otherwise the regulator will preside over progressive deteroration in the transmission system. What transmission investors need is a real incentive to be there for the long term.

Planning and performance standards

Before carrying out the main steps of the preparation—the planning coordination process—we need to:

- review the existing planning and operating standards and the regulatory performance targets
- look closely at the trading arrangements and market rules
- decide who should set and enforce the standards. The expertise that resides within the industry can be used to assist in the design of the standards, but the ultimate responsibility for setting and enforcing the standards should reside with the regulator, because of its impact on the prices to consumers and permitted revenues. The system operator can be used as a front line monitor
- an attempt should be made, wherever possible, to replace detailed technical standards with broadly-based economic standards (but with adequate performance targets to ensure that quality of service does not deteriorate)

The planning coordination process

Central coordination of the planning process is desirable, involving as many stakeholders as possible in an effort to build consensus. We need to:

- decide who should coordinate. The decision of who should carry out the planning coordination lies between the system operator (RTO) and a transmission company owning a significant part of the existing infrastructure. FERC has already determined that the RTO/ITP should have the role of planning coordinator in the U.S., and this makes sense because of the multiple transmission ownership. In other jurisdictions, where there is a transco and a single transmission system, the role should be given to the transco. However, where there is a multiple ownership of transmission and new entry is to be facilitated, the system operator would be the best option
- utilize the recent developments in planning tools, such as the application of real options and game theory, the Cigré stakeholder methodology, and the reference network concept
- ensure that the planning coordination process is carried out on an annual or biennial cycle, looking over a five- to seven-year time horizon. It would involve a number of steps, including establishing and maintaining an accurate planning database; collecting information for demand forecasts and inputs from all stakeholders; assessing the degree of likely generation and demand-side responses; the publication of information as to the performance of the transmission system, its adequacy and reliability, and the current level of congestion costs; obtaining wide stakeholder input as to needs; local community and environmental sensitivities; and taking into account committed generation, demand, transmission (regulated and merchant), and interconnection projects and long-term transmission interconnection and service requests
- require the planning coordinator to publish an assessment of the transmission performance and capacity enhancement needed,

but without designing specific projects. The input of stakeholders must be taken into account in a process that endeavors to reach full consensus and acceptance by a substantial majority of the stakeholders that their views have been heard
- require the planning coordinator to invite the incumbent regulated transmission owner (or owners) to design projects to meet the capacity and performance enhancements that have been identified and to file complete applications with the relevant regulatory, siting, and permitting agencies by a specified date
- enable the planning coordinator to issue a request for proposals (RFP) if applications covering all needs are not filed by the specified date. The RFP would be open to regulated and merchant developers for projects to assist in meeting the needed performance and capacity enhancements described in the assessment, inviting generation, transmission, and demand-side solutions

Permitting and siting processes

We have looked at a number of different issues that explain why the permitting process takes a very long time. If we could start with a blank sheet of paper, the recipe for the ideal process to reconcile national or regional interests with the impact on the environment and local communities, would look something like this:

- Transmission would be taken into account in regional and local development plans
- The regulator would decide the issues relating to the need for a regulated project and whether, in principle, it should proceed (because it will need to do so anyway for the purposes of cost recovery through regulated charges to users and consumers)
- There would be no determination of need with respect to merchant transmission projects
- A special purpose, multidisciplinary body established on a regional basis, presided over by a judge or a regulator, would provide all other permits on a "one-stop shopping" basis

- The process would be non-adversarial, problem-solving, and consensus-building in character, and a real effort would be made to take into account the impact on the local communities and the environment
- Politicians would not make or second-guess decisions, and national, state, and local government would participate in the process in the same way as any other party
- The process would be handled efficiently through good coordination and case management, rules that were rational and consistent with timetables for the procedural steps to be carried out, and for decision-making requiring strict compliance
- The multidisciplinary body would encourage early engagement of local stakeholders, publish guidelines for participation in the process, employ trained mediators, encourage settlements of disputes, and maximize the transparency of the process in order to create more certainty as to outcome, without applying a rigid system of precedent

However, we do not live in an ideal world and do not have a blank sheet of paper. We cannot assume that there is much political will to recognize that things are broken and to take active steps to fix them. We can, however, try to achieve improvements through small steps. Our conclusions are as follows:

- Dealing with multiple agencies increases cost, creates uncertainty, and has the potential for delay. There is scope for creating really effective coordination among multiple agencies, if, out of necessity, they must all retain some jurisdiction over aspects of the project
- Public hearings or inquiries are not the best forum for solving problems. Structured consultations with stakeholders, assisted negotiation, and consensus building work much better
- Compliance with rules with respect to timetables for decision-making would be extremely useful, because controversial decisions will always be delayed (especially if they are made by politicians)

- The system for expropriation of rights-of-way should be coordinated with the permitting process (if the powers are not otherwise conferred), and private sector, non-public utility, and merchant developers must be able to obtain the necessary powers for a project
- It is important to put together a team of people who are effective communicators with landowners and objectors and who have credibility and the ability to create good relations with them throughout the process (continuity being important)
- Ensuring that the concerns of landowners and other objectors are taken into account is of paramount importance. The best means to achieve this is through a consensus-building process, convened at an early stage. Trained mediators should be used if the conflict assessment process and report indicate that consensus-building is likely to achieve a better outcome than the conventional permitting process

The "Facility Siting Credo" provides a useful summary of the recipe for success in siting and permitting a transmission facility. Its salient points are:

- Institute a broad-based participatory process
- Achieve agreement that the *status quo* is unacceptable
- Seek consensus
- Work to develop trust and constructive (non-adversarial) dialogue
- Choose the solutions that best address the problems
- Fully address all of the negative aspects of the facility
- Make the host community better off

The Recipes

We are now in a position to put the transmission expansion recipes together for each class of potential developer—the merchant investor, a regulated incumbent or a regulated non-incumbent transmission owner. We shall assume that all merchant developers can be new entrants or incumbent merchant investors. Where we refer to incumbent and non-incumbent transmission owners, we assume that they are or will be regulated.

Small-scale controllable projects

Cook: Merchant developer (most likely) or non-incumbent transmission owner (incumbent transmission owners not excluded)

Ingredients: Identifiable beneficiaries (unless the developer intends to and is permitted to exploit all of the capacity of the project itself)
Technologies that reach economies of scale plateau
Market-based prices (if carried out on a merchant basis)
Locational marginal pricing
Property rights (financial or physical)
No regulatory hurdles or restrictions

Preparation: Preparation is best left to the skills and good entrepreneurial sense of the developer (in coordination with the relevant system operator)

Presentation: Early, constructive (non-adversarial) dialogue with stakeholders is desirable (if there is an impact on the environment or local communities), using the same team of skilled communicators. Stakeholder input must be taken into account. Consensus-building and mediation may be useful

Taste: Useful small-scale capacity additions, mostly around the edges of the system or between systems, that may reduce congestion costs and improve reliability

Large-scale controllable projects and AC radial lines

Cook: Merchant developer, non-incumbent transmission owner, or incumbent transmission owner

Ingredients: Identifiable beneficiaries (probably investor coalitions)
Technologies that reach economies of scale plateau or ability to declare a lower rating (*i.e.*, withhold capacity from transmission rights market)
Market-based prices (if carried out on a merchant basis)
Locational marginal pricing
Property rights (preferably financial, not to interfere with dispatch)
No bias in favor of affiliated users
Regulatory oversight of open access, rights auctions, tradability of rights, market power, reserve prices, transparency, etc.
Posting an up-to-date list of transmission rights holders and the rights held to facilitate secondary trading
Coordination with the system operator in meeting its reliability requirements
No impact on pre-existing property rights
If the project is regulated:

- incentives set on a comprehensive basis (performance standards and targets) to maximize availability and flexibility of existing system
- enhanced price cap regulation, using a menu of options, if feasible, for the developer to choose (see chapter 7)
- flexibility to achieve efficient pricing
- clear rules on asset revaluation for imprudent investment
- real upsides for the investor
- regulatory scrutiny of investment plans
- accelerated depreciation

- an implicit rate of return that realistically reflects the risks involved that is sufficient to attract investment

If the project is market-based, a level playing field with the incumbent transmission owners (discussed in chapter 5)
Expropriation or eminent domain powers if land and rights-of-way must be acquired

Preparation: Review planning standards, rules, and conditions of the transmission systems to which the project will interconnect, arrange system studies in conjunction with relevant system operator and transmission owners, and resolve issues with respect to system control and reinforcement at an early stage. Participate in the regional planning coordination process. Otherwise, leave to the skills and good sense of the developer

Presentation: Early, constructive (non-adversarial) dialogue with stakeholders is necessary. Stakeholder input using the same team of skilled communicators must be taken into account. Consensus-building and mediation will be necessary

Taste: Major contribution to needed capacity additions that can reduce congestion costs and increase reliability. Mainly radial lines and interconnectors that may bring the added value of increased market liquidity as well as reduced market power and price volatility. They may also improve reliability and reduce ancillary services costs

Small-scale AC network projects

Cook: Merchant developer, incumbent transmission owner, or non-incumbent transmission owner

It's All in the Taste and the Presentation

Ingredients: Identifiable beneficiaries
Technologies that reach economies of scale plateau
Market-based prices (if carried out on a merchant basis)
Locational marginal pricing
Property rights—financial (not to interfere with dispatch) or lease of the facilities to the interconnected transmission owner
No bias in favor of affiliated users
Regulatory oversight of open access, rights auction (open season), tradability of rights, market power, reserve prices, transparency, etc.
Posting an up-to-date list of FTR holders and the rights held to facilitate secondary trading
Coordination with the system operator and meeting its reliability requirements
No impact on pre-existing property rights
If the project is regulated:

- incentives set on a comprehensive basis (performance standards and targets) to maximize availability and flexibility of existing system
- enhanced price cap regulation, using a menu of options, if feasible, for the developer to choose (see chapter 7)
- flexibility to achieve efficient pricing
- clear rules on asset revaluation for imprudent investment
- real upsides for the investor
- regulatory scrutiny of investment plans
- accelerated depreciation
- an implicit rate of return that realistically reflects the risks involved that is sufficient to attract investment

If the project is market-based, a level playing field with the incumbent transmission owners (discussed in chapter 5)

Expropriation or eminent domain powers if land and rights-of-way must be acquired

Preparation: Review planning standards, rules, and conditions of the transmission systems to which the project will interconnect, arrange system studies in conjunction with the relevant system operator and transmission owners, and resolve issues with respect to system control and reinforcement at an early stage. Participate in the regional planning coordination process. Otherwise, leave to the skills and good sense of the developer

Presentation: Early dialogue with stakeholders is desirable (if there is an impact on the environment and local communities) using the same team of skilled communicators, and stakeholder input must be taken into account. Consensus-building and mediation may be helpful

Taste: Useful small-scale capacity additions that may unlock bottlenecks on a large system, reduce local market power, reduce "must run" generation and ancillary services costs and improve reliability

Large-scale AC network projects

Cook: Incumbent transmission owner or non-incumbent transmission owner, where the incumbent is unable to raise the finance (because it is likely to be difficult to develop this kind of project on a merchant basis, although, in theory, it could be done)

Ingredients: Identifiable beneficiaries (unless the benefits really are system-wide)

Market-based prices (if on a merchant basis)
No bias in favor of affiliated users
Property rights—financial (not to interfere with dispatch)
Regulatory oversight of open access, rights auction (open season), tradability of rights, market power, reserve prices, transparency, etc.
Posting an up-to-date list of FTR holders and rights held to facilitate secondary trading
Coordination with the system operator and meeting its reliability requirements
No impact on pre-existing property rights
If expansion is regulated:

- incentives set on a comprehensive basis (performance standards and targets) to maximize availability and flexibility of existing system
- enhanced price cap regulation, using a menu of options, if feasible, for the developer to choose (see chapter 7)
- flexibility to achieve efficient pricing
- clear rules on asset revaluation for imprudent investment
- real upsides for the investor
- regulatory scrutiny of investment plans
- accelerated depreciation
- an implicit rate of return that realistically reflects the risks involved that is sufficient to attract investment

If project is market-based, a level playing field with incumbent transmission owners (discussed in chapter 5)
Expropriation or eminent domain powers if land and rights-of-way must be acquired

Preparation: Review and, if necessary, revise planning standards, rules, and conditions of the transmission systems to which the project will interconnect, arrange system studies in conjunction with the relevant system operator and transmission owners, and resolve issues with respect to system control and reinforcement at an early stage. Participate in the regional planning coordination process. Otherwise, leave to the skills and good sense of the developer

Presentation: Early, constructive (non-adversarial) dialogue with stakeholders is necessary, using the same team of skilled communicators. Stakeholder input must be taken into account. Consensus building and mediation will be necessary

Taste: Major contribution to needed capacity additions that will reduce congestion costs and increase reliability. The project is likely to exhibit strong economies of scale and likely to be paid for by all beneficiaries on a regulated basis. It should mitigate local market power and "must run" generation costs. It should also bring the added value of increased market liquidity as well as reduced market power and price volatility

The right recipes will produce results

The recipes show that we can expect both merchant and regulated developers, whether incumbents or new transmission entrants, to play important roles in delivering needed capacity additions, resulting in reliability and performance enhancements. We need to put in place the key ingredients to ensure that they can all flourish. As both the theory and the experience that we have discussed in this book have shown, if we use the right recipes, we can expect that:

- Merchant developers will deliver small and larger controllable facilities and interconnectors and also AC radial lines quickly, using innovative technology and financing techniques. Interconnectors developed in this way will be a real bonus because the degree of coordination between regulatory agencies that is necessary to enable them to proceed makes it harder to carry them out on a regulated basis

- New entrants will be able to respond to all opportunities, on a regulated and a merchant basis, which will be a great advantage in countries where the incumbent transmission owner cannot raise the finance for expansion
- Regulated, incumbent transmission owners will respond to market-based opportunities, create some real competitive pressures, and act as a backstop if there is a market failure. They will also have a key role to perform in maximizing the availability and flexibility of the existing systems, and in building the larger projects that exhibit strong economies of scale. Because their costs of entry are already sunk, they will be in a strong position to compete with new entrants and use their existing assets to expand the system at least cost to the consumer. However, they need good regulatory incentives to do this

Competition between merchant and regulated transmission owners

There is likely to be plenty of opportunity around the world for both regulated and merchant developers. They need not be afraid of each other. Each of them can wear the other's hat in order to create a market in transmission expansion. They may compete for projects, but this will play to their strengths in designing projects and innovative financing schemes. The regulated developers should not fear "cherry picking" of the best project by merchant developers. They can "cherry pick" as well. This will act as an incentive to build expansion that obviously has a value quickly. Society will benefit from these forces.

Preemptive planning

In chapter 8, we discussed a coordinated planning process that should help to reconcile the issues at the interface between merchant and regulated expansion. It is difficult to argue that the plan, once in place, should be binding and that no market-based project that is then proposed can proceed if it conflicts with a planned, regulated project. If a merchant developer is prepared to put its own money at risk in a competing project, a regulator would be bound to look at it, in the interests of saving the consumer money. If the merchant project made a competing regulated project redundant, in whole or in part, the regulator should order cancellation or redesign of the project, but allow the regulated developer to recover the costs that it has incurred in developing the project and redesigning it (if appropriate). This will look like chaos to a central planner. An independent planning coordination process along the lines discussed in chapter 8 is the best compromise to mitigate the disorganization.

More controllable projects

The majority of the world's transmission systems are AC networks. Clearly, controllable technologies do make expansion easier, particularly in terms of identifying beneficiaries and deciding who must pay. In the past, engineers did not value controllability in the same way as we do now. Perhaps, we shall now see a move to the addition of controllable facilities that will enable more and more parts of existing systems to be controlled. The transmission system was never conceived as the marketplace or race track it has increasingly become. However, we cannot really expect it to be completely redesigned as a controllable system over time, even though the value that controllability adds will be better understood.

Exploiting the public benefits of transmission to achieve expansion

We have observed several times that transmission expansion adds more than just reduction in congestion costs and access to markets for generators. It can improve reliability, reduce ancillary services and "must-run" generation costs, and reduce price volatility. It can also help to alleviate market power. Merchant developers, for the most part, are only remunerated for the reduction in congestion costs, access to markets, and price volatility hedging through the sale of transmission rights (although there is no reason that loads should not be prepared to pay them to reduce market power). We should look for ways to incentivize both merchant and regulated developers to find projects that achieve these wider benefits. Should the system operator run a solicitation process for a pure reliability project and recover the cost on a socialized basis through some form of uplift? Should a regulator offer an incentive or subsidy to provide these wider benefits for consumers under a solicitation process inviting both market-based and regulated projects? Can we really rely on merchant investment for public policy needs? The answer is "why not give it a try and see what happens?" It may need regulatory oversight, but regulators are used to regulating transmission. The idea of hybrid or public-private partnerships in infrastructure projects is appealing and fashionable (and used in other infrastructure development, such as water and transport). Why not give them a try? If the result is unsatisfactory, we know that we can improve the regulatory regime to create incentives for incumbent transmission owners to invest in transmission.

Regulating transmission for the long haul

There is consensus that performance-based or incentive regulation has superior properties to rate-of-return regulation. Transmission owners have been asking for it in the U.S., because it will incentivize them to manage an increasing level of risk, using their core competencies, in return for greater reward for the benefit of the consumer. Regulators have found performance-based regulation attractive as a means of creating effective incentives for more efficient use of existing assets and greater operational efficiency, cost saving, and productivity gain. It creates the flexibility for more efficient pricing. The price cap formula using a negative "X" factor or a cost pass-through factor has been effective to incentivize long-term investment, but it has some difficulties:

- The explicit review period is short by comparison to the lives of the assets, and the uncertainties surrounding the review do not encourage investors to be there for the long term

- The transmission owners may focus on short-term efficiency gain to the exclusion of longer-term investment
- Performance standards need to be put in place to ensure that cost cutting is not at the expense of the quality of service to users
- It seems that there will always be a need for some degree of regulatory micromanagement in the form of a review of expansion plans (despite the difficulties of conduct regulation)

It is clear that incentives work better than rules and statutory duties to expand the system. The cost of the incentives to the consumer will be far outweighed by the benefits. Put another way, the consumer will be better off if the regulator runs the risk of allowing large profits to be made in the short term. The alternative—where the regulator tries to guess what an efficient level of costs would be in advance with no incentive on the part of the transmission owner to beat that level—is much worse.

Regulatory objectives and asymmetry of information

The regulator will always be at a disadvantage, because of the asymmetry of information available to it to regulate transmission. Menu regulation, if it can be well done, is an appealing mechanism to create incentives for the transmission owner to choose a higher-powered incentive (that it can expect to beat), than would be the case if the regulator had chosen it. There is not much experience of menu regulation in the field of transmission, and it is clearly challenging to get the menu of options right. It is worth making the attempt to use this approach, because it can provide win-win solutions for the transmission owner and the consumer.

Get the prices right

The life of transmission developers and of regulators will be much easier if the prices are right. Locational marginal pricing is gaining acceptance and is already regarded as "standard" in good market design in the U.S.[1] We need to make sure that we minimize the need to expand transmission by making the most efficient use of it. We need to create signals as to when to invest and who the beneficiaries are likely to be. Locational marginal pricing is the best tool for this.

Transmission rights

Market participants have learned and are continuing to learn how to value and trade FTRs. Progress is being made with simultaneous feasibility tests for option FTRs and with loss FTRs. We do know that merchant projects have been developed on the basis of FTRs.

Permitting and siting

There is no doubt that the permitting and siting process deters investment and that there is little political appetite to change or streamline it. We are grappling here with one of the most fundamental questions that society must tackle. How are decisions made with respect to infrastructure development in the national or regional interests and how can the impact on the environment and local communities best be mitigated? We have looked at an ideal solution in chapter 9 and also small steps that could be taken to bring about some efficiency improvements. The easiest solution is to use consensus-building and mediation to provide better answers. It avoids the much harder task of changing the whole process.

Transition

To use the automobile analogy, there are a lot of moving parts in transmission expansion or, to use the recipe metaphor, a lot of ingredients and flavors. It is difficult to move to a new paradigm for transmission expansion in a single step. We have looked at many small changes that could be made in a pragmatic manner. Some changes will be easier to make than others. It is always simpler to work on a blank sheet of paper, but the transition in the U.S. and in many other countries will not be on a blank sheet of paper. It is particularly hard to see how the regulatory reforms toward performance-based regulation will take place. The transition will take considerable political and regulatory will.

Transmission developers and owners are not the problem. They are not the dinosaur monopolies that everyone likes to criticize. They are ready to create vibrant, stand-alone transmission businesses and able to manage assets and expansion risks for the benefit of the consumer.

It is a matter of taste

We have found ourselves in a number of ideological debates over the key ingredients of transmission expansion. In the absence of long experience, we can only conclude that it is a matter of taste as to which ideology to choose. The most fundamental of the competing ideologies involves deciding whether regulation, imperfect as it is, is better or worse than imperfect markets for transmission expansion. It is easy to understand why some economists do not like the large, regulated, "black box" transmission monopoly and prefer market solutions. As William Hogan reminds us:

> *This approach results in a powerful monopoly with the familiar problems of finding the right level and form of incentive regulation.*[2]

It is also easy to understand why some engineers feel that the market will fail to deliver optimal projects that deliver reliability and all the other public benefits that transmission can provide. They question whether merchant projects should be allowed to destroy a regulated project that was developed as a backstop when the market appeared to have failed. They worry about a series of small projects that might have less impact on the environment and local communities than a larger one that has greater impact, but will cost the consumer less, because of strong economies of scale.

The answer is that it is all a matter of taste and ideology. Paradoxically, what you see and what you taste can be different. We should also remember that tastes have a tendency to change. NGC, a regulated transmission owner, became highly innovative and market-orientated, well before merchant transmission was a reality.

We should allow both merchant and regulated expansion to succeed. Our recipes indicate that they will probably provide different forms of expansion, all of which society will need. Between them, regulated and merchant developers will create a market for transmission expansion that will benefit the consumer. The imperfections of the market should reduce over time. Regulatory oversight will always be needed, but that comes as no surprise. Regulators will simply need to regulate transmission differently and better. We can make course corrections in the light of operational experience. If the result is some marginal over-investment in transmission, it will be at a cost to the consumer that is insignificant by comparison to the cost of under-investment. As Paul Joskow has pointed out to FERC:

> *The Commission should be much more concerned about underinvestment and should recognize that the societal costs of underinvestment in transmission are much larger than the societal costs of overinvestment. External factors such as licensing requirements, the need for rights-of-way, and NIMBY opposition to transmission infrastructure already places significant constraints on overinvestment in major new transmission projects.*[3]

NOTES

1. FERC Notice of Proposed Rulemaking, Docket No. RM01-12-000 (March 15, 2002)

2. William W. Hogan, "Market Based Transmission Investments and Competitive Electricity Markets," John F. Kennedy School of Government, Harvard University (August, 1999), p. 1, available on Professor Hogan's website, www.whogan.com

3. Regional Transmission Organizations, Notice of Proposed Rulemaking: Comments of Professor Paul L. Joskow, Docket No. RM99-2-000 (August 13, 1999), p. 5

GLOSSARY OF KEY TERMS

Some of the following terms and definitions were taken or adapted from the August 1996 NERC Glossary of Terms and from the Glossary of Terms prepared by the National Council on Competition and the Electric Industry. Some were developed by the U.S. Department of Energy Task Force on electric system reliability and appear in the report "Maintaining Reliability in a Competitive U.S. Electricity Industry." Many were prepared for this book.

access or use of system charges. Charges levied by a transmission owner or an ISO/RTO for access to and use of the transmission system, not including usage (congestion) charges. Where locational marginal pricing is in place, these charges are usually referred to as access charges, and where it is not in place, they may simply be referred to as use of system charges.

adequacy. The ability of the electric system to supply the aggregate electrical demand and energy requirements of the customers at all times, taking into account scheduled and reasonably expected unscheduled outages of system elements.

affiliate. A corporation or business under the same ownership or voting control as the industry participant in question.

ancillary services. Services such as reactive power, voltage control, regulation, reserves, load following, frequency control, black-start capability, and network stability services that the system operator requires in order to maintain system security and reliability.

availability. A measure of the time that a generating unit, transmission line, or other facility is capable of providing service, whether or not it actually is in service.

available transfer capability. A measure of the transfer capability remaining in the physical transmission network for further use or for power transfers over and above already-committed uses.

balancing. The balancing of scheduled power flows and contracted or nominated transactions with actual power flows to ensure that supply and demand for electricity match in real time. It is also the service that provides energy correction for any hourly mismatch between supply and demand. *Synonym*: energy imbalance service.

binding constraint. A constraint, limitation, or bottleneck between two points on the transmission system that limits the amount of power that can be transferred between buyers and sellers. A constraint may be caused by the volume of transactions scheduled across that path exceeding the rated capacity or transfer capability of the transmission facilities along that path or by a planned or unplanned reduction in the transfer capability of those transmission facilities (*e.g.*, for the reasons that one or more of the facilities has been taken out of service for maintenance). *Synonym:* constraint

black-start capability. The ability of a generating unit or station to go from a shutdown condition to an operating condition and start delivering power without assistance from the electric system.

bus. A busbar or conductor forming a terminal or junction point in a power system.

capability. The ability of any part of the transmission system to transmit or transform power (which may depend upon the ambient conditions, the configuration of the transmission system, the pattern of power flows, and transmission system usage).

capacity. The continuous production, transmission, or transformation ability of a facility, expressed in megawatts (MW), megavolt-amperes (MVA), or megavolt-amperes-reactive (MVAR) of generation, transmission, or other electrical equipment.

congestion. A state on the transmission grid that restricts or constrains the ability to add or substitute one source of power for another.

congestion costs. The difference in cost between the cost of energy in the wholesale market calculated on the basis that there are no binding constraints on the transmission system that affect scheduled spot market and balancing transactions and the cost of energy calculated on the basis of the actual binding constraints at the relevant time.

connection. The equipment and facilities that connect generators and loads to the transmission system. *Synonym*: interconnection.

connection charges. Charges levied by a transmission owner for equipment that needs to be installed to connect generators, transmission facilities, distributors, or loads to the transmission system. In some jurisdictions, these charges may be called interconnection charges.

consumer. A customer, person, or entity that purchases electricity for its, his, or her own consumption.

contingency. The unexpected failure or outage of a system component, such as a generator, transmission line, transformer circuit breaker, switch, or other electrical element. A contingency also may include multiple components, which are related by situations leading to simultaneous component outages.

contract path. A notional contiguous electrical path from a source to a sink for which transfer rights have been contracted.

contracts for differences (CfD). An option contract designed to hedge the volatility of energy prices in a spot market or pool.

control. The process of deciding the switching and reconfiguration of the transmission system and giving operating instructions for this purpose. The system operator carries out transmission control. There is a subtle difference between transmission control and other control responsibilities of the system operator in giving operating instructions to generators and other users of the transmission system.

control area. An electric system or systems, bounded by interconnection metering and telemetry, capable of controlling generation to maintain its interchange schedule with other control areas and contributing to frequency regulation of the system or systems.

curtailment. A reduction in scheduled capacity or energy delivery.

demand-side solutions. Consumer behavior such as changing the patterns of consumption of electricity (both as to timing and amount) and technologies that can reduce power consumption.

deregulation. The reform, liberalization, unbundling, restructuring, or change of an electricity trading system in order to create (or simulate) a market or to increase the capture of market forces.

direct costs. The costs associated with the ownership of the transmission system, such as the costs of maintenance, transmission operations (switching), and the construction of new transmission facilities.

direct current. A flow of electricity that is controllable as to direction and amount.

dispatch. The process of determining which generators should be operated each day to meet the daily demand of the system. *Synonym*: unit commitment.

distributed generation. A distributed generation system involves relatively small amounts of generation located on a utility's distribution system for the purpose of meeting local (substation level) peak loads and/or displacing the need to build additional (or to upgrade) local distribution or transmission lines, such as modular electric generation or storage located close to the point of use, including microturbines, fuel cells, photovoltaics, and small wind turbines.

distribution. The process of moving power from transmission lines to consumers over low-voltage lines and other equipment.

distribution system. The portion of an electric system that "transports" electricity from the bulk-power transmission system to consumers, consisting primarily of low-voltage lines and transformers.

distributor. The owner and/or operator of a distribution system.

disturbance. An unplanned event that produces an abnormal condition on a transmission or distribution system.

economic dispatch. The process under which a transmission system operator orders sources of generation to inject electricity into the transmission system, in order of least to greatest cost or price. *Synonyms*: merit order dispatch or dispatch in merit.

economies of scale. Circumstances in which costs of production or transportation decline as plant or equipment size and the amount of goods or services produced or supplied increases.

economies of scope. Circumstances in which costs of production or transportation decline as the nature or types of goods or services produced or supplied increases.

electrical energy. The generation or use of electric power by a device over a period of time, expressed in kilowatt-hours (kWh), megawatt-hours (MWh), or gigawatt-hours (GWh).

embedded cost. The original or historic cost of a transmission facility (less depreciation and including operation and maintenance expenses and taxes).

embedded generation. Generation connected to (or interconnected with) a distribution system.

emergency. Any abnormal system condition requiring automatic or immediate manual action to prevent or limit loss of transmission facilities or generation output that could adversely affect the reliability of the electric system and continuity of service to consumers.

eminent domain. The power to expropriate or acquire compulsorily land or rights-of-way.

expansion. Expansion is a multi-part process. It is the process of evaluating the need for new facilities, deciding where and when to build new assets, dealing with the necessary regulatory, siting, and environmental approvals, and physically building and commissioning the facilities.

expropriation. The compulsory acquisition or taking of rights-of-way and land (usually in accordance with law), subject to the payment of compensation to the relevant landowner.

Federal Energy Regulatory Commission (FERC). An independent federal agency within the U.S. Department of Energy that, among other responsibilities, regulates the transmission and wholesale sales of electricity in interstate commerce.

financial transmission rights (FTR). Rights granted under a financial instrument representing the financial value of congestion costs and sometimes also losses between two points. Transmission congestion contracts (TCC) grant FTRs representing the financial value of congestion costs only. The rights are sometimes referred to as "point-to-point" rights. FTR sometimes means "fixed transmission right." *Synonym:* congestion revenue right (CRR)

firm. A service is firm when it is not liable to interruption.

fixed cost. A cost that does not vary with the scale of production or the scale of service provided.

flexible alternating current transmission systems (FACTS). These systems involve the use of high-speed, solid-state technologies to control transmission equipment and power flows in order to improve reliability and increase capacity.

flowgate. Transmission paths or elements that are likely to be congested. Flowgate modeling of the transmission system is essentially a simplification based on relevant paths or "flowgates." Flowgates are defined for NERC's transmission loading relief (TLR) processes. Flowgate rights (or flowgates) are rights to the congestion revenues across a particular congested physical transmission element.

for-profit. A business model that is characterized by the contribution of investment capital and the freedom to take management and entrepreneurial actions to manage revenue, usage, and other risks in return for the opportunity to retain a significant share of the value created in excess of the costs involved in providing goods and services to customers.

frequency. The rate, in cycles per second (Hertz, Hz) at which voltage and current alternate in electric-power systems.

grid. A reticulated or meshed transmission network or system.

impedance. The vector relationship of the resistance and the reactance of an electrical line.

independent system operator (ISO). A system operator that is not an affiliate of an entity that owns or operates generation, transmission, or distribution assets and has no interests in trading that is carried out in wholesale or retail electricity markets.

independent transmission company (ITC). A for-profit company that owns transmission assets that is not a system operator and is not affiliated with or in common ownership with generation, distribution, or retail businesses or facilities.

independent transmission provider (ITP). Any public utility that owns, controls, or operates facilities used for the transmission of electric energy in interstate commerce that administers the day-ahead and real-time energy and ancillary services markets in connection with its provision of transmission services pursuant to the FERC standard market design tarriff, and that is independent (i.e., has no financial interest, either directly or through an affiliate, in any market participant in the region in which it provides transmission services or in neighboring regions).

indirect costs. The costs of transmission associated with the role of the system operator, such as the costs of losses; out-of-merit dispatch of generating plants to manage congestion and to maintain network frequency, stability, and voltage criteria; some of the costs of ancillary services; the societal costs of local market power including must-run (constrained-on) generation and the costs of regulatory mechanisms aimed at mitigating these costs; the costs associated with poor generation location decisions; and excessive costs and delays in connecting new generators to the system.

interchange. Electric power or energy that flows from one entity to another.

interconnected system. A system consisting of two or more individual transmission systems that are normally synchronized and have connecting tie lines.

interconnector or interconnection. The facilities or tie lines that connect two systems (transmission or distribution) or control areas. An interconnection can also refer to the facilities that connect a generator to a control area or system.

interface. The specific set of transmission elements between two areas comprising one or more electrical systems.

kilowatt hour (kWh). A measure of electrical energy.

least-cost planning. The traditional process for planning generation, transmission, and distribution on an integrated basis in order to serve native load or captive customers at least cost.

license. A regulatory instrument defining the detailed rules and incentives that apply to a particular regulated entity.

load. A consumer of electric energy; also the amount of power (sometimes called demand) consumed by a utility system, individual consumer, or electrical device.

load shedding. The process of deliberately removing (either manually or automatically) preselected consumer demand or load from a power system in response to an abnormal condition in order to maintain the reliability or integrity of the system and minimize overall customer outages.

locational marginal pricing or location-based marginal pricing (LMP). Electricity pricing that reflects the marginal cost of transmission losses and congestion as well as the local cost of generation at every point or node on the transmission system.

loop flow. The characteristic (reflecting Kirschoff's Laws) that electricity flows on all available routes and parallel paths from one point to another. It can be calculated as the difference between the scheduled and actual power flow, assuming zero inadvertent interchange, on a given transmission path. *Synonyms*: unscheduled power flow, circulating power flow, and inadvertant transfer.

losses. Total electric energy losses in the electric system. The losses consist of transmission, transformation, and distribution losses between generation sources and delivery points. Electric energy is lost primarily because of heating of transmission and distribution facilities.

lumpiness. The characteristic of a transmission project that adds facilities in standard sizes and capacity in discrete blocks or "lumps", rather than continually increasing amounts. New investment could create a relatively sharp increase or "lump" in the progression of price increases to consumers over time.

maintenance. The process of keeping the transmission system in good working order and repair, including checking for wear and tear, replacing defective or worn parts, and lopping trees and plants that might interfere with or touch the equipment. Normally, elements of the transmission system are taken out of service for maintenance by the transmission owner at times approved by the system operator.

margin. The difference between net production capacity resources and net demand of consumers. Margin is usually expressed in megawatts (MW) for operating reserves and as a percentage of either system load or installed generating capacity for planning reserves.

marginal cost. The cost of providing one more unit of capacity or of producing one more unit of energy (or the cost that would be saved by providing or producing one less unit).

marketers. Commercial entities that trade in electricity at wholesale or retail or act as brokers or agents for market participants and consumers in the purchase or sale of electricity.

market participant. Any agent, company, entity, or person that trades in the wholesale electricity market, including generators, consumers, marketers, retailers, and, in certain circumstances, distribution and transmission facility owners.

market power. The ability of a firm to charge a price above marginal cost.

megawatt (MW). A measure of capacity or availability.

merchant. An investor or investment that is made on the basis that the investor takes the full usage and revenue risk of the project.

must-run generation. Generation (that tends to have local market power) required to run or to be constrained on to the system in order to enable the system operator to maintain reliability and system security. *Synonym*: constrained-on generation.

National Grid Company (NGC). The system operator, balancing mechanism operator, and owner of the transmission system (a transco) in England and Wales.

native load. The collective electric power and energy demands of retail and wholesale requirements contract customers of the electric utility providing such electric service. Such customers typically are located within the service territory of the utility. Because of this and the fact that the utility has a long-term contractual obligation to serve them without a specified maximum demand, the loads or consumers are considered "native" or "captive" to the utility. Other loads served (*e.g.*, through economy interchange or short-term market transactions) and loads served in a competitive retail market normally would not be considered native load.

natural monopoly. A monopoly providing a service at a price that is lower (or likely to be lower) than would be the case if the service were provided by two or more firms.

n-1 standard or criterion. A design rule intended to ensure that a power system will continue to function without interruption of service to consumers even with the loss of any single component (such as a generator, transmission line, or substation). If there are n elements in the transmission system, it should continue to function with any set of n-1 elements. If the standard were n-2, the system should continue to function with any set of n-2 elements.

nominal. A figure expressed in money terms, not adjusted for inflation.

non-firm. A service is non-firm when it is liable to interruption.

nodes. Grid locations where spot prices of energy differ as a result of congestion and losses. The differences may be positive or negative.

North American Electric Reliability Council (NERC). A not-for-profit company formed by the electric utility industry in 1968 to promote the reliability of electricity supply in North America.

open access same-time information system (OASIS). An electronic posting system for transmission access data that allows all transmission customers to view the data simultaneously.

opportunity cost. In general terms, the cost of the lost opportunity to make different investments. In corporate finance terms, it is the expected rate of return offered to investors by the equivalent-risk investments traded in capital markets.

outage. The removal of a facility from service or a reduction in the capacity of a facility.

pancaking. The payment of multiple charges for transmission service to more than one utility or transmission owner, with an additional charge applied every time a transaction crosses a transmission owner boundary.

parallel paths. The paths of an AC power grid where the impedance is lowest, over which electricity flows in a manner that prevents electricity from being directed to flow through specific lines and in specific directions.

permitting. The obtaining of siting authority, planning permission, zoning permits, governmental and regulatory consents, approvals and certificates, environmental consents, approvals, and permits, and where the processes are integrated or coordinated, obtaining powers for the expropriation or compulsory acquisition of land and rights-of-way (eminent domain).

planning reserve. The difference between an electric utility's generating capacity (usually expressed in megawatts) and the anticipated peak load. Operating reserve is a subset of planning reserve.

Pool. The power pool that was in place in England and Wales from 1990 to 2001.

postage stamp pricing. A transmission pricing method that is not sensitive to distance and allows a user to use the whole of the interconnected transmission system.

Powergrid. The Indian national transmission company.

Powerlink. The state-owned system operator and owner of the transmission system in Queensland, Australia.

power pool. A collective wholesale electricity trading arrangement. *Synonym*: pool.

power transfer distribution factor (PTDF). A PTDF is defined for every part of a node and an element (link or flowgate) on a transmission system relative to a slack bus (node). It is the fraction of the power injected at the node and taken out at the slack bus that will flow through the particular link.

public good. An economic good, such that the value of which to other consumers is not diminished by its consumption by any customer. Although transmission is not strictly a public good, it does have some aspects of a public good. Examples include the additional choices afforded to consumers by virtue of economic access to additional suppliers and the price benefits of a reduction of market power. More generally, examples of public goods would be lighthouses to guide ships, clean air, and green space in cities.

radial. A radial line or system is one that is not part of a meshed grid, or system that may radiate from a generating source to a load center or a neighboring system.

rates. The price or charges for electricity or interconnection, connection, or access to and/or usage of the transmission system.

rate case. A proceeding before a regulator that results in the determination, approval, or setting of rates (prices, tariffs).

rate of return. The rate of return (profit) to a firm calculated by reference to its assets, investment, or equity.

rate of return regulation. A regulatory model or regime linking a firm's costs to the firm's rate of return. *Synonym*: cost of service regulation.

reactive power. The portion of electricity that establishes and sustains the electric and magnetic fields of alternating-current equipment. Reactive power must be supplied to most types of magnetic equipment, such as motors and transformers. It is also needed to make up the reactive losses on transmission facilities. Reactive power is provided by generators, synchronous condensers, or electrostatic equipment such as capacitors, and it directly influences electric system voltage. It is usually expressed in kilovars (VAR) or megavars (MVAR).

real. A figure adjusted for inflation.

regional transmission organization (RTO). An entity responsible for certain system operations, market administration, and transmission functions, and meeting specified criteria in accordance with FERC orders.

regulation and frequency response service. The ancillary service that provides for following the moment-to-moment variations in demand or supply in a control area and maintaining frequency.

regulator. An individual, group of individuals, commission, agency, government department, governmental body, or authority empowered (usually by law) to oversee, monitor, control, make, and enforce certain rules and decisions with respect to the electricity industry or specified parts of, functions in, or transactions within the electricity industry or energy sector, with a view to protecting consumers or the public as to prices, quality of service, abuse of monopoly power, anticompetitive practices, environmental impacts, and the adverse effects of certain types of land use.

regulatory lag. The period of time between regulatory approvals, reviews, or rate cases.

reliability. The degree of performance of the elements of the bulk-power system resulting in electricity being delivered to consumers within accepted standards and in the amount desired. Reliability may be measured by the frequency, duration, and magnitude of the adverse effects on the electric supply.

rent. A revenue or profit (typically in excess of marginal cost).

retail. Sales to final or end consumers.

retailer. An entity that buys electricity at wholesale and sells it at retail to consumers or other retailers.

rights-of-way. The rights granted to a transmission owner to place transmission facilities on, under, or over the land belonging to a third party and includes easements, wayleaves, licenses, and servitudes (whether permanent, terminable, registered, or unregistered).

schedule. An agreed transaction size (megawatts), start and end time, beginning and ending generator output ramp times, rate, and type required for delivery and receipt of energy between the contracting parties and the control areas involved in the transaction.

Scottish Power. The utility that serves the southern region of Scotland and is a participant in the England and Wales wholesale electricity market.

seam. The interface between two adjacent electrical systems or markets, such as control areas or jurisdictions of system operators. Because the rules and operating practices may differ across these interfaces, problems can arise at the seams.

security. The ability of the electric system to withstand sudden disturbances such as electric short circuits or unanticipated loss of system elements.

security-constrained economic dispatch. The process of economic dispatch that takes into account binding constraints.

security coordinator. One of 23 entities established by NERC with the responsibility and authority to direct actions aimed at maintaining real-time security for a control area, group of control areas, NERC subregion, or NERC region.

sink. A point of consumption.

siting authority. The court, individual, government minister or department, authority, agency, or commission empowered (usually by law) to decide the need for a generation, transmission, or distribution expansion project or to approve the project and to grant the necessary planning or zoning permits and environmental consents. There may be more than one siting authority that has jurisdiction over a single project in a single jurisdiction.

socialized. The allocation of costs to all buyers in a market according to a methodology usually determined or approved by the regulator. *Synonym*: rolled-in.

source. A point of production.

stability. The ability of an electric system to maintain a state of equilibrium during normal and abnormal system conditions or disturbances.

Statnett. Statnett A.S., the Norwegian state-owned system operator and transmission company. It has a part interest in the market administrator of the Scandinavian power pool, Nordpool. It is the owner and lessee of the transmission system in Norway.

stranded. An asset is said to be stranded when it is unable to earn revenues sufficient to cover its cost.

supply. Generation or production of electricity (*e.g.*, in the U.S.) and retail sales to final consumers (*e.g.*, in the U.K. and Australia).

switching. The physical opening and closing of circuit breakers and switchgear in order to reconfigure the transmission system.

synchronize. The process of connecting two previously-separated alternating current apparatuses after matching frequency, voltage, and phase angles.

system operator. An operator or control unit responsible for maintaining the generation-load balance of the system and the security and reliability of the transmission system in real time. The system operator performs its function by monitoring and controlling the transmission system and certain generating units to ensure that generation matches demand in real time and that system reliability is maintained.

Terna. The owner of the transmission system in Italy.

thermal limit. The temperature-based, power-transfer limit of an item of electrical equipment. This limit is a consequence of the electrical resistance of the equipment, which causes the equipment to heat when power flows through it.

transco. An independent organization that owns and operates a transmission system. A transco differs from an ISO in that an ISO does not own the transmission resources. A transco combines the functions of system operations and transmission ownership, asset management, and maintenance.

Transener. Transener S.A., the concessionaire of the main transmission system in Argentina.

TransÉnergie. The U.S. affiliate of Hydro Québec that owns and develops transmission facilities and interconnectors. The name also applies to other affiliates of Hydro Québec that own or develop transmission facilities and interconnectors in other parts of the world.

transfer capability. The ability of any part of a transmission system at any moment in time to transmit or transform power, taking into account the prevailing system conditions, ambient temperature, and patterns of power flows or system usage.

TransGrid. The state-owned system operator and owner of the transmission system of New South Wales, Australia.

transmission or transmission system. An interconnected group of high-voltage lines, transformers, switchgear, and associated equipment for the movement or transfer of electric energy between points of supply and points at which it is transformed for delivery to consumers or is delivered to other electric systems. *Synonym*: grid.

transmission administrator. An entity that administers transmission system operations prior to real time or prior to a short period before real time. *Synonym*: transmission system administrator.

transmission loading relief (TLR). TLR is a procedure defined by NERC for relieving overloads that could cause a transmission line to malfunction. The procedure has various steps and can call for the reduction or curtailment, on a *pro rata* basis, of all transactions. TLR is a congestion management methodology that uses curtailments instead of a pricing mechanism.

transmission operations. The process of changing the physical configuration of the transmission system by switching to take parts of the network into or out of service and to activate transmission equipment. The transmission owner normally carries out operations, although several operations can be carried out by the system operator directly or automatically. Where the system operator and the transmission owner are the same company (*i.e.*, a transco), there is less of a distinction.

transmission owner. The owner, concessionaire, or lessee of a transmission system or transmission facilities that may also be responsible for certain transmission operations not carried out by the system operator, such as switching. If it is also the system operator, it will be a transco. *Synonyms*: transmission utility, transmission company, network service provider, transmission facility owner.

Transpower. Transpower New Zealand Limited, the state-owned system operator in New Zealand and the owner of the New Zealand transmission system.

unbundle. The separation (disaggregation) of the electric supply system into its component parts (such as generation, system operations, transmission, distribution, and retail sales).

uplift. The difference between the wholesale market price of electricity paid to sellers and the wholesale market price charged to buyers. The difference typically covers the costs to the system operator of operating the transmission system and administering the market that are not allocated directly to market participants. The uplift is usually charged on a $/kWh basis on the volume purchased by each wholesale buyer.

usage charges. The charge for using a congested path between the nodes (under locational marginal pricing), being the difference in energy prices at the nodes at either end of the path.

user. Any entity or person that uses and/or pays for access to or use of transmission facilities that may or may not be directly connected to a transmission system. A user could be a generator, distributor, retailer, marketer, or consumer. *Synonym*: transmission customer.

utility. A corporation, person, agency, authority, or other legal entity or instrumentality that owns or operates facilities for the generation, transmission, or distribution of electric energy or sells electricity to consumers.

vertically-integrated electric utility. An electric utility that owns and controls more than one electric utility function, *i.e.*, generation, transmission and distribution, and sale of electric energy.

volt-amperes reactive (VAR). The unit of measure of the power that maintains the constantly varying electric and magnetic fields associated with alternating-current circuits. See reactive power.

voltage. The unit of measure of electric potential.

voltage control. The control of transmission voltage through adjustments in generator reactive output and transformer taps, and by switching capacitors and inductors on the transmission and distribution systems.

wholesale. Sales to other market participants or firms rather than to final or end consumers.

INDEX

A

B

C

Index

D

E

Index

F

G

H

I

Index

Index

N

Index

O

P

R

Index

S

U

V

W

X

Y

Z